THE VIEW *from the* CENTER *of the* UNIVERSE

DISCOVERING OUR EXTRAORDINARY PLACE IN THE COSMOS

JOEL R. PRIMACK

and

NANCY ELLEN ABRAMS

RIVERHEAD BOOKS
Published by the Penguin Group
Penguin Group (USA) Inc., 375 Hudson Street, New York, New York 10014, USA • Penguin Group (Canada), 90 Eglinton Avenue East, Suite 700, Toronto, Ontario M4P 2Y3, Canada (a division of Pearson Penguin Canada Inc.) • Penguin Books Ltd, 80 Strand, London WC2R 0RL, England • Penguin Ireland, 25 St Stephen's Green, Dublin 2, Ireland (a division of Penguin Books Ltd) • Penguin Group (Australia), 250 Camberwell Road, Camberwell, Victoria 3124, Australia (a division of Pearson Australia Group Pty Ltd) • Penguin Books India Pvt Ltd, 11 Community Centre, Panchsheel Park, New Delhi–110 017, India • Penguin Group (NZ), Cnr Airborne and Rosedale Roads, Albany, Auckland 1310, New Zealand (a division of Pearson New Zealand Ltd) • Penguin Books (South Africa) (Pty) Ltd, 24 Sturdee Avenue, Rosebank, Johannesburg 2196, South Africa
Penguin Books Ltd, Registered Offices: 80 Strand, London WC2R 0RL, England

Published simultaneously in Canada

Library of Congress Cataloging-in-Publication Data

Primack, J. R. (Joel R.)
The view from the center of the universe : discovering our extraordinary place in the cosmos / by Joel R. Primack and Nancy Ellen Abrams.
p. cm.
Includes bibliographical references and index.
ISBN 1-59448-914-9
1. Cosmology—History. 2. Physics—Philosophy. I. Abrams, Nancy Ellen. II. Title.
QB981.P85 2006 2005055262
523.1—dc22

Printed in the United States of America
1 3 5 7 9 10 8 6 4 2

Book design by Stephanie Huntwork
Illustrations by Nicolle Rager Fuller unless noted otherwise

To our daughter Samara, our shining star,

and the galaxy of other young people

here and to come

who deserve the best chance we can give them.

And to our parents,

Rena and Larry Abrams and Loretta and Roy Primack,

who gave us that chance.

Contents

THE VIEW *from the* CENTER *of the* UNIVERSE

Preface

This is a pivotal moment in human history. New technologies and theories are allowing us to see a universe no one ever could have imagined. The two of us feel extremely fortunate to have witnessed firsthand, and in some cases participated in, these discoveries. The universe has filled much of our professional lives, through Joel's work as an astrophysicist and Nancy's work as a writer and a lawyer working in science policy. It has helped inspire our personal lives and our marriage as well. We have collaborated in developing new ideas, in teaching, traveling, writing—in life itself since 1977. While cooking or sitting around the dining room table, and in late-night talks with many of the world's leading scientists, we have been able to explore the ideas presented in this book. We have repeatedly found in the other's viewpoint the insight we needed. Over decades we have created for ourselves a sense of the universe that is much richer than either of us could have conceived alone.

We have done this because the nature of the universe matters to us—and, we believe, to millions of others as well. Where do we really come from? What are we made of? Are we alone? And, perhaps most important, *do we matter?* The answer to this last question is yes: we humans are significant and central to the universe in unexpected and important ways. We are discovering this fact at a moment in history when so much is at stake. It is our hope that this new picture of the universe will help convey the preciousness of the cosmic experiment on planet Earth. An understanding of our universe and our extraordinary place in it may reveal solutions to the problems that confront us personally and globally.

We invite you to join the conversation and explore the view from the center of the universe.

Nancy Ellen Abrams and Joel R. Primack
Santa Cruz, California
September 10, 2005

Introduction

In their hearts, most people are still living in an imagined universe, where space is simply emptiness, stars are scattered randomly, and common sense is a reliable guide. In this imagined universe, we humans have no special place and often feel insignificant. But today's golden age of astronomy is revealing that this lonely understanding of the universe is misguided. Our universe is rich, fascinating, and meaningful, and in it we humans occupy an extraordinary place.

If you close your eyes and try to picture the universe as a whole, what do you see with your mind's eye? Shooting stars, spiral galaxies, an ember-red moon rising over an unknown planet? Images like these can evoke the strangeness that lies beyond the earth, but they don't represent the universe *as a whole* any better than a single atom would, or your own face. The strange fact is that in this information age, when powerful and fast-paced images are our currency of communication, most of us have no idea how to picture the universe. But every prescientific culture did, and in their own cosmos they had a central and significant place.

Prescientific people had believable answers to big questions that became impossible to answer once we started to demand scientific accuracy. Does time run in one direction, or is it cyclical? Has the universe always existed, or did it come into being? If it had a beginning, how did it start? What is it made of? How does it work? *How do we humans fit in?* People hardly even ask such fundamental questions anymore, or appreciate that the answers affect not only how we live but what we believe is *possible*—including all our goals and plans. Ours is probably the first major culture in human history with no shared picture of reality.

Many of humanity's most dangerous problems arise from our seventeenth-century way of looking at the universe, which is at odds with the principles of science that we blithely use in countless technologies. The main threats to our survival result from the almost total disjunction between the power of our technologies and the wisdom required to use them over the long period during which their effects will last. A wise perspective must include a *cosmological* time-scale and be based in a reality that includes quantum physics, relativity, evolution, and other scientific theories that underlie technologies like computers and cell phones, global positioning systems, and genetic engineering. You don't have to understand those theories, any more than you have to be an expert in automotive mechanics to drive a car. But you can't drive a car safely if you are blind to other cars and expect only horses and buggies to come around the next bend. As a society, we have been exploiting the powers of a universe to whose existence we are blind. Now we finally have the opportunity to end this alienation: the modern science of cosmology is discovering the universal reality in which we are all immersed.

Cosmology is a branch of astronomy and astrophysics that studies the origin and nature of the universe, and it is in the midst of a scientific revolution that is establishing its lasting foundations. What is emerging is humanity's first picture of the universe as a whole *that might actually be true*. There have been countless myths of the origin of the universe, but this is the first one that no storyteller made up—we are all witnesses on the edges of our seats.

The last time Western culture shared a coherent understanding of the universe as a comforting cosmic dwelling place was in the Middle Ages. For a thousand years, Christians, Jews, and Muslims believed that the earth was the immovable center of the universe and all the planets and stars revolved on crystal spheres around it. The idea that God had chosen a place for every person, animal, and thing in the Great Chain of Being made sense of the rigid medieval social hierarchy. But this picture was destroyed by early scientists like Galileo, who discovered about four hundred years ago that the earth is not the center of the universe after all. The idea of the cosmic hierarchy lost its credibility as the organizing principle of the universe, but those early scientists couldn't replace it. Instead, for centuries they were able to say with authority what the universe is not, but not what it is.

With little data from beyond the solar system, early scientists extrapolated a bleak picture of the universe as endless emptiness randomly scattered with stars. When the French physicist-mathematician-philosopher-monk Blaise Pascal (1623–1662) absorbed this post-Galileo picture, he experienced a cosmic malaise that no one had ever before expressed in literature: "I feel engulfed in the infinite immensity of spaces whereof I know nothing and which know nothing of me. I am terrified. . . . The eternal silence of these infinite spaces alarms me."[1] From the medieval universe, which had felt like a magnificent, high-ceilinged cathedral, Pascal felt tossed into a "scientific" universe that was cold, shapeless, and incomprehensibly huge, in which humans are rootless and insignificant. This impression of the universe has lasted to this day.

Cosmically homeless, our culture over the centuries downgraded the importance of having a cosmic home; today "the universe" in the popular mind has become little more than a shapeless space or a fantasy setting for science fiction, neither of which appears to matter much in what people call "the real world." In a reversal of all historic and even prehistoric precedent, it is normal today to consider people who are more concerned with cosmic reality than with making money to be out of touch and *un*realistic. As a people, we now have the scientific ability to see so much more deeply into the universe than ancient people, yet we experience it so much less and connect with it almost not at all. This widespread cultural *indifference to the universe* is a staggering reality of our time—and possibly our biggest mental handicap in solving global problems.

The modern world has so deeply absorbed the four-hundred-year-old picture of a universe in which we have no special place that it seems like reality itself. But actually this Newtonian picture is founded on physics that explains accurately the motions of a single star's planetary entourage, but not the entire universe. Until the late twentieth century, there was virtually no reliable information about the universe as a whole. That has now changed. New kinds of powerful telescopes including the Cosmic Background Explorer, the Hubble Space Telescope and other satellite observatories, and the Keck Telescopes in Hawaii and other huge new ground-based telescopes have begun to provide the first reliable data, not just about the nearby universe but also about the early universe and the most distant galaxies. Astronomers can now observe every bright galaxy in the visible universe and can even see back to the cosmic "Dark Ages" before galaxies formed. Voluminous data from the never-before-seen universe are supporting exciting and counterintuitive directions of thinking. The great movie of the evolution of the universe is coming into clearer focus: we now know that throughout expanding space, as the universe evolved, vast clouds of invisible, mysterious non-atomic particles called "dark matter" collapsed under the force of their own gravity. In the process they pulled ordinary matter together to form galaxies. In these galaxies generations of stars arose, whose explosive deaths spread complex atoms from which planets could form around new stars, providing a home for life such as ours to evolve. Clusters, long filaments, and huge sheetlike superclusters built of galaxies have formed along wrinkles in spacetime that were apparently generated before the Big Bang and etched into our universe forever. There may also have been different Big Bangs creating different kinds of universes beyond our own. In fact, instead of starting our origin stories "In the beginning . . ." we may need to humble the phrase to "In a beginning . . . ," which is an equally accurate translation of the Hebrew word *bereshit* at the beginning of the Bible. Or we can simply say, "In our beginning."

The possession of this new story is a gift so extraordinary that most of us don't know what to do with it. We have been living for centuries in a black-and-white film. There were no obvious gaps in the scenes before us, so we didn't notice that anything was missing. Becoming aware of the universe is like suddenly seeing in color, and that changes not just what's far away but what's right here. The universe is *here*, and it's more coherent and potentially meaningful for our lives than anyone imagined.

Most of us have grown up thinking that there is no basis for our feeling central or even important to the cosmos. But with the new evidence it turns out that this perspective is nothing but a prejudice. There is no geographic center to an expanding universe, but we are central in several unexpected ways that derive directly from physics and cosmology—for example, we are in the center of all possible sizes in the universe, we are made of the rarest material, and we are living at the midpoint of time for both the universe and the earth. These and other forms of centrality have each been a scientific discovery, not an anthropocentric way of reading the data. Prescientific people always saw themselves at the center of the world, whatever their world was. They were wrong on the details, but they were right on a deep level: the human instinct to experience ourselves as central reflects something real about the universe, something independent of our viewpoint.

Working from the assumption of their own centrality, the ancients took the cosmos—as they understood it—as the model for their lives and their religions. This book argues that we should, too. The big difference today is that science is finding out how the cosmos really works, and therefore *we are the first generation that can know what the universe may really be saying.* The universe is speaking and always has been, only now we humans have both the technological tools and the intellectual capability to hear and understand a lot of it.

The discovery of our universe challenges us to reframe everything, and this is truly difficult. For the vast majority of busy people, there's little point in learning a lot of science unless you can do something valuable in life with that knowledge. We want to show you that you can. The aim of this book is not only to help people understand the universe intellectually, but also to develop imagery that we can all use to grasp this new reality more fully and to open our minds to what it may mean for our lives and the lives of our descendants.

Unlike earlier cosmologies, the new scientific story is not intuitively simple. It can't be, because both intuition and common sense are always based on the assumption that we're on Earth. There's no way to have intuition about things one has never experienced, and most of the universe fits into that category. The new cosmology can't be intuitively simple for a second reason as well: it is based on unfamiliar concepts including relativity and quantum physics. But these theories can be translated into ordinary language and compelling images. This book suggests mythic images that we can adapt to a modern understanding of the universe. This is essential, because just as scientific cosmology cannot be explained in numbers alone, neither can it be adequately explained in everyday language. Many religions have concepts that resonate harmoniously with aspects of the new scientific picture—concepts that can, in fact, help us tremendously to appreciate the depth and meaning of the universe—but all religions also have concepts that don't. An attempt to explain the modern universe in terms only of a favorite religion would result in scientific ideas being crushed and distorted to fit narrow preconceptions, while beautiful and apt imagery would be dismissed. We need to find those concepts that work, and only those, borrowing from many religions as well as other sources.

A famous photograph taken from the Apollo spacecraft by the first human beings to orbit the moon shows Earth as a sparkling blue-and-white ball, suspended in blackness, with a sterile lunar landscape in the foreground. This photo jolted many people into realizing in their hearts what of course they knew intellectually: that maps and globes have imprinted a false picture of reality on our minds. The photo showed no countries on our planet, only land masses, oceans, and clouds. The endless preoccupation with nations and racial or ethnic groups has completely misled our intuitions. This modern icon of our gorgeous, undivided home planet shows the power of a new image to alter perceptions and attitudes. In this book we try to portray the universe in similarly meaningful pictures.

We use the word "picture" metaphorically, however. Any picture of the expanding universe can only be symbolic, because the universe can't directly be seen. No one can step outside of it to look at it, no one can see all times, and over 99 percent of its contents are invisible. Symbols are the way to grasp the universe. They free us from the limits of our five senses, which evolved to work in our earthly environment. Symbols let us see the universal with our mind, which doesn't have any obvious limits. And symbols are far easier to remember than a long, logical argument or a mathematical equation. Carl Jung wrote, "A symbolic work is a perpetual challenge to our thoughts and feelings because even if we know what the symbols are, they do not refer to a given thing, like a sign, but are 'bridges thrown out towards an unseen shore.' "[2] Each of the symbols in this book represents a fundamental but incomplete insight about the universe, our unseen shore. No single symbol can ever represent the universe completely. To get a sense of the whole, we have to somehow absorb the meanings of all the symbols *together*, and this takes imagination. It's ironic that seeing reality takes a lot of imagination.

This book places the scientific meat at the center, sandwiched between the human past and human future. In the first part, we look at how earlier cultures saw the universe and how their cosmologies shaped people's sense of what they were and could be. Ancient cosmologies created not only a meaningful mental homeland in the cosmos for their own members but also much of the mythological language, imagery, and questions that still matter today and that continue to inspire artists and thinkers. Although fascinating origin stories have been told around the world, here we focus on those cosmologies in the line of development toward Western scientific culture and on the two great cosmological revolutions that marked the shifts from one universe picture to another. In investigating these early cosmologies, we're looking for time-tested, powerful symbols and other forms of expression and inspiration that touched our ancestors but that also resonate with essential modern cosmological ideas. We will in later chapters reinterpret them in light of modern cosmology so that these new-ancient symbols can represent the mythic power of the new cosmology.

In the central part of the book, we present the new scientific picture by focusing on five aspects of reality that are universal. They are, in essence, the answers to these profound and timeless questions:

What is the universe made of?

How did it get this way?

How big is it?

Where did it come from, and where is it going?

Are we alone in it?

But the big question science does not answer or even ask is: What difference does this all make to *me*? This book will try to answer all these questions, including the last, but the real answer to what difference the new cosmology will make will only become clear as more people open their eyes to the possibilities.

In the final part of the book, we ask what the future implications of the new big picture may be for our planet, for the human race as a whole, and for each of us personally. In a world as complicated and volatile as today's has become, in order to raise our probability of success, it's essential to narrow the field of contending political and other ideologies to those that have some chance of succeeding in the real universe, and this book gives examples of how people might start thinking from a cosmic perspective about global concerns. A cosmic level of awareness is no longer a luxury. As Einstein is supposed to have said, "Problems cannot be solved at the same level of awareness that created them."

In the last chapter, we explore how thinking cosmically might help us experience what it means to be the human part of the universe. Though people tend to focus on their differences—classifying others into *us* and *them*—when we humans confront the universe all differences among us become trivial: we vary no more than pearls on a string compared to what's out there. And we pearls may be far more cosmically rare and precious than most people realize. It's only because humans are all bunched together on one planet that we fail to see how extraordinary we are. Science is allowing us to start sketching the outlines of what it takes for a planet to produce intelligent life, and this is beginning to tell us what aliens *can* be like. In this way we can begin to situate ourselves among all intelligent life and see ourselves as others might see us. Intelligent life is neither incidental nor insignificant but has a place in the universe so special it could not even have been imagined before the invention of modern cosmological concepts. By understanding the universe, we begin to understand ourselves.

Traditional religious stories can still arouse a sense of contact with something greater than we are—but that "something" is nothing like what is really out there. We don't have to pretend to live in some traditional picture of the universe just to reap the benefits of the mythic language popularly associated with that traditional picture. People around the world should be able to portray our universe with all the power and majesty that earlier peoples evoked in expressing their own cosmologies. Mythic language is not the possession of any specific religion but is a human tool, and we need it today to talk about the meaning of our universe. Big changes are happening on our planet, and shepherding ourselves through them successfully is going to require tremendous creativity. An essential ingredient may be a cosmic perspective, and such a perspective is just becoming available. Not a moment too soon.

This book presents various kinds of explanations, including contemplations, symbols, metaphors, and for those who desire them, further details in the endnotes with data and graphs in key cases. (If the endnote number appears in ordinary typeface, it's simply a reference; if the number appears in boldface, it provides explanation.) We also have a website for this book: TheViewfromtheCenteroftheUniverse.com. Please skip the explanations that don't work for you, and don't feel obliged to read the endnotes. What matters above all is not the details but the overarching realization that we are living at the center of a new universe at a pivotal time.

Part One

COSMOLOGICAL REVOLUTIONS

ONE

Wrapping Your Mind Around the Universe

THERE ARE FEW FAMILIAR WORDS that can describe the universe as a whole. Those who study the universe use physics and mathematics, but what are their equations *about*? Cosmologists toss ideas back and forth until they either find or create some meaning for mathematical expressions, some way of tying them together into a theory. When applied to aspects of the universe far beyond the ordinary conditions on Earth, almost every word is a metaphor. Science is both a consumer and creator of metaphors and is meaningless without *thousands* of them.

Not only is the universe a challenge to words—so, to a great extent, is the way of thinking that we have to adopt to do cosmology. Cosmology is undergoing a scientific revolution, producing the first theory of the universe that may actually be true, but all the key words in this assertion—"cosmology," "scientific revolution," "theory," "universe," and "true"—have meanings beyond those used in ordinary speech. These words encapsulate a tremendously liberating attitude toward knowledge that makes it possible to discuss the universe with both ambition and humility. In this chapter we will try to crack the words open and let them express these larger possibilities. Other relevant concepts like "common sense" and "myth" we must also redefine, since common sense normally carries a positive connotation and myth a negative one, which we propose to reverse for the purpose of understanding the universe.

What Is Scientific Cosmology?

"Cosmology" means two very different things. For anthropologists, who study human cultures, "cosmology" means a culture's Big Picture, its shared view of how human life, the natural world, and God or the gods fit together. The Big Picture almost always portrayed everydayness as being embedded within an awe-inspiring but mostly invisible and therefore spiritual cosmos. The stories that together created the Big Picture in people's minds were told metaphorically in imagery derived from the local environment—imagery of people, relationships, animals, plants, landscape, sun, moon, and stars. This picture of reality was constructed through a lifetime of hearing such stories and witnessing or performing rituals that were often incomprehensible or fantastic to outsiders. It was this Big Picture that made sense of the world. Cosmological stories establish a context for life—they create what we might call a "cosmic dwelling place" in which human affairs acquire meaning on a larger scale.

But for astronomers and physicists, the word "cosmology" means something quite different: it is the branch of astrophysics that studies the origin and nature of the universe as a whole by developing theories and testing them against observational evidence to support or rule them out. Traditional cultures' cosmologies were not factually correct, but they offered guidance about how to live with a sense of belonging in the world. Modern scientific cosmology says nothing about human beings or how we should live. It aims to provide scientific accuracy, not meaning. This book seeks to connect these two different understandings of cosmology by offering a science-based explanation of our human place in the universe. It is by no means the only possible explanation; we hope creative people everywhere will interpret the findings of scientific cosmology in new ways that are valuable to participants in the emerging global culture.[1] This book assumes that a global culture is developing, and that it is up to us who are alive today to make it as good as we can. "Good" is admittedly a word that is both too small and too big—it can be a tepid grade halfway between fair and excellent, or it can be God smiling on creation, but the important thing is that it is a direction, a commitment. Scientific cosmology can serve the global good.

Astrophysical cosmology is one of the "historical sciences"—that is, one of the storytelling sciences, not a laboratory science based on experiments.[2] Experimental sciences like physics, chemistry, and much of biology seek to understand the underlying principles of their field, but historical sciences like archaeology, paleontology, evolutionary biology, geology, and cosmology seek to understand not only timeless principles but also the story of what actually happened. Scientific predictions in the historical sciences are not about what will happen but about what will be discovered about what has already happened. Cosmology also predicts the future of the universe, but we can only trust predictions about the future if our understanding of the past is repeatedly confirmed with new instruments that are ever more ingenious and powerful.

How do we know what we know in scientific cosmology? A torrent of data from new telescopes has washed away most old cosmological theories and suggested counterintuitive new theories. As recently as 1998, convincing data was discovered that there really is a cosmic repulsion of space by space, something first envisioned by Einstein in 1916 and worked out theoretically by modern cosmologists. This repulsion is causing the expansion of the universe to accelerate. Although until recently there appeared to be inconsistencies—for example, stars that seemed older than the universe—by the beginning of the twenty-first century all the data fit together beautifully. As we will explain, the key facts of cosmology are now all cross-checked several different ways. The result is a single theory that has so far passed all tests.

To construct a view of the very distant past, say twelve billion years ago, we have a direct source of evidence: light that has happened to fall upon a few telescopes and other detectors on or near Earth after traveling across the universe for twelve billion years. The light from such distant galaxies is proof of the past—but its sources are inferred by complex and possibly fallible chains of human reasoning. There are many kinds of light, most of which are invisible to the human eye but not to our instruments.[3] Using this almost infinitesimal sample of the universe's radiation plus constantly honed lines of reasoning and the willingness (not always totally successful) to subordinate intuition to the data, cosmologists investigate extreme physical conditions, including those at the beginning of the universe that can never be reproduced.

Our goal is ambitious: we are trying to reason out the laws of the universe and piece together its unfolding story. Although this may seem outrageously presumptuous to some people, in fact our problem is more often that we are not daring enough. The heat radiation from the Big Bang was discovered in 1965, for example, but it could probably have been detected almost twenty years earlier. The technology for detecting it had been invented during World War II, and relevant theoretical papers had been published by the late 1940s. As the physicist and Nobel laureate Steven Weinberg said, "Our mistake is not that we take our theories too seriously, but that we do not take them seriously enough. It is always hard to realize that these numbers and equations that we play with at our desks have something to do with the real world."[4]

Cosmology addresses some of the deepest questions that people are capable of asking. Odd though it may sound, faith motivates cosmologists to pursue these questions—faith that we humans can get close enough to some aspect of the real universe to uncover a secret. We test our theories mercilessly, pushing them as far as we possibly can to see where, or whether, they fail—but that this is worth doing is not tested: for us it's self-evident. Our faith is confirmed on those rare but wonderful occasions when our predictions are confirmed.

Ask cosmologists why they do what they do, and almost invariably you will hear, "I want to know how the universe really is." In this we are like the ancients, who never questioned the value of knowing the Big Picture; but modern people do question it, perhaps partly because they no longer think that such knowledge is possible. One thing we hope to show in this book is that it is, and that the value of knowing the universe is as great as ever. The faith of active research cosmologists—a faith shared with the ancients—is that human beings can personally connect in a meaningful way to the real cosmos. This faith is possible because scientific cosmology also shares an essential, liberating tool with the ancients: we have developed a language and a set of ideas and images that let us discuss *the universe we actually believe in*. One purpose of this book is to help more people join in this conversation.

Is the Universe Something?

"The universe" in common speech is a basket term: it's just a container for everything else. But the universe of modern cosmology is not just a container—it's a dynamic, evolving being. Its initial explosive expansion slowed down as it made most galaxies and stars, but about five billion years ago it began expanding faster and faster. The universe exists in different ways on every size scale from the largest to the smallest, and all times are within it. What scientific cosmology does is to put a mental frame, so to speak, around the universe. A frame gives its contents an identity, and until something has an identity, we can't think about it; we can't distinguish it from what is not-it.

The universe as understood by any given civilization, however, is actually a *social consensus* on how to think about whatever is out there. This consensus is "a mask fitted on the face of the unknown Universe."[5] To many people today the word "consensus" suggests an unexciting compromise that's not really true or fully satisfactory but simply the expedient solution of the moment. In fact, consensus on reality is what has made cooperative societies possible, but today it's shaky. A clear consensus on the reality of the universe would be an enormously valuable achievement, but consensus on anything gets harder to maintain as communications become more global. Any story of the universe will do if everyone around you shares it. But if different people you meet have different stories, no story will do, because there is always the chance that it is only a story. Consensus has to be based on something impartial, and science is the closest the modern world can come to that. There already exists a global unspoken consensus on science: no matter where people fall along a political or religious spectrum, and no matter what they may claim, in practice they trust their *lives* to airplanes, computers, and other technological products based on modern science.

But isn't there something *really* out there, the Universe with a capital U that all masks are hung upon and that created us? No doubt. But the moment we start to say anything about it that is not supported by scientific evidence, we can only fall back into remixes of old metaphors, and thus drag out yet another mask. *What our species needs is a general consensus on the universe that lets us operate by making and testing assumptions while remaining open to future evidence.* Scientific cosmology has given us a goose that lays golden eggs. If in our impatience for Ultimate Truth we rip open the goose to discover its secret, we will find only meaningless entrails. Treasure the golden eggs—treasure the solid discoveries that redirect our theories ever closer to reality.

Just Theories?

What modern cosmology is offering is a theory of the universe as a whole. Just a theory? Anybody can have a *theory*; wake me up when you find the *truth*. The tendency of many people to dismiss evolution as "just a theory" shows how poorly they understand what a scientific theory is. In ordinary language, "theory" is used in many ways: a belief, an ideal, a hunch, even a put-down ("Yeah, in theory"). People often say they have a theory when all they really mean is, "I would guess" or "I have a possible explanation." They're not betting their life on it. But a scientific theory has to stake its life. To be scientific, a theory has to make predictions that are testable—the kind that can rule the theory out, end of argument. Any theory flexible enough to be able to explain all possible data is not scientific because it can't be ruled out no matter what data are discovered. A scientific theory is a working hypothesis—a proposed picture of reality, to help us think about reality. So we keep on testing it and improving our understanding of it in order to get closer to reality. One way we have tested cosmological theories, for example, is by creating theoretical universes in supercomputers. The universes are allowed to evolve from the Big Bang to today according to each theory, and the results are compared with actual observations of the universe to see whether they are the same—that is, to see whether the theory's predictions can survive being compared to the reality of the data. The point of a scientific theory is not to have an agreeable explanation; the point is to get it right.

A theory gets tested until it fails. Ever since 1916, scientists have been testing Einstein's theory of general relativity, our modern way of understanding gravity, space, and time. The theory of relativity has never failed, but scientists are still doggedly testing it.[6] This process inspires confidence! The more intense the testing that relativity survives, the more confidence we have in it. A scientific theory can be disproved by a single counterexample, but it can never be proved true because that would mean it couldn't be refuted; and if it can't be refuted, by definition it's not a scientific theory—it's faith, not science.[7] So the great theories get tested the most, because it would be so cool to be the person who disproved one of them—and there is always that chance. In fact, at the frontiers of research it is well known that general relativity can't be *completely* right, because it is at some point inconsistent with the theory on which the entire electronics industry is based: quantum mechanics. Quantum mechanics describes how elementary particles behave on the atomic scale—for example, how electrons move in computer chips.

Whoa—so even the theory of relativity is still just a temporary understanding? If that's true, why should I bother to read a whole book about these "theories" of the universe?

This kind of reaction is entirely justified when it comes to the kind of "scientific discovery" that newspapers report on every day—what foods some new study indicates are healthy, what medical treatment works best, how safe your local industry's waste stream is—because a year later another study often seems to find the opposite. But although olive oil may be bad for you last year and good for you today, the fundamental theory of biology—evolution—is not going to change; these minor discoveries don't even challenge it. That fundamental level is what we're discussing in this book. Like physics after Newton or biology after Darwin, cosmology is now undergoing the transformation from a loose collection of observations and speculative ideas to a discipline with immense amounts of reliable data, organized by an overarching theory that may last forever.

The only way for humans as a whole to get closer to the truth of the universe is through the partnership between scientists who are improving their theories and observations and a public that not only supports them to do this work but pays attention, asks what those theories may mean, and uses them to enrich and strengthen the larger culture. There is no truer guide than our scientific theories, and if they turn out to be wrong, they have raised us up and we have lived at their level and are ready for wider understanding. Current cosmological theories have already broken through to a new world, and there will be no going back.

Anyone who has ever rented an apartment or bought a house knows that no matter how carefully you inspect the place beforehand, you never discover all of its problems until you actually live there. Nevertheless, it is much better to move in than to live on the street. The same is true of theories. It is crucial to commit to your theories, even while knowing they might be wrong, because it is only the act of moving into the theory with all your intellectual furniture and living there that lets you find its problems and limitations—or its secret passageways. If its problems turn out to be insuperable, we will find out much faster, and with luck the solution may be down one of the secret passageways. But we would know neither the problem nor the solution if we had been afraid to move into the theory in the first place. Our theories are our way of understanding what is real. We hope that this book conveys not only intellectual ideas but a felt understanding that the universe is *real*, and that this has personal implications for every one of us. To demand Ultimate Truth rather than scientific theory is to choose homelessness because no house is perfect.

The cosmological theory emerging today explains the Big Bang and the evolution of galaxies and even gives us a glimpse of the state of being just before the Big Bang. Beyond that we have no evidence, but we do have fascinating and important theories that make good mathematical sense even though they try to explain events beyond the limits of actual testability. In this book we will try to make clear where science gives way to the kind of speculative theory that verges on metaphysics, but we need to present speculative untested theories such as "eternal inflation" because what they attempt to explain is essential to a satisfying picture of the universe. Ultimately, only a picture of the universe that can be challenged and overturned or else encompassed by new data will ever feel honest in an age of science.

What Happens in a Scientific Revolution?

Will what we now know about the universe today still be correct in a thousand years? While these ideas will remain theories, tested and transformed by further data, we now have a picture of the universe that may very well stand the test of time—and further discoveries. Most people think this is impossible, because scientific theories always overthrow the previous theories, and this new one will also someday be overthrown. This misunderstanding is due to a famous book of the 1960s, *The Structure of Scientific Revolutions* by Thomas Kuhn,[8] which gave many people a profoundly mistaken impression of how science works. Kuhn basically argued that scientific research proceeds for long periods of time within a certain manner of thinking (a "paradigm") until too many pieces of evidence have turned up that are unexplainable and even paradoxical, and then suddenly there is a great leap (a "paradigm shift") and the old theory is abandoned for a new theory that explains much more, and in which the old paradoxes disappear. The concepts of the new theory, according to Kuhn, are so different from the concepts of the old that they are "incommensurable" because the implicit assumptions have changed. For example, in the Copernican Revolution of the sixteenth and seventeenth centuries, when scientists abandoned the idea that the earth was the immovable center of the universe, the earth became a planet and the sun and moon lost that status. Furthermore, even the scientists who adopt the new theory are often a different generation. The old theory is overthrown and never again taught as science.

This was in fact what happened in the Copernican Revolution, and Kuhn, whose first book was on the Copernican Revolution,[9] effectively assumed that all scientific revolutions are like that one. His argument implies that no scientific theory can ever be considered true, because it will eventually be overthrown by a bigger and better theory. Since the new theory will eventually be overthrown in its turn, it is ultimately no truer than the old one (even though it's temporarily more useful), so it is questionable whether science actually progresses or just keeps changing. Following what they take to be Kuhn's views, a substantial number of "postmodern" scholars have argued that a scientific theory is nothing but an opinion about reality, even if it is an educated opinion.[10]

Kuhn was wrong. There have been several revolutions in physics since Newton, but none of these has overthrown the previous theory.[11] Revolutionary scientific theories do not have to overthrow their predecessors except in the earliest stage of a science when a scientific theory is replacing earlier ideas that were not well supported by evidence. Once a field of science undergoes the revolution that creates for it a solid intellectual foundation—like the one Newtonian mechanics gave physics, or Darwinian evolution gave biology—that foundational theory can stand forever. Science then progresses by encompassing the foundational theory in a new and larger theory that explains things beyond the ken of the older theory. But unlike Kuhn's description, an encompassing theory does not overthrow the older theory—instead, it defines the limits within which the older theory is reliably true. It puts the old theory in a box and tells us where the walls are.

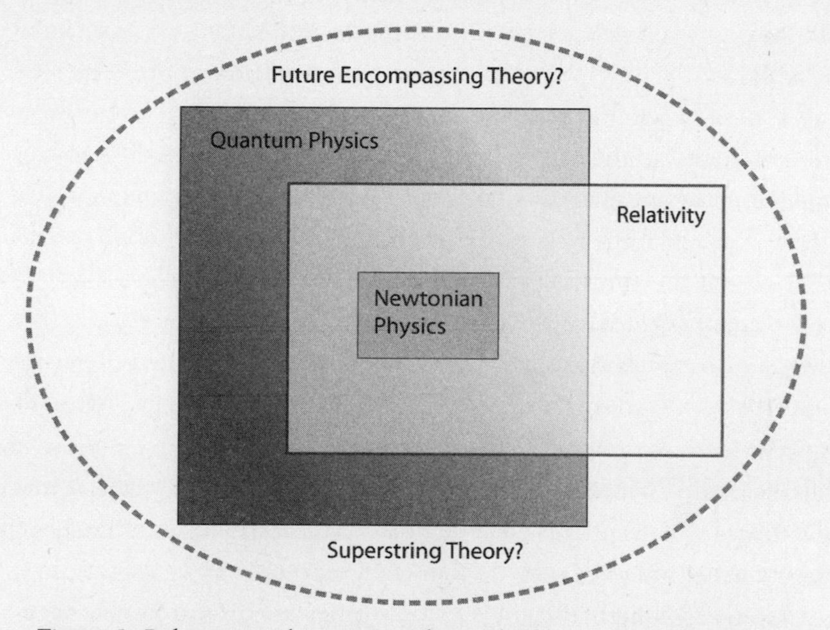

Figure 1. Relativity and quantum physics both encompass Newtonian physics, but not each other. They may in the future be encompassed in turn by superstring theory.

Einstein's theory of relativity did not overthrow classical Newtonian physics. During the centuries after Newton but before Einstein, scientists tried to explain the entire universe with Newtonian gravity. What they did was extrapolate Newtonian theory endlessly—they assumed that the whole universe worked like the solar system and was basically just a lot bigger. They were doomed to fail, because this assumption is wrong. They did not know that relativity is necessary to make accurate predictions about the larger universe, where some objects are traveling at speeds approaching the speed of light, or moving in extreme gravitational fields near black holes. However, when relativity is applied to the slow speeds and comparatively weak gravity we're familiar with on and near Earth, it makes essentially *the same predictions* as Newtonian theory, even though conceptually it explains differently what is happening. The conditions under which the new and old theories make the same predictions define the region inside the box.[12] In Newton's case the box contains the solar system. On the size scale of the earth and solar system Newton's laws of motion are essentially right. They accurately predict the planetary motions, with only tiny errors even for the planets closest to the sun, Mercury and Venus (which feel the strongest gravitational force and move the fastest). Those tiny deviations, however, were predicted by general relativity and provided the first observational tests of that theory. *Inside its box, classical Newtonian mechanics can forever be considered true.* Outside the box, relativity is required, and thus, despite its revolutionary transformation of physics, Einstein's relativity *encompassed* Newton's mechanics without overthrowing it. This is the kind of truth the physicist Charles Misner meant when he wryly said, "The only sort of theory we can know to be true is one that has been shown to be false"[13]—by which he meant limited. Such an encompassed theory is the highest grade of scientific truth that can be achieved.

Relativity in turn will almost certainly be encompassed in the future by some theory that tells us the *limits* within which relativity can forever be trusted. We will then be able to say with complete confidence that relativity is true—within its truth box. And thus science does not simply toss one theory out for another: it makes real progress toward ever-larger truths. There is a built-in humility enforcer in science: we can never call something true until we know about something bigger. The theories of modern cosmology that we discuss in this book may someday be encompassed, but it is increasingly unlikely that they will be overthrown.

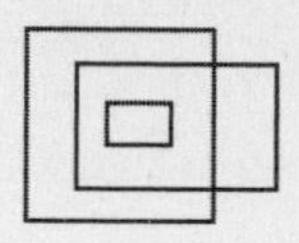

Think of this miniature version of Figure 1 as a kind of abbreviation or mini-icon for the concept "encompassing, not overthrowing." We will use this and other mini-icons occasionally as reminders.

Is the New Theory of the Universe True?

If modern cosmology's overarching theory is likely someday to be clapped into a box that cuts off its edges, exactly how true is it *now*? "Truth" in science is different from truth in religious belief or truth as accepted by a court of law. For some people truth in religion is a quiet certainty that is the bedrock of their lives, and no external fact can ever shake it. For others, religious truth is some kind of cosmic obligation, worth killing or dying for—a certainty so obvious that for other people to believe otherwise is an offense against God. Both soft and loud versions of religious truth illustrate a single understanding: that truth is by definition unquestionable. Law has yet another take on truth: truth, as a judge or jury finds it after due deliberation, is a set of facts upon which the court must base its sentence or order. Our society agrees that the whole power of the state can be marshaled in support of that sentence or order. The facts "found" by the judge or jury may or may not be what actually happened, but if they arrived at their findings *by following the right procedures*, then those findings become officially true. This allows the disputed matter to be closed in an orderly way without violence, and the parties to move on with their lives. Even if new information later reveals that those "facts" were wrong, the case will rarely be reopened except in special circumstances. In the judicial system, the nonviolent closure of a dispute is generally a higher value than factual truth. It needs to be this way. If law used the scientific standard of truth, no case would ever close. If law used the religious standard—well, those countries where it does so are proof enough that such a course is disastrous. Unlike truth in law, scientific truth lasts until and only until there is contradictory evidence, however many centuries it may take for that evidence to surface. Unlike truth in religion, scientific truth is always open to question. Scientists, however, try to find the most fertile questions.

Many people assume that if something is true, it has to stay true. Truth is presumed to be unchanging and eternal. People who crave this kind of Ultimate Truth rarely consider that they themselves are at only an intermediate stage of evolution and therefore in no position to understand anything ultimately. They are hoping for truth to be a kind of bedrock that they can build a house upon in complete security. But in a globally wired world where ideas are always clashing and cross-fertilizing, the claim that any of them is bedrock truth leads to a lot of rock-throwing. The interplay between cosmology and culture goes on all the time, and wisdom is possible without omniscience. How much of the nature of reality science has figured out is astonishing, but most of it is working knowledge, not Ultimate Truth. Working knowledge has turned out to be a far more fruitful goal for beings like us humans who keep evolving and learning. Science tells us to let go and flow with the data. As long as humans are creative, there can't be a final truth. To insist on Ultimate Truth is to live in an all-too-human fantasy and miss those wondrous aspects of the universe we can actually verify. "Ultimate" doesn't mean highest; it just means last.

Are Truth and Beauty Really Inseparable?

There is a romantic notion that the true scientific theory is always the most beautiful one. John Keats's "Ode on a Grecian Urn" ends: "Beauty is truth, truth beauty,—that is all / ye know on earth, and all ye need to know." However, truth may be beautiful, but beauty is not always truth. A theory appears beautiful generally because it appeals to some deep human preference like simplicity (meaning we can understand it) or symmetry,[14] but the universe exists on size scales, such as the size of atoms, to which humans have no conscious connection. What happens on these size scales is beyond human experience—and maybe even beyond our imagination. If quantum physics, relativity, and modern cosmology have taught us anything, it is that *things are not always the way they seem*. The universe is under no obligation to be the way our aesthetic sensibilities might wish or expect. To assume that we earthlings can accurately judge cosmic truth by what seems beautiful to us here and now is really hubris, and using beauty this way as a criterion for truth can be a prejudice if it keeps a scientist from finding a successful theory that looks very different from expectations.[15]

Here is a personal example. In 1972 as a young scientist Joel, along with two colleagues, was the first to calculate the mass of what was then a hypothetical particle called the "charmed quark." It seemed to him at the time that hypothesizing this particle was an ugly idea, and that if it really had the relatively small mass he and his collaborators had just calculated, it would probably have been discovered already—so it probably didn't exist. After publishing the results, he dropped the subject and moved on to other topics that he felt had a better chance of panning out, while his colleagues and others continued to write papers developing the idea. Two years later, the charmed quark was discovered at particle accelerators, and it does indeed have a mass consistent with the prediction in Joel's paper. He learned that it is not wise to rely on intuition when several possibilities are open, and he was determined not to repeat the same mistake.

He had the chance to make that mistake again in the 1980s, when he was helping to create the Cold Dark Matter theory.[16] Although this has since become the standard picture of the evolution of galaxies and larger structures in the universe, in the 1980s there were huge uncertainties about the existence and quantity of the various constituents of the universe, so the number of possible variations on the theme was daunting. This time, rather than follow his intuition, Joel with his graduate student Jon Holtzman worked out in detail ninety-six alternative variants of the theory. Three years later, in 1992, NASA's Cosmic Background Explorer (COBE) satellite discovered tiny differences in the temperature of the heat radiation from the Big Bang in different directions, just as the Cold Dark Matter theory predicted—but which variant of the theory was right? Because Joel and Jon had worked out so many possibilities in advance, they and the COBE project scientists quickly realized that only two versions had any chance of matching the data: one variant that many scientists (including Joel) found attractive, and one that was widely considered too weird to be true because it requires that most of the density of the universe is not matter at all but instead a hypothetical "dark energy" that makes empty space repel itself. Once again ignoring his intuitive preferences, Joel and several colleagues worked out both variants in equal detail. In 1998 new data decided the matter: it was the weird one. In 2003 another NASA satellite and much data from other sources confirmed this new theory of a counterintuitive universe. This was a vindication for resisting the temptation to prejudge the universe on the basis of feelings about beauty or symmetry. Eventually when we finally do understand a new theory, we may learn to appreciate its beauty. Keats's deepest insight may be that he limited his simple equation of beauty and truth to "on earth."

Common Sense and Intuition Are Earthbound

It is absolutely essential that we use common sense in many situations, particularly when trying to understand and deal with other human beings. But this book is about situations where common sense and intuition are not reliable guides. What people call common sense is the ability to use good judgment and reasoning based on unexamined but widely shared assumptions that arise from collective human experience. Common sense has a key unexamined assumption: it assumes that the way things work on Earth is the way things work, period. Intuition doesn't involve reasoning at all, but is something like the experience of knowing something without having thought about it. Intuition, too, assumes we are on Earth.

Since the sun, moon, planets, and stars have been part of human experience for thousands of years, we all have a kind of commonsense understanding about how the sky works, be it right or wrong. But no one has any personal experience—and therefore no possibility of a commonsense understanding—of anything bigger than the region near Earth. We particularly delude ourselves if we assume we have some kind of intuition about how the whole universe is put together. To believe something about the universe because it's described in ancient texts, it feels right, or it's beautiful—handicaps us. Feelings are bound into intuition and common sense. When we extrapolate such feelings about how things work to the universe, what we are actually imagining is how the universe would work in miniature *if it existed on the size scale of our experience*. But miniatures never work like the real thing. A toy car doesn't run with a combustion engine, an atom is not like the solar system, and Earth doesn't work like the larger universe.

People often think of common sense as innate and unchanging, but despite its default-setting of "on-Earth," it can be educated. Eventually it can be reset at a different level—it will have to be if humanity is ever going to function consciously as part of the expanding universe. The scientists who work with cosmological ideas on a daily basis gradually retrain their intuitions, but non-experts as well can do this if they move into these ideas with an open mind. If cosmological concepts at first "make no sense," this is simply to say they are not common sense. That's the point: *beyond Earth, we need scientific guidance.* A coherent picture of the universe as a whole is now possible, but the catch is that unlike earlier cosmologies it is not intuitively simple. Cosmology is challenging us to expand beyond our common sense, but this can happen. The obviously flat earth became the obviously spherical Earth.

There is a popular idea that scientists get stuck in a paradigm and persist in their favorite (even if wrong) theories until they die. This has convinced many people that once scientists begin to think about something a certain way, they won't change. Although this has been the case for a few scientists with too much ego invested in their theories, good scientists change their minds in the face of convincing data, and the best scientists try their hardest to shoot down their own theories. Today getting stuck in a paradigm is actually more likely among non-scientists. A personal, intuitive view of the universe can often last a lifetime unacknowledged and thus safe from challenge, unaffected by any amount of formal education. Schooling rarely teaches us how to integrate new ideas into our thinking. Instead it lets us store them in the mental rooms where we first learned them, so that while intellectually we may juggle new cosmological ideas like erudite entertainment, our intuition reclines in an old, overstuffed, and unaffected image of reality. Everyone knows, for example, that the earth goes around the sun, but we still talk about sunrise and sunset. Language is powerful and it can reinforce obsolete intuitions, as those words do—or it can challenge them.

One approach to challenging intuition that we will take sporadically throughout the book is what we call "contemplations." These are brief meditations on what it might feel like to experience the scientific ideas you're reading about. Here is one example to help get a feeling of how the planet turns.

Imagine it is late afternoon and you are lying on your back in soft grass, looking up at the sky, your feet toward the south. You lazily spread out your arms and legs and feel the warm earth below you. There is nowhere you need to go. You are just a part of the earth. Slowly turn your head to the right and look west toward the orange sun. You can feel your patch of earth turning away from the sun, heading into night. Just as the horizon on your right rises to meet the sun, you turn your head all the way left, and there you see on the eastern horizon the moon appearing at the same moment. Tonight is the night of the full moon, when sun and moon are opposite and in perfect balance. Now feel your patch of earth moving slowly toward the huge, orange moon. The moon appears to rise and become whiter because you are seeing it through less and less atmosphere. It seems smaller, but only because it is no longer near the horizon where your mind compares it with familiar objects in the earthly landscape. Time passes. The earth carries you around until the last glow of sunlight disappears and stops masking the stars. It is cold now. You have traveled into night on your planet. You are the face of the planet that is traveling into night.

No one now doubts that the earth turns, but our cosmological theories can be a lot harder to swallow. A contemplation like this one is an exercise in suspending disbelief. Ironically, it can be far easier to suspend disbelief in a theater piece that we know is not true than in a theory we know may very well be true. Refusal to believe the evidence is what holds people in mindsets that don't work. Only when we've stepped out of our comfort zone can we reap the rewards of expanding intuition.

Myth Is Cosmology's Native Language

Myth is a way human beings *relate* to their universe. What exactly is a myth? In his last book, *The Inner Reaches of Outer Space*, mythologist Joseph Campbell made the passionate argument that what our society most desperately needs is a new story of reality for all of us—not just some chosen group. The story must demonstrate humanity's connection to all there is, yet be consistent with all we know scientifically. What he was longing for was a new myth, but he knew that no one can simply create a myth, any more than they can "predict tonight's dream."[17] A myth, he said, must develop from the life of a community. He hoped inspiration for such a story might come from physics. When he died in 1986, the revolution in cosmology was just beginning.

Unlike Campbell, many people use the word myth to mean "the opposite of reality"—that is, a notion held by other people who are delusional or at best quaint, but this is a very unfortunate understanding of myth that limits our ability to draw on its power and its importance to the human mind. Myths don't just *represent* a reality people already know about: myths *present* reality. They define it. In this book, when we use the word "myth," we mean what Campbell meant: the highest-order explanations, the stories that people of any culture, at any time including today, communally believe and use to explain the larger reality, including their own place in it. Other people's myths always appear to be arbitrary, but your own myths bring you and your family into the story. Religions offer many benefits in addition to a mythology, such as comforting practices, holidays, and social networks, but these bonding activities are justified by an underlying mythology. Without the story of Moses going up Mount Sinai to talk to God, for example, the Ten Commandments are just a random list.

But let's look at a myth of a people very different from most of us, the Huichol (pronounced WEE-chol) Indians of Mexico, an indigenous people who in large part maintain the shamanic worldview prevalent before the Spanish reached the New World.[18] Most indigenous people feel they have been here since the creation of the world (however their mythology may portray that creation as having happened). They are earth people, not sky people. They don't see themselves as placed on this earth by a formless divine being in the sky but as rooted in the earth, drawing their strength from it as the plants and animals do. In the creation myths of many such people, the first humans emerged from within the earth as a child emerges from its mother.

Grandfather Fire is the original light, the original wisdom, the universe's own memory. In the beginning he took the raw energy of creation and transformed it into vision by creating colors and images and into sound by singing. In this way he gave us human knowledge, and we are forever grateful. Grandfather Fire is alive in every flame and spark, and fire is to be treated as an honored being.

Is this Huichol myth merely quaint? In actual fact, the insights this myth presents are profound. The point we want to communicate here is not that this story is somehow true (although we will return to its parallels with modern cosmology), but that it viscerally connects the people who believe it to their universe. The Huichols not only call fire "Grandfather"—they experience it as one of their ancestors; in fact, they experience all the primal forces of nature (Grandmother Growth, Grandmother Ocean, Mother Earth, Father Sun) not as gods whom they have to worship but as ancestors to whom they have always been related. They have no word for "god." They honor these ancestors all the way back to the beginning (as they understand the beginning), and by doing so they keep a mental window open to their common origins.[19] This expands their consciousness of time far beyond their own lifetimes, back into the mists of myth, and their sense of self and of cosmic connection grows proportionately. The Huichols know perfectly well that Fire, Ocean, Earth, and Sun are not ordinary people, but by thinking of them with the titles of respected relatives they cultivate a sense of kinship, or organic connection, with the universe itself. Naturally, the universe to which they connect is a different universe from that of most modern people. The scientific revolution in cosmology is the discovery of *our* universe. We need to connect to *that* to discover our extraordinary place in the cosmos. The question is how. How can we use the power of mythic thinking to connect to an accurate picture of our universe? We will return to the Huichols to understand better how they do it, but only after coming to know our own universe.

Everyone, especially children, wonders whether there is a reality beyond what we see. Cave paintings and other archaeological evidence show that tens of thousands of years ago people were already seeking to understand a reality beyond the visible.[20] By endlessly creative means, including prayer, alcohol and other drugs, meditation, music, study, contemplation, sexual practices, shamans, priests, rituals, dancing, drama, art, and now science, people have sought to connect to the invisible at a level deep enough to trigger in themselves a sense of awe. A drive like this that appears across time and cultures says something undeniable about human nature. We are no different today—we all need a meaningful explanation of the invisible that includes our place in it and makes it possible for us to experience connection with it.[21] In this book we call our overall interpretation of scientific cosmology and its implications for our outlook on life "the meaningful universe."

The narrow, local kind of mythic explanation that sufficed when cultures rarely mixed will never work in the emerging global culture. We now need myths that are not only scientifically believable but allow us to participate—all of us. To experience the human meaning of modern scientific cosmology, and to turn it into a working cosmology—a meaningful universe—in which we feel like participants, our culture will gradually have to transform it into myth. However, mythmaking is no longer a purely imaginative, spiritual endeavor. Today the leeway for speculation about the nature of time, space, and matter has narrowed. Now that we have data, whole classes of possibilities have been ruled out, and science is closing in on the class of myths that could actually be true.

This book is not presenting the next myth. No matter what these chapters may say about the universe, by coming to you in the form of silent paper rather than stories told to you as an infant and communal activities and holidays raising spiritual consciousness and promoting learning and discussion, the heart of it will feel missing. This book is instead an invitation to take part in the creation of the next myth. Stephen Hawking, at the end of his landmark book *A Brief History of Time*, wondered why the universe goes to the bother of existing. "What breathes fire into the equations?" he asked.[22] We creative humans, to give one answer to his famous question, are what breathe the fire.

Both scientific and traditional cosmologies are in some sense searching for the same thing: a voice through which the universe can talk to us. The universe is endlessly speaking. People ask the universe the questions they are capable of conceiving and hear the answers they are capable of hearing. Traditional peoples each discerned a kind of voice, and they described it mythologically and symbolically. They honored it as a sacred standard and tried to harmonize their lives and societies with it. Science has now discovered whole new languages through which the universe is chatting all the time, and what we are hearing today is expanding the class of questions we ask and radically revising traditional views of reality. But as humans, we need to express this cosmic conversation mythologically and symbolically.

Without modern scientific cosmology, no people, no matter how wise, creative, and good, can create a mythic language through which the universe can speak to our global, science-based culture. We need to work together to achieve a cosmic perspective that can inspire a vision powerful enough to master the technological forces that threaten our survival. Whatever myth might emerge, if it is science-based, it won't stand still. As long as the universe expands, the myth must absorb, be tossed out by, or else be enfolded in larger understandings. No myth is for all time, but mythmaking is.

Our Big Picture

Today Western culture is ripe for an explanation of the universe that makes sense of our full existence, but most members of our culture have no faith that such a thing is possible. Education in the United States now strives for "diversity" rather than coherence, as though diversity were an end in itself. Diversity is simply a fact. The goal of education is not just to be educated, but to *do* something with it. Respect for diversity is like education: it nurtures the widest possible range of abilities and creates the conditions for cooperation, but that is not the end. The real goal of a diverse community is to do something big together. *For example, save the earth.*

Many people assume that saving the earth requires focusing on the earth. Why waste time worrying about the universe when we have poverty and injustice and environmental problems at home? But there is no dividing line between the globe and the universe. From a Darwinian point of view, it may seem inexplicable that humans should be able to decode the origin and nature of the universe, since this kind of knowledge seems to have no practical consequences and thus no survival value—and survival is the bottom line for evolution. But survival value may lie in the wider perspective that cosmology gives us on reality. "Encompassing, not overthrowing," may be the way we need to proceed. When a larger theory encompasses a narrower one, the paradoxes of the narrow theory disappear. From a cosmic perspective, it is possible that some of the paradoxes of our world may also disappear, the problems of our time may make a new kind of sense, and solutions may at last become visible.

To get these practical results, we need to move into the expanding universe as citizens—as real participants, not visitors or voyeurs. Refusing to accept the best theory we have of the universe because it might turn out wrong in the future is like never letting yourself fall in love because you want to avoid possible disappointment. There is of course danger in risking your heart, but there are greater benefits. To embrace the possibilities of this new universe is truly a leap of faith—but not the kind of faith that asks you to believe something absurd despite the evidence. This is a leap of faith in yourself that you *can* believe the evidence. In the long term it may be possible to turn what is now a mostly scientific cosmology into a personal cosmology of the gut-level, water-to-a-fish kind that finally makes sense of modern life on the global scale and binds us to our tribe—humanity—in such a way that we overcome the catastrophes waiting in the wings for their cues. This book's attempt at seeking meaning through history, symbols, imagery, metaphors, and contemplation, as well as straight scientific explanation, is not entirely cosmology, but perhaps it is the point of cosmology.

TWO

From the Flat Earth to the Heavenly Spheres

EVERY COSMOLOGY IS A LENS through which people see a unique image of the universe. Literally, a lens is a transparent material shaped in such a way that it bends incoming light to form the kind of image that is sought. Metaphorically, each cosmological lens is partially blocked by filters, which correspond to limitations on experience. Thus in Egyptian cosmology the desert played a major role, but the snow and ice central to the Norse creation stories were never mentioned. By expanding our experience and understanding of the natural world and our imaginative use of a wider selection of metaphors, we can remove filters and immeasurably improve our view, but we can never simply see the universe "as it is." Lenses are how humans biologically see—we even have them in our eyes—and the filters represent culture. The lens of one's cosmology is thus a limitation only to the extent that having a unique viewpoint is a limitation. The filters are limitations only in the sense that speaking a particular language—and not others—is a limitation. A viewpoint and a language are liberating in the same way that choosing among life options lets you move ahead, although every choice also eliminates alternative possibilities. Today science is removing filters and widening our world, and with new observations and theories it is helping us to expand our imagination.

Traditional cosmologies worked by "centering" their members around a common core of shared beliefs, expectations, view of the past—and assumptions about the natural world. To a great extent what defined the group, whether a tribe or a civilization, was such a centering core. This chapter shows what a "centering cosmology" can be by presenting examples of past ones. In the telling of these cosmological stories, we hope to give some sense of what it felt like to live within a centering cosmology. Since the modern West doesn't have such a cosmology and hasn't for centuries, it is hard for people today even to imagine what the words "centering cosmology" might mean. However, it is essential to imagine precisely that, because then we begin to understand what is missing in our lives. These ancient cosmologies hint at what a modern cosmology would need *beyond science* to become indispensable and transformative. The expanding universe cannot be fully appreciated through the intellect alone but also needs, like every earlier cosmology, to be taken as real. A new science-based cosmology could be capable of centering a global community around a common consensus on what reality is and how it all came about. Science speaks for itself as to whether it is convincing or not. But as to whether the universe it describes feels real, the examples that follow help us see that what feels real is subjective and depends on your cosmology.

The cosmologies we have chosen as examples contributed essential ideas to the Western picture of reality.[1] Each cosmology historically piggybacked on previous ones, but there have also been two complete upheavals: from the flat earth to the spherical cosmos, and from that to the Newtonian picture. Another cosmological revolution is underway now. We focus here not only on the cosmological pictures that preceded ours but also on the cosmological upheavals themselves—the great mental and social transformations as one picture was replaced by another. Now that we have a new scientific picture, our society is primed to re-envision once again our relationship with the larger reality.

We begin with the flat earth, focusing on how this picture of reality was interpreted in ancient Egypt and then in Israel in the period of the Hebrew Bible. Egypt and Israel were very different from each other, indicating that the physical picture of the universe alone by no means determines what the culture as a whole will be like. Every cosmology is based on a physical picture, but the rest of the story arises from the character of the people and how they live and think. Modern science is presenting a new physical picture, but what it will mean is not determined by the science. This book proposes a possible interpretation—an optimistic and generous one. There is plenty of room for creative interpretations, but no room for any interpretation that claims to be *inherent* in the science.

The geocentric picture arose in ancient Greece and prevailed in all of Europe and the Mediterranean through the Middle Ages until it was overthrown by early modern science in the seventeenth century. Neither the flat earth nor the geocentric cosmos is remotely suggestive of the modern picture, yet both will be valuable for helping us to approach the new picture, because they put humanity at the center of the story. Modern scientific cosmology doesn't even discuss us, and it is a simple fact that if science has nothing to say *about* human beings, it will have little to say *to* most human beings. This book is committed to figuring out how we humans might fit into the story.

This chapter whips through three millennia of cultural developments.[2] To stay focused on the narrow purposes of this journey, we have some guiding questions of each cosmology: What was the story of the origin of everything, and what was the mental picture of the universe? How did people relate to the sky? How did it feel to live within this cosmology? What did their cosmology help them accomplish? Our goal here is more inspirational than educational. You do not need to remember the historical details of the cosmic stories, but instead let them blend into a vivid picture of each total, believed-in cosmology. Getting the overall feeling of these early cosmological pictures is what matters, because experiencing the contrast between that feeling and today can make clear what has been missing from modern life, and can also suggest ideas and imagery to help us think more visually about the fathomless reality that scientific cosmology is opening to us.

The Cosmology of Ancient Egypt

Imagine that you are an ancient Egyptian, sitting on a hilltop in total darkness on a warm, moonless night. There are no fires tonight, not even in the valley you know is below but cannot see. You look upward, and the sky is sparkling with stars. A great swath of pale white light crosses the sky: the Winding Way. That is the real Nile. The river below you is merely its reflection.

You lie back and relax into the beauty of the stars. You can feel the presence of their mother, the great goddess Nut, arching over you, and her brother the earth god, Geb, whose warm, sweet body is the ground you lie on. Nut is as huge as the sky. You can feel her protecting you from the forces of chaos beyond. If you look deeply into her, between the stars, you can almost see the world of spirit, the Duat, ruled by Nut's son, Osiris. In the perpetual darkness within the goddess Nut, the sun is traveling on his nightly journey, stripped of his brilliant outward form, being renewed and prepared to be reborn tomorrow. You too were born from the Duat, and at death you hope to return.

The story of creation surrounds you—how Nut and Geb, heaven and earth, were born in loving embrace and how their father Shu, the air, aided by the gods of wind, split apart his children to create the space between heaven and earth. Even in the deepest darkness, you know that Shu is standing firmly upon Geb, supporting Nut with his life force to keep her from falling back into the embrace of Geb. This arrangement was implicit in Ma'at, the Order/Harmony/Truth of the universe, which is older than heaven and earth, and even older than their father Shu. Harmony is the Order of the Universe; this is the Truth, and the Truth is eternal. You wonder which of the countless stars above are actually past

Pharaohs who have risen after death, to serve with the god Osiris in protecting the Order of the universe. Could all the stars be long-dead Pharaohs? Can Egypt possibly be that old? All you really know is that as long as the Pharaoh sits upon the throne and upholds the Truth, the Order of the universe will be maintained, and the world is safe. "Oh, my Mother Nut," you pray, using well-known words carved into the stone of the pyramids, "spread yourself above me so that I can be placed among the unchanging stars and never die."[3]

Figure 1. Shu (space), supported by Geb (earth), upholding Nut (the heavens).[4]

The ancient Egyptian cosmology derived from careful observation of the natural world, but its goal was not merely to explain facts but to nurture in human beings *the experience of being* the cosmos, of belonging to the structure and drama of creation. The inhabited part of ancient Egypt was a narrow strip of land on either side of the Nile, which was the only source of fertility and life in the midst of desert. The Nile comes out of the south and runs almost directly north to the Mediterranean Sea, dividing Egypt in halves, while the sun rises each morning in the east and descends each evening in the west, moving perpendicular to the Nile in many places. The world as the Egyptians saw it reflected the shape of a human being, arms outstretched, facing the source of the Nile, and this image was embedded in their language. The word for "west" was "right," while the word for "east" was "left"; a word for "south" meant "front," and the usual word for "north" is related to "rear" or "back of the head." The word for "land of Egypt" also meant "earth."[5]

Egyptians saw the sun reborn each morning in the east, and so the eastern shore of the river was the land of rebirth, while the west represented the gate to the underworld and became the location for tombs and pyramids. Every year at about the same time, the Nile flooded, and for a period towns became islands and the fields were obliterated. As the floodwaters receded, the fields were left soaked and fertilized with nutrient-laden silt, and the cycle of life could begin anew. This benign, predictable, ordered geography was seen as part of cosmic Order. In the seasonal cycle of floods, a submerged Egypt returned to the formlessness of the beginning, and the reemergence of land reenacted the creation of the world. By experiencing the invisible forces of the natural world as gods with characters and histories, the Egyptians saw through the ordinary world to a meaning behind or within it.

The underlying physical assumptions the Egyptians made about the universe as a whole are revealed through their stories of creation. Unlike the creation stories of monotheistic religions, the first thing a polytheistic religion usually explains is the origin of the gods. The oldest gods are the most remote and abstract, while the youngest gods in the family tree are usually the most humanlike and important in governing everyday life. Here we are interested in the earliest, remotest gods. Egyptian divine genealogy was a theory of the origin of the world; gods represented the primal forces of nature. The world the theory was explaining was the flat earth. Far from being quaint, the process described in the following myth is not unlike our best scientific theory of the origin of the universe, as we will see.

In the beginning, Egyptians believed, there was Primeval Water—not a sea with a surface but a black and formless Watery Abyss in all directions. The Abyss was called "Nun." It had no characteristics and was

eternal and did nothing. Sometimes it was artistically symbolized as an old man with jagged waves drawn all over his body, but sometimes as a great serpent that contained—but also entrapped—the life energies of all future existence within its coils. From the Abyss the Creative Principle itself, called Atum, pushed out the sun.[6]

The creation of light was thus the first phase in the transition from the Primeval Watery Abyss to the existence of heaven and earth—the crucial turning point between eternal nothingness and the beginning of everything. The god Atum, who was both male and female, now gave birth to "Order/Harmony/Truth," personified as the goddess Ma'at. Inspired by his daughter Ma'at (literally—he breathed her in), Atum then gave birth to Shu, who represented the space held open in the midst of the Watery Abyss for the world to come.

The Egyptian creation stories seamlessly connected the origin of the gods and the creation of the universe to the founding of Egypt and the institution of the Pharaoh. From Shu descended the main gods down to Horus, who was believed to be incarnated in each Pharaoh. The Pharaoh was thus not just a king. He brought the power of the gods to earth. When the Pharaoh died, his job was not finished: he gave up his identity of Horus and could choose to rise into the heavens to join Osiris and become a living star, continuing to administer the cosmic Order on which the calendar and the seasons depended. The newly crowned Pharaoh would then become Horus. There was no clear line between gods and humans.

There was no clear separation in general between the spiritual and the physical. The waters of the Nile flowed from the spiritual realm onto farms. People came from the spiritual realm and returned there after death. There was no clear line between the time of the gods and the time of humans, either. Myths were literally about the doings of the gods at the beginning of time, but they were always symbolic of what was going on in the present. The visceral feeling of these cosmic connections was maintained through public ritual: every public act of the Pharaoh harkened back through language and symbolism to the "First Time"—the mythic golden age when the god Osiris had ruled on earth. The landscape of Egypt was dangerous and beautiful, pitiless desert and fertility, drought and flood, famine and plenty—personified as the treacherous god Seth and the good god Horus in endless battle, but also as opposites in balance.[7] Maintaining this cosmic balance was the job of the Pharaoh. The Pharaoh's ritual acts repeatedly evoked the consciousness of mythic time, creating the potential for people to experience living both now and then, to renew their awareness of themselves as *participants* in the sweep of cosmic history as they understood it. Ritual was (and can still be) a kind of time travel. It was as if the present act of the Pharaoh or priest were an *echo*—not just a remembering or reenactment—of the past. An echo is a physical, factual connection, fading perhaps, but unquestionably set in motion by an actual event in the past. For thousands of years the Egyptians never just lived in the present and took the world for granted; they were always ritually protecting their world from the forces of Chaos that according to their cosmology were held at bay but never conquered by the Order/Harmony/Truth of the goddess Ma'at.

This experience of connection across mythic time is not just a primitive stage of religion; it serves a crucial purpose in expanding human consciousness beyond the deadly narrowness of the everyday. Today's commonsense idea of time makes it difficult for many people to appreciate in what way the ancient Egyptians were right. The sense of echo between past and present is an important concept because in modern astronomy, as we will explain, the past in certain well-defined ways continues to exist: every atom in our bodies is a living relic; all space is still filled with the heat radiation from the Big Bang, which carries detailed information about the period right after the Big Bang; and astronomers literally look into the past whenever they observe distant galaxies. Maintaining a felt connection with our own distant cosmological origins is a possibility that we may begin to appreciate from ancient Egypt.

Egypt is often seen as morbidly obsessed with death, mummification, and the afterlife. But the Egyptians were not obsessed with the afterlife *just to save themselves*—they were obsessed with the longevity of their *civilization*, which for them was literally identical with the longevity of the cosmos itself. The creation stories were about Egypt, not a large earth on which Egypt was a mere country.[8] People outside Egypt were not considered real people because they were not descended from the (Egyptian) gods.[9] Whatever one may think of Egypt's practices, Egyptians succeeded in maintaining their civilization for three thousand years.

We don't want to return to the magic, superstition, and social horrors of the ancient world but to mine the good as best we can. What can we mine from ancient Egypt? An attitude that could be of tremendous value today. To understand it, let's go back to the creation story. In addition to the one that started with the creation of light, there was a second seemingly contradictory version. In this version, out of the Watery Abyss a mound of land arose. This Primeval Mound was the "First Place," and upon it Order and Mind came into existence.[10] Here the god Atum was the Primeval Mound—in other words, the first thing to come into existence was earth. Land rising above waters was a form of creation that people saw every time the Nile Valley reemerged after the annual floods, so of course this makes sense. But what matters for our purposes is that *there was no single canonical story*, even though Egypt was a unified country, ruled by a single king from about 3100 B.C.E. onward. Both stories we have told belonged to the city of Heliopolis, but other great cities had different stories. Over the millennia countless stories and beliefs arose and were reinterpreted and mixed with others. Egyptians were not a "people of the book" but a people of books, actively encouraging speculation and recombination of ideas. The very concept "Ancient Egypt" is a collage of myths and ideas from many different times, and there is often no definitive way to tell how far back in time we are looking when we consider any given one.[11] As one Egyptologist put it, "The way of the Egyptian was to accept innovations and to incorporate them into his thought, without discarding the old . . . like some surrealist picture of youth and age on a single face."[12] Looking back at ancient Egypt is a little like looking into the night sky: the eye sees the stars all at once, but in fact since light travels at a constant speed and the stars are all different distances away, we actually see each star as it was at a different time in the past (when it emitted the light reaching us now). The night sky too is a surrealist picture of youth and age on a single face.

The Egyptian mixing and melding of imagery and myths had the tremendous benefit of preventing the dogmatic imposition of myths. It was completely unnecessary that myths be logically consistent. The Nile, the land, the sun, and the stars were the fundamental observational realities, and all myths were *consistent with these observations*, but that left plenty of room for interpretation.

What we need to take from this is the attitude that *multiple non-dogmatic interpretations* of the cosmos are more inspirational than a single arbitrary story—but not as good as the true story, to the extent that it can be found. It is essential for creativity and the cultivation of the human spirit that dogma—that is, beliefs imposed by tradition or authority—be seen as a matter of personal choice, but that facts be accepted. This requires that there be a clear method for distinguishing dogma from facts. The Egyptians knew perfectly well that there was no way *for them* to say that one myth was truer than another. Without knowing it, they obeyed the maxim proposed thousands of years later by one of the founders of quantum mechanics, Niels Bohr: never speak more clearly than you can think.

Egyptian cosmology provided the purpose of life: to take care of the creation (entrusted to humans by the gods) by preserving and upholding the Order of the Universe. Ma'at was the Order/Harmony/Truth of both the cosmos and also moral behavior, and behaving morally helped uphold the cosmos. The Egyptians believed that maintenance of the entire cosmos depended on them, and thus *they mattered.*

Egyptian cosmology went unchallenged for some 1,700 years, but when it was, the challenge lasted a mere seventeen years, the reign of a single Pharaoh named Akhenaten. What Akhenaten did revealed a fundamental truth of centering cosmologies: nobody can create one intentionally. Akhenaten imposed by fiat a sort of monotheism in which the sun was the only god.[13] He suppressed the myths of the other gods, thus erasing the only explanations people had for natural phenomena. For example, every year people experienced the parched land as the terrifying god Seth winning. Just before the annual flood, they saw with tremendous relief Osiris, the god of fertility, reemerge from below the southern horizon in the form of the constellation that we call Orion. Shortly after Orion's rising, Isis would return as the brightest star in the sky, Sirius, bringing the flood, which was caused by her tears for the murder of her husband by Seth. Without Isis and Osiris, why should the Nile flood?

Chaos in the spiritual world let loose chaos in everyday life. Egyptians believed that good, moral people would be reborn after death. There was tremendous inequality of property, but morality, not wealth, was the reason for rebirth. This moral code was a major factor in the acceptability and therefore longevity of Egyptian civilization.[14] But under Akhenaten, rebirth itself depended on doing good for the king. *Under the cloak of cosmology*, the king grabbed power not only over life and death but even rebirth. Before long his queen and then he disappeared under suspicious circumstances. The old religion was restored, and he was referred to as "the criminal" and sometimes left off the lists of the Pharaohs. Myths reflect the consciousness of the culture, and when they are suppressed, access to that collective consciousness is shut off. The rebellion against the artificial cosmology was "the revenge of myth."[15]

Not only Osiris (Orion) and Isis (Sirius) but other gods were also identified with stars or constellations, and myths about the behavior of those gods often reflected the motions of those stars across the sky from season to season. The Great Pyramid of Giza, the only wonder of the ancient world still standing, was designed as a connection to the stars.[16] It contains more than four million tons of rock so precisely oriented north to south and east to west that Egyptian engineers had to have done it by aligning it with the stars. Among its inner chambers, hidden doors, and passageways are two narrow shafts that emerge from opposite sides of the room called the King's Chamber and go deep into the walls, then bend so that they cannot be seen through. For many years scholars assumed they were simply air shafts, but in 1964 it was discovered that they had nothing to do with ventilation—they had a spiritual purpose connected with the stars: in 2500 B.C.E., when the pyramid was built, the north shaft pointed toward what was then the North Star, not Polaris as today but Thuban.[17] Thuban had a special meaning for Egyptians: it was the leader of what they called the "Undying Stars." These are the stars that never set below the horizon and are therefore visible all year from Egypt. These circumpolar stars were the faithful, celestial army perpetually guarding the harmony of the cosmos. The dead Pharaoh was expected to fly from his tomb in the pyramid to these stars to take command as he had on earth. Meanwhile the southern shaft pointed toward the belt stars of Orion. Since the Pharaoh at death was expected to join the god Osiris in Orion, it was now clear that the purpose of the two shafts was to launch the Pharaoh's soul by targeting his special stars.[18]

Egypt's commitment to make sense of the starry cosmos led to a flowering of imagination and creativity and a level of culture the human species had never before seen. Egypt invested huge resources and energy into symbolic art and ritual, and this riot of imagery and ideas created a *mental homeland* for its people. The Great Pyramid became a symbol of Egypt to the ancient Egyptians themselves. A thousand years after it was built to echo the mythic Golden Age, its own time was seen as a Golden Age. The pyramid in its very shape was echoing the

Primeval Mound rising out of watery chaos, and promising the dead Pharaoh buried within that he would rise again into new life. For the ancient Egyptian, the cosmic symbolism embodied in the pyramid was not "just" symbolism but active participation in the ideas symbolized.

This discussion may appear to idealize ancient Egypt. However, we are not trying here to make a sober judgment of that complex civilization but to find some of its high points of consciousness and borrow the ones that resonate with modern cosmology. Great imagery is an essential ingredient of cosmology. Cosmology is always based on symbolism—there has never been and can never be any way of describing the whole universe in straightforward language. The rich symbolism of Egyptian belief helped lead to a mental integration of life, spirit, and cosmos that is alien to us today—yet such integration was once widespread, and it could be again. Of course, the elevated state of mind we have described could never have been universal. But something awoke in people five thousand years ago and burst out through complex metaphors so artistically and spiritually inspiring that they transformed a group of small, neolithic settlements along a river into the high civilization of ancient Egypt.

The Egyptian choice of tying religion to the stars had the ultimate effect of stabilizing the culture. The effort to orient the pyramids and other religious buildings according to the stars helped sink foundations for the ages into the most stable material anyone could possibly know at the time: the cosmos. On a practical level, the mobilization of resources and people necessary to build the Great Pyramid created a concentration of economic and political power that helped stabilize the institutions of Egypt and strengthen it against domestic disorder and potential invaders. The Great Pyramid of Giza was a solid connection that the Egyptians built between the earth and the heavens. When the Great Pyramid launched the Pharaoh's soul, it in some sense also launched the civilization.

Stability for its own sake is, of course, not enough since it can be repressive, but a civilization without stability can do nothing of lasting importance—it can build neither pyramids nor cathedrals, nor send machines or humans into space on an extended voyage. It cannot invest in the future because it doesn't believe in the future. One of the gifts of modern cosmology is a new way to think about the future—a way to grasp the immensity of time and space, and to carry that consciousness into our understanding of our own lives and futures.

The Cosmology of the Bible

Imagine that you are a shepherd in ancient Israel. It's a warm clear night, completely quiet, and as usual you're alone on a mountainside with sheep. They are dozing and occasionally nuzzling each other. You lie on your back under the stars, so many stars you could never count them, and you pick up the ongoing conversation with your constant companion. "God, I'm wondering, where did the first light come from three days before You made the blessed sun, moon, and stars? And if these stars, may they shine forever, are exactly where you put them on the fourth day, how did you open the windows of heaven in Noah's time and let down the cosmic water? Not that I'm questioning your wisdom, God forbid, but really, how? Not that I expect an answer. I'm not Moses. But it seems to me, if I may say so, that they look like holes with shiny covers that only You know how to open. Am I right, God? Why choose us to tell the whole world about You and then make us guess at everything?"

Contemplating the conundrums of monotheism was a major pastime of the Hebrews in the period when intellectual and political power was shifting from the great civilizations of Egypt and Mesopotamia to the Greek world. Unlike their predecessors and all their neighbors, the Hebrews started their creation story with God already on the scene. The Book of Genesis skipped not only God's genealogy but also all the divine battles and victories that for other traditions established the primacy of the creator god. This left a lot of unanswered questions. But unlike the Egyptians hundreds of years earlier under Pharaoh Akhenaten's enforced monotheism, the Hebrews, while accepting the bottom line that God was One, filled in the gaps by borrowing what they could from the many cultures with which they had contact, and speculated about the rest, establishing a habit of debate that continues among Jews to this day.

In the beginning, according to the story from Genesis that this shepherd puzzled over, there was darkness, the earth was formless and void, and God hovered over the Primeval Waters, which were called "the deep." Suddenly God created light, and this was called the first day. Then on the second day He performed an act that is a dead giveaway that this was a flat-earth picture: He made a space in the middle of the Primeval Waters by *dividing them in two*, pushing half the water up and half down. The upper water He held in place with a "firm thing" (*raqi'a* in Hebrew), translated millennia later in the King James Bible as "firmament." The firmament was probably understood originally as a kind of invisible dome, perhaps with a curvature suggested by the rainbow. This dome covered the entire flat earth and was strong enough to hold up inconceivable amounts of water. On the third day God laid dry ground over the lower water and put in plants. On the fourth day He created the sun, moon, and stars for illumination and for purposes of keeping time and the calendar. On the fifth day He made the creatures of the sea and air. On the sixth day He made all the land animals; then He created man and woman together equally in His own image, and handed them dominion over the plants and animals. He regretted nothing; everything was good. On the seventh day He rested from creating and in doing so created the Sabbath.

What this story explained was the creation of the same flat earth as in Egypt, with a bubble of air above it, topped with an arching sky, all surrounded by water. There are many passages in the Hebrew Bible that reflect its flat-earth cosmology—for example, "You stretch out the heavens like a tent and build your palace on the waters above."[19] The firmament has much the same job as the goddess Nut as she arches her back over the flat earth: the job of holding the celestial bodies and protecting the world from the watery chaos beyond. But heaven, earth, and air—which were the gods Nut, Geb, and Shu in Egypt—are in Genesis inanimate. Monotheism required that no part of the natural world be personified.

Whether or not the idea of monotheism originally came from Abraham and Mesopotamia, or from Moses and Egypt, or later, is unknown. But for the Hebrews it became a new ethical stance based on the radical conception that the same God that created and sustained their world created and sustained *everyone's* world, whether others knew it or not. The language and imagery the Hebrews used to describe God, however, was not always the transcendent, abstract imagery that would seem logically consistent with such a total conception; instead they often used language that portrayed God as emotional, full of desires, occasionally regretful, and greatly preoccupied with humans. This was the language of mythology, because, as we have said, mythology is a way of *participating* in something larger than everyday life, and when the Hebrews emphasized God's relationship to human beings, describing Him in human terms, they were making it possible for themselves to feel they were participating in the same reality as He was.

At the end of the story of creation in six days, Genesis immediately tells a second, very different creation story: Adam and Eve in the Garden of Eden. Genesis thus opens with two creation stories in which, like the Egyptian stories of Atum (which were already two thousand years old when the Bible was compiled), light is created first in one version and earth is created first in the second.

But the borrowing had just begun. The *sequence* in which everything is created in the six-days story is taken from the Babylonian creation story.[20] In both the Bible and the Babylonian creation myth there is watery chaos in the beginning, then the creation of the firmament, dry land, celestial objects, and humanity—in precisely that order—before the creator rests. Many later commentators have proposed religious reasons why God created plants before the sun, moon, and stars, when everyone knows plants need sunshine. But this was probably not a religious or logical choice. The likeliest explanation is that this was the order of the planetary gods in the Babylonian week, and it was simply adopted by the Hebrews along with the Babylonian calendar. The pastoral god came third in the Babylonian week, while the god of astronomy came fourth. By adopting this order the Hebrews were emphasizing that their God claimed the planetary powers for Himself.[21] Many leading Hebrews were influenced by Babylon because they had been held captive there for two generations. Individuals who had grown up in Babylon were among those who selected and edited the stories that became part of the Bible.[22]

Some Hebrews felt an emotional need to see God as the pagans did—not as having simply created the world with a few words, alone and unopposed, but as having instead fought hard and earned His victory over the forces of Chaos. Several Biblical passages suggest that the Hebrew God did not just order inanimate water to divide on day two but in fact had won a great battle with the Deep and her monsters—as the Babylonian creator god had done. For example, "Thou didst divide the sea by thy might; thou didst break the heads of the dragons on the waters. Thou didst crush the heads of Leviathan . . ."[23] This battle imagery was the language of myth, continuing to bubble up despite valiant efforts to suppress such ideas by the priests of monotheism.

The Babylonian penchant for astronomy, however, was not adopted by the Hebrews. By the time the Hebrews lived among them, the Babylonians had accumulated hundreds of years of careful astronomical observations.[24] However, in Babylon astronomy was not a source of comforting symbolic connection to the gods, as it was in Egypt. Babylonians didn't consider the stars to be gods, but they relied on the appearance of the heavens to predict the future, or at least to reveal portents of good or evil.[25] All celestial behavior was assumed to be portentous even if the astrologer was unclear what it portended. Babylonian astrologers worked almost exclusively for kings. The power to affect the decision of a king is seductive, and ominous portents were sometimes censored or else invented to achieve a political result. A prudent king would use several astrologers in order to get second and third opinions, but even then he had no way to figure out which one, if any, was right. Despite these major shortcomings, palace astrological predictions were treated as state secrets.

The Hebrews, however, despised worship of heavenly bodies and astrology.[26] Their Book of Daniel is set during the Babylonian captivity and mocks astrologers, showing them repeatedly failing to see in the stars what Daniel's faith in God reveals to him through revelation. It portrays the Babylonian king Nebuchadnezzar as so infuriated with his astrologers for their incompetence and dishonesty that he orders all these "wise men" killed.[27] The Babylonian wise men may have been incompetent fortune-tellers, but their astronomical calculations were based on sophisticated mathematics. Babylonians invented the place system for writing large numbers, and later the concept of the zero and simple algebra. They knew the Pythagorean theorem a thousand years before Pythagoras,[28] and they were the first to describe celestial phenomena mathematically, accurately predicting eclipses. Babylonian astronomy was the most advanced in the ancient world, but by keeping it arithmetical, without asking *why* the stars and planets moved as they did, they divorced the celestial bodies from reality. They never tried to *visualize* what was going on behind the numbers, and thus they never developed any geometric picture of the heavens, as the Greeks later did. Had they tried, they would have run into extremely unsettling inconsistencies with their religious worldview, since the flat-earth assumption simply could not coexist with that kind of thinking. Furthermore, by using astronomy mainly for secret prophecies for the powerful, they limited its cultural effect and limited their own imaginations.

Perhaps the Babylonian entanglement of astronomy with astrology turned the Hebrews away from the whole subject, but the Hebrews probably could not have focused on the stars and planets without risking some kind of worship of them, which would have weakened their burgeoning monotheism. In the end, the Hebrews limited their understanding of stars to timekeeping and evidence of God's glory. For them *whatever* existed, God had created it. They simply made assumptions about the physical facts of the natural world, which they saw no reason to question, in order to focus on what they saw as the big question—the implications for their own lives of a single, all-powerful, yet unvisualizable God, in a world where everyone around them was and always had been polytheistic. They gave to the question of One God tremendous original thought. We modern people, however, have continued in their footsteps to give much thought to the question of God without reconsidering their fundamental—wrong—assumptions about the "physical" universe, the creation of which partly *defines* this God, who is after all called the "Creator."

There was never any specifically monotheistic revelation as to the nature of the physical world. The early Christians were not very interested in it. There is almost nothing about creation in the Gospels. When twenty-first-century biblical literalists fight science on the grounds that the creation stories in the Bible are the last word on the subject, they are not defending ideas integral to the moral truths of their religion, but instead ideas picked up from the pagan, polytheistic cultures of ancient Mesopotamia and Egypt and passed on down for thousands of years, unexamined.[29]

Poets, artists, prophets, and other thinkers through history have shaped cosmologies with words and images, borrowing, often unconsciously, from earlier times and neighboring peoples. Countless people contributed to biblical cosmology, not only those who cut and pasted, but also all the earlier polytheistic people whose sophisticated cultures created the stories that fed into it. The complications of their backgrounds, the many ideas to which they were in turn exposed, their personal agendas, dreams, and fears, and their widely varying artistic abilities all combined to create stories with enough complexity and synergy to keep people debating their possible meanings for thousands of years. The liveliness and richness of the Hebrew Bible's often inconsistent borrowings may have helped save Hebrew monotheism from the fate of Pharaoh Akhenaten's minimalist monotheism. Acknowledging the Bible's history of multicultural fusion is important because our book follows loosely in that ancient tradition, cutting and pasting valuable icons and ideas from many cultures in order to produce a new meaning.

Every religion is a development from older beliefs and practices. Borrowing stories and poetic descriptions of gods, while changing their names, was common among the many peoples of the Middle East, and the key question was *in whose honor* the stories were told and the songs were sung. The Bible may call worship of Baal an abomination, but beautiful poetic phrases in praise of Baal, which have been found on Ugaritic tablets, can also be found almost word for word in the Bible adapted as praises of God.[30] Science also began by reinterpreting concepts and imagery from religion and mythology. This is how culture advances. We all stand "on the shoulders of giants."[31] There is no value in getting back down on the ground and rejecting the far-reaching perspective those shoulders give us in order to owe nothing to the "wrong people." History's most powerful cosmological images are not just arbitrary inventions—*they may be discoveries about human nature*. Images and ideas that expanded people's cosmic sense of themselves in earlier cosmologies may therefore still be useful, as long as they can be reinterpreted to carry a scientifically accurate message. The next centering cosmology must incorporate the best science of our time, welcoming future corrections, but it will inevitably involve the same patient and loving work of cultural cutting and pasting that brought to life the great cosmologies of the past.

Ancient Greece Invents the Geocentric Universe

At the same time that the compilers of the Bible were working, a mere eight hundred miles away Greek thinkers were teaching that the earth was not flat but spherical, and that the sun, moon, and stars rotated about the earth every day. Over the next thousand years this Greek picture became the cosmology of Jews, Christians, and Moslems alike, until by the Middle Ages a spherical universe, which had once been radical, was unquestioned. Most people may not have understood the technicalities underlying the picture, but their very intuition and common sense had changed by the integration of new knowledge.

How did the Greeks make this enormous imaginative leap beyond the flat-earth picture to reimagine the universe? The Greek city-states were just being consolidated in the seventh and sixth centuries B.C.E., and several of them went through periods of tyranny, oligarchy, and democracy, with men arguing endlessly as to what was the best system of government. In the Athenian democracy men made decisions in meetings of thousands, and a great orator could sway the crowd. Rhetoric became a high art, as did the ability to quickly grasp the weak points in the other man's argument. Mythical explanations had been based on the authority of priests, storytellers, and tradition, and they legitimized and reinforced those authorities, but the kind of debate happening in the political sphere took place within a framework of law, where actions were determined by reason and persuasion. Eventually philosophers arose who questioned the value of powerful rhetoric designed to convince a crowd in a public square. Rhetoric was not the way to reach truth, the philosopher Socrates insisted, but just a way of manipulating emotions. These philosophers were products of a democratic atmosphere, and they were able to criticize its faults because they enjoyed a freedom of spiritual self-development that may never have existed before.

According to Aristotle, the philosopher Thales of Miletus invented the founding idea of science: *that the seemingly infinite complexity of the world can be explained by means of a small number of hypotheses*. Thales introduced the study of geometry to Greece after learning it on a visit to Egypt.[32] He was also the first to propose a *theory* of matter. This was his theory: there is a substance from which everything is made, and the substance is conserved; it is not created and cannot be destroyed. Thales' principle of conservation was a limitation on what even gods could do. Thales realized that it was possible to think about the world in such a way that one could make progress in understanding—without offending the gods—by pursuing the logical consequences of what everyone already believed. His work launched centuries of brilliant Greek efforts to rationalize nature and extend the concept of law to all phenomena. Like the peoples of the Middle East, Thales still assumed the flat earth rested on water; it was his successors who would develop the idea of the earth as spherical. But when Thales proposed that earthquakes are caused when the (flat) earth is somehow rocked by waves in the water below it,[33] he was declaring that a naturalistic explanation of the Primeval Water need not mention the god (Poseidon in the Greek pantheon) usually associated with the sea.

The idea of "physics" came from Thales' student Anaximander, who taught that everything in nature had an inherent character, which made it what it was. This character he called "physis," and it was incorruptible, immortal, and eternal. His followers, who constructed explanatory models for why things have the characters (and characteristic behavior) they do, acquired the name "physikoi," or physicists (although not quite in the modern sense).[34] "Incorruptible, immortal, and eternal" were qualities that no one before Anaximander had ever ascribed to nature, only to the gods. But in his understanding, although the gods existed, they were not the cause of everything. Things had their own inherent nature that even the gods could not change.

The birth of scientific thinking took place when people stopped seeing the Primeval Water as ever-present Chaos held in place by a god and began thinking of it as a natural substance that could be understood logically. Thales and Anaximander never questioned the existence of the gods. They made science possible *within* a universe of willful gods, not by denying the supernatural but by inventing that very concept, because in doing so they also invented its opposite, the natural. Natural phenomena, they could then claim, are those which are not the products of willful divine influences but are regular and governed by cause and effect.[35] But their earth was still flat.

Now came the great turning point—the break from the flat earth. Pythagoras or his followers envisioned the entire universe anew. They proposed that the earth, sun, moon, and planets all turned around a central fire, and their distances from the fire corresponded to the intervals of the notes in the musical scale.[36] This was a breakthrough, because they were theorizing a new universe to satisfy a new standard: they wanted it to be consistent with astronomical observations and based on mathematics. This was the beginning of cosmology. The word "cosmology" is Greek and meant reasoning about or explaining the cosmos, but "cosmos" did not originally mean the universe but rather "a crafted, composed, beauty-enhancing order."[37] The word was widely used in moral, military, political, domestic, and architectural contexts.[38] Either Pythagoras or Heraclitus, just a generation or two after Thales, first used "cosmos" to mean the principles behind the entire universe, and this was a new meaning—a challenge to all thinkers to try to explain the universe as a whole.

The central-fire theory was soon followed by an equally shocking and drastic idea: the philosopher Heraclitus proposed that the existence of the universe could be explained without any story of a beginning. "Cosmology," for Heraclitus, meant not an explanation of where the current organization of the universe came from but an explanation of what *powers* it all. Heraclitus focused not on the large-scale geometry, as Pythagoras did, but on the smaller-scale dynamism of the universe, and he saw implicit in everything an active principle—fire—that can burst out, and that, when not bursting out, is a kind of hidden soul of all matter.[39] That was what powered the universe. It's important to remember that this was a time when divine intervention was the preferred explanation, and people commonly believed that the gods could not only change the course of external events but reach into men's minds and change the way they thought and felt, for example punishing a man by making him act foolishly or against his own interests. To Heraclitus, this popular view of reality was a random and lonely nightmare. If people only awoke to his theory of the reality of nature—that it could be explained by principles, not miracles—their lives would be changed forever.

All these daring thoughts emerged not in the great empires of Egypt or Persia but in the little city-states of ancient Greece. By the fourth and third centuries B.C.E., the question of what the universe was like had become a topic of lively philosophical discussion and there were competing schools of thought. In Egypt, as we have said, there coexisted multiple stories of creation, and special myths associated with each city. A myth would never be criticized for being inconsistent with some other myth. But among the Greek thinkers, it was understood that different cosmological theories were in competition with each other. *Not all of them could be true.* Without any way of testing their theories, however, the Greeks had no reliable way of choosing among them.

In the fourth century B.C.E., Plato, perhaps the greatest philosopher in history, set the agenda not only for much of Western philosophy but also for astronomy for the next two thousand years. In his time there was no clear distinction between philosophers, cosmologists, and physicists, nor any profession equivalent to "scientist." Plato's two key ideas relevant to our discussion were first, that only Ideas (or Forms) are real, true, and eternal—the material world around us appears real, but all material things decay and disappear, and whatever decays is not eternal but temporary and is thus ultimately illusory. Thus for him mathematics and the ideas of philosophy were more real than the world itself, which was only matter slapped together in an imperfect copy of Ideas. Every story about the changing and decaying material world—including cosmology—could at best be a likely story, never the truth.[40] The second relevant idea was that reason is the only path to truth. Plato proposed a radical way of doing astronomy: by reasoning out the underlying mathematical regularities, rather than observing the sky. This had serious ramifications, because science is founded on the willingness to let go of wrong ideas, no matter how reasonable and beautiful they appear, if the data don't support them.

Plato posed a famous question about "the planets" to his students at his Academy, and the question ended up shaping the future of astronomy. The word "planets" in Greek meant "wanderers," since it was known that they don't just cross the sky in a simple way like the stars but move at different speeds and even seem to stop and go backward with respect to the stars for a while (retrograde motion). This cried out for an explanation. Plato asked his students: *"By the assumption of what uniform and orderly motions can the apparent motions of the planets be accounted for?"*[41] By asking *what* uniform motions, Plato imposed on his students as an absolute requirement that the motions of the planets *had* to be uniform and orderly, in short, mathematically ideal. In response, Plato's student Eudoxus—who had apparently mastered much of geometry sixty years before Euclid—came up with an ingenious geometric picture involving twenty-seven separate spheres centered on the earth, carrying the planets, sun, moon, and stars.[42] Eudoxus's system was adopted by Plato's student Aristotle, who would become the most respected authority of the ancient world and would enshrine Eudoxus's scheme under his own imprimatur. The Platonic prejudice in favor of spheres or circles persisted until the early seventeenth century, when Johannes Kepler discovered how the planets really move by daring to consider the heretical idea that they do not move on perfect circles, but rather that their orbits have some other shape, namely ellipses.[43]

Today many theoretical physicists are philosophically Platonists in the sense that we believe that the world can be understood mathematically, and that the physical laws (in the form of mathematical equations) are more fundamental than any particular phenomenon that they describe. But we part company with Plato when it comes to his doctrine of strict separation between Ideas and matter. Material objects are not merely imperfect copies of ideal Platonic forms.[44] According to modern quantum physics, only certain electron patterns are possible in atoms, and these are actually determined by geometry. The possible structures of atoms (which underlie all of chemistry), and indeed the elementary particles out of which the atoms themselves are made, are apparently *perfectly* described as representations of purely mathematical possibilities.[45] Thus at the quantum level not only does perfection exist, but there is no meaningful distinction between Form and matter. At that level, matter *is* geometry.

Greek astronomers and physicists pursued their research mainly for the pure value of knowledge. There were a few Greek astronomers who were influenced by Babylon and studied the planets in order to discern their influence on human destiny, but at least through the time of Aristotle astrology was not a significant motivation for doing astronomy. Athenian democracy was not perfect—it excluded women and slaves, engaged in an ultimately suicidal war against Sparta, and put Socrates to death—but some of the most brilliant and wealthy people at the time considered life's highest goal to be the reasoned search for truth. Again, as we suggested in the discussion of Egypt, we are not trying to paint a balanced and sober history of Greece, but to borrow some of its high points of consciousness for ourselves. Surely the reasoned search for truth was one, and another was the fact that to these Greek thinkers the *value* of knowing the universe was self-evident. The ideas that developed in this atmosphere transformed reality itself for Western civilization.

The Alexandrian Synthesis

The spread of Greek culture was partly due to its intrinsic brilliance, but also to the historical twist that Philip of Macedon, the conqueror of Greece, hired Aristotle to tutor his son. The son, who would become Alexander the Great and conquer much of the known world, was introduced by Aristotle to the highest intellectual life of the time. Then he became king, won major battles in Persia, and headed for Egypt. The meeting of a young man of Alexander's education and self-confidence with the culture of ancient Egypt was electric, and there he decided to establish a new city in his name. The city of Alexandria on the Mediterranean coast of Egypt became a cultural center and the site of the greatest library of antiquity and the first great research institute and center for advanced studies in the sciences, literature, philosophy, and the arts, including agricultural and other technologies. One of its first directors, Eratosthenes, accurately measured the circumference of the earth.[46] Archimedes, perhaps the greatest mathematician and physicist of antiquity, studied there with Euclid, the great geometer.

In Alexandria, Greeks encountered not only Egyptians but also Jews, many of whom were educated in Greek schools. It was these Hellenized Jews who translated the Bible from Hebrew into Greek, making it available for the first time to non-Jews.[47] Although Jewish writings of the time were usually religious and Greek writings secular, one thing Jews and Greeks had in common that distinguished them from the Egyptians and Mesopotamians was that their learned writings were available to anyone and not limited to the powerful monarchs and their clergy.[48] Alexandria in the first century C.E. was also a center for the leaders of the early Christian Church, who mostly spoke Greek, and the development of Christian dogma was from the start influenced by controversies over how to interpret and evaluate Greek ideas.

In this cultural ferment in the first to second century C.E. lived the culminating astronomer of the ancient world, Claudius Ptolemy. He worked out a geometric theory of planetary motion that could explain with impressive accuracy all the astronomical observations that had accumulated over many centuries. He did it within the assumptions of Plato and Aristotle that the orbits of the planets were perfect circles by proposing that the planets move on circles which in turn ride on other circles.[49] Ptolemy's view would last almost 1,500 years, although for most people only in a simplified, intuitive version.

A new picture of reality began to take shape in Alexandria in the early centuries of the first millennium C.E., as Romans,[50] Egyptians, Greeks, Jews, Christians, and others all interacted. It combined, among other things, the Hebrew story of creation with the Greek (now Ptolemaic) physical picture of the spherical cosmos and still-developing Christian concepts of God.[51] But this was the end of the ancient world. The Library of Alexandria was destroyed, Rome fell, Greek science and other learning were lost to most of Europe, and the Dark Ages intervened from about the fifth century C.E. to the eleventh. When we pick up our story in the next chapter, those centuries will have passed, and we will find ourselves in the high Middle Ages. From Persia and North Africa to Scandinavia there will be one widespread cosmology—the "medieval picture"—synthesized from the vastly disparate cultures that mixed in ancient Alexandria. Though the word "medieval" today sounds very distant and long-surpassed, the European cosmology of that time lies at the root of many of our current intuitions as well as misconceptions. Religious views of reality in the West today are mainly based on this medieval synthesis, not on the cosmology of the Hebrew Bible or the early Christians. The medieval picture will probably be the first in our history of cosmologies to seem comfortable to people today, because this was the beginning of the modern world. These people thought much like us.

THREE

From the Center of the Universe to No Place Special

Imagine that it is the year 1200 C.E., *and you are a monk in a monastery somewhere in Europe. You have just awakened in your cell. It is pitch-black and very cold. You wrap yourself tightly in your woolen habit and fling open the window. The moon has not yet set. The world outside is silent and the sky sparkles with stars. You shiver, not only with the cold but because of the awesome beauty above.*

Everything in creation has a place that God has decreed for it, and it moves toward that place by its own desire, because it loves God and wants to fulfill His will. God has given every object this tendency, to keep the universe orderly. Earth is at the center of the universe because it is the heaviest of all elements. The waters lie above it, the air above that, and fire soars upward to the heavens. How nice a fire would be now. You try to imagine the heat and the reddish glow of earthly impurities burning off. Real fire, of course, is invisible and flies always toward its proper home just beneath the sphere of the moon. But beyond the moon—there lies perfection! The crystal spheres, made of quintessence, all turning at different speeds, not a hairsbreadth of empty space between them yet absolutely frictionless—only God could have engineered such precision. You rarely have time to contemplate the stars, but right now you hold your breath and listen very carefully—perhaps you can just make out the ethereal music of the spheres as they revolve!

You look up at Jupiter, gleaming brightly and riding ever so regally upon its sphere below the stars. Everyone knows kingly Jupiter brings great fortune, but he is following on the heels of Saturn, who brings great misfortune. What can this portend? Of course, the planets are not gods. The Romans were unredeemed pagans, and their gods' names on the planets mean nothing. The planets simply ride their crystal spheres, tracing perfect circles forever around the earth. Still, their influence is powerful.

The stars are bright and clear on their crystal sphere. You know the sun right now is still on the other side of the earth, so that the monastery is in earth's shadow. You try to imagine the way it looks from outside, and you can visualize the cone of darkness extending out from earth opposite the sun. Your monastery and the whole village are in that cone of darkness. But high up beyond earth's shadow, all is bright, all is lit by the sun.

Soon it will be time to awaken the other monks for morning prayers. If you all pray at exactly the same time as all the other monks at all the other monasteries, God will surely hear you. It's a good thing, too, because if you had to send prayers to God by mule, and the mule walked forty miles a day straight up, right past Saturn, the Seventh Heaven, it wouldn't even get to the sphere of the fixed stars for eight thousand years. What a huge sphere—so big, in fact, that you can hardly imagine how it can turn all the way around the earth every day, even for the love of God. And God is even farther away! The thought of being that high makes you dizzy. Whenever you walk into the Cathedral, though, and look way, way up at the angels on the ceiling, you feel grandeur and you do understand, at least a little. No matter how high the sphere of the heavens is, God and the angels are always there, looking down at you.

The heavens are so much more beautiful than the corrupt earth. But we each have to live out our time here first, in whatever place God put us, and he has placed us here in this cold world full of suffering, far from heaven. You look longingly at the realms of the heavenly spheres. They are filled with angels and other ethereal beings, all overflowing with divine love. The moon is just setting. It is time to awaken the monks for morning prayers. This is your lot in life. Yours is an important role, and you are grateful for it. You close the window and begin your day's work.

The multicultural synthesis that had begun in Alexandria had transformed the secular, geometric Greek picture into a spiritually centering cosmology with a sustaining and profoundly religious significance and a place for everyone. The freedom of thought of ancient Greece was long gone, and although practical new technologies spread quickly in the Middle Ages,[1] the nature of the universe was no longer a topic of debate. The idea that the earth was a sphere at the center of the universe was unquestioned, eternal reality for Christians, Jews, and Muslims alike. Medieval cosmology, however, was not just an image of the earth and celestial bodies—it was a whole way of reasoning, of defining truth, and of valuing and categorizing knowledge, and this way of thinking and the image of the cosmos supported each other.

The Medieval Cosmology of the Heavenly Spheres

"Medieval" to the modern mind may evoke images of knights and chivalry, damsels, dragons, and very poor hygiene, but the real preoccupation of medieval man was organizing, defining, and tabulating. Everything had its place.[2] A fundamental assumption of medieval thinking was that all beings—the worm in the soil, the lowliest serf, and the king himself—had been placed by God exactly where they belonged in a Great Chain of Being, and that this hierarchy was divine.[3] Therefore, it would be blasphemous to question the hierarchy or the place God had chosen for anyone in it. When a king claimed to rule by divine right, he wasn't just saying that God had selected him personally to rule as opposed to somebody else; he was upholding a cosmology in which kings had no more say over whether they would be kings than dogs had over whether they would be dogs. Like the royal families, the Church justified its ecclesiastical hierarchy as the divinely ordained downward continuation of the cosmic hierarchy. Wars were fought according to rules, courtship followed rigid rules, reasoning followed rigid dialectical rules, and people played the roles into which they were born.[4]

Putting everything in its proper place was a challenge, because in the Middle Ages written authority was accepted as the source of knowledge, and since literature was available from pagan, Platonic, Stoic, Jewish, Christian, Coptic, and many other sources, medieval thinkers felt an urgent need to reconcile all the apparently contradictory truths.[5] Since they took their understanding of science from all the ancients, not only from the natural philosophers and physicists but from poets, there was even more inconsistency. Consequently, their model of the universe had to be very complicated.

"Everything in its proper place" was not only a way of organizing society and the heavens—it also explained physics. With no concept of gravity, why did people think things fell down? Because, as Chaucer explained, every "kindly thing" that exists has a "kindly place" reserved for it, and to this place it "kindly enclynes" or goes.[6] Hidden in this explanation is the medieval assumption that an object has a spirit or soul that makes it move to its proper place.

What kept the heavenly spheres turning? A thousand years earlier, Aristotle had said Love turned the spheres but that it was meaningless to ask what lay outside the largest sphere, because that was the edge of space and time. For medieval Christian thinkers, however, what lay beyond the outermost sphere was God, angels, and divine light.[7] In the early 1300s, the Italian poet Dante described that realm as "pure light, intellectual light, full of love."[8] But medieval theologians worried, how could crystal spheres love? Their answer was that the spheres were beings. The Christian theologians accepted that the spheres had souls that forever sought God's love. Therefore, the entire universe was literally immersed inside God, literally bathed in divine love, and divine love powered the universe.

Although the Church tried, it could not blot out the pagan sense of power that emanated from planets still bearing the names (as they do today) of Roman gods.[9] In a world where the ancient writers were an accepted source of truth, the gods of those writers still influenced the thinking even of people who didn't believe in those gods. Between planets that might be exercising "influences," the intelligent spheres eternally loving and making music, and angels flitting up and down among the spheres and waiting upon God, it was a religious but not a very monotheistic cosmology. As a result medieval Christian writers, with the exception of Dante, largely avoided the topic of cosmology.

But this was not so among the Jews. Members of the secret Jewish mystical movement called Kabbalah[10] focused intensely on the actual process of how the universe was created. They strove to read what they believed was the hidden story between the lines of Genesis. The reason we bring up this small group is that in their quest—the goal of which was not to explain the universe but to know God—they developed a unique story of creation strikingly similar to some of our best current theories, although it was the medieval concentric-sphere universe whose creation they were describing. We will return to that story when we explain what may have happened before the Big Bang, because the Kabbalistic terms are a surprising fit.

The medieval picture was the last great centering cosmology in Western history, and it made sense, given the limited knowledge of the time. There was comfort in belonging to the cosmic hierarchy, even though most people were close to the bottom. But it reveals the dangers of unquestioned hierarchies, and above all the danger of justifying power relations on Earth as determined by a view of the heavens. For all its spiritually comforting aspects, medieval cosmology was dogmatic and socially repressive. One should not imagine that because a cosmology is long-lived and centering and gives everyone a role, it is necessarily good. Cosmologies are thought-control systems[11] that can have a dark side of limiting both imagination and membership. Today most people in the world have either tasted or at least seen freedom, and we don't want a cosmology at all, shared, centering, or otherwise, if it means dogma and social repression. Meaninglessness is better than dogma; that is pretty much the choice the secular West has made for the past few centuries, and will no doubt continue to make until presented with a nonrepressive alternative.

The rigidity of medieval thinking eventually cracked and then shattered. The Crusades of the 1100s–1300s exposed Europeans to a high and initially open-minded Arabic culture. A wealth of works by the ancient Greeks, which had been lost to Europe during the Dark Ages, were preserved by the Arabs and widely read in Arabic translation.[12] Starting in the 1200s, Europeans began translating them into Latin and circulating them throughout Europe. Early-fifteenth-century Italian painters developed perspective, and Michelangelo and other Renaissance artists opened people's eyes to the beauty of the human body again, as it had been appreciated in ancient Greece and Rome. Gutenberg invented the printing press, and soon books were widespread. Martin Luther in 1517 challenged the authority of the Church, leading to the Protestant Reformation, and all these factors contributed to the weakening of the medieval worldview. But it was the scientific revolution of the sixteenth and seventeenth centuries that brought down the medieval cosmology. When the heavenly spheres were shown to be an illusion, the entire hierarchy of the cosmos became a fantasy, and suddenly there was no convincing cosmic justification for the hierarchy of the Church or the divine right of kings—and before long kings of England and France lost their heads.[13]

Science, however, shot down the heavenly spheres centuries before it could offer an alternative picture, and how this happened and what it has meant to us to live those centuries without a shared cosmology is the focus of the rest of this chapter. Science has often eliminated the impossible more easily than it could tell us what is likely to be true. But the simple fact that a picture of the universe gets scientifically ruled out does not erase it from anyone's mind, or there would be an unthinkable gap of the type that never arises in a normal brain. A universe picture can only be pushed out by a more satisfying universe picture.

The lens through which most educated people today see the universe is the seventeenth-century Newtonian picture that replaced the medieval one—a universe in which space is shapeless, endless, cold, and empty except for scattered stars and other celestial bodies. When we look through this lens, Earth is not at the center. There is no center. There is no special place for God. Humans are not essential and seem as random an occurrence as the arrangement of the stars. Objects do not move because of their own souls but simply because of the laws of physics. But through this basically inanimate picture of reality, the West found technological power and mental freedom no one had ever dreamed possible before.

The Copernican Revolution Leads to the Newtonian Universe

And behold, in the midst of all resides the sun. For who, in this most beautiful temple, would set this lamp in another or a better place, whence to illuminate all things at once? . . . Truly indeed does the sun, as if seated upon a royal throne, govern his family of planets as they circle about him.[14]

This hymn to the sun was written by Nicolaus Copernicus, a Polish churchman and astronomer who in 1543 published the earthshaking idea that Earth is not the center of the universe but one of the planets, and that it, like all the others, revolves around the sun. The word "helios" is Greek for sun, so this was called the heliocentric model.[15] Copernicus arrived at his insight while seeking a way of predicting the motions of the planets that was simpler than Ptolemy's. For the sake of geometrical elegance Copernicus accepted the huge—and soon to be extremely controversial—consequence that the earth must move.[16]

Copernicus's book was prefaced by a disclaimer (inserted anonymously by a Lutheran theologian working with his publisher) that the scheme was intended merely as a mathematical shortcut to make astronomical calculations easier, and no one should believe that it actually described reality.[17] These precautions were wise, since a more outspoken advocate of the Copernican picture, Giordano Bruno, was burned at the stake in 1600 for challenging Church teachings. Among many other topics, Bruno had written of an infinite universe containing the possibility of other worlds.[18]

This was a different but equally controversial challenge because—since the Church held that God surrounded the outermost universe—it would push God infinitely far away. The possibility of other worlds also offended the Church because it challenged the uniqueness of Jesus: if earth is a planet like the other planets, then there could be people on the other planets. If so, how could they have descended from Adam and Eve and know the Savior? How can earth be a sink of iniquity and the heavens be perfect if there are people up there? Catholics were led by the rising Protestant Reformation to become less tolerant, while in earlier centuries such challenging ideas had been proposed by scholarly priests.[19]

The main reason Copernicus favored the heliocentric picture was that it naturally explained why the inner planets Mercury and Venus were always near the sun in the sky, appearing only as morning or evening stars, while also explaining retrograde motion. Copernicus felt his sun-centered theory was more elegant than Ptolemy's, but he also knew that he could present no definitive evidence for it. The Copernican theory profoundly influenced Galileo, Kepler, and other astronomical experts and unsettled a few theologians and thinkers, but by Copernicus's own choice his impact was mainly on experts. He launched what is called the Copernican Revolution, but its completion took a hundred years.

The idea that the earth goes around the sun seemed absurd in the 1500s. Today it seems elementary, but people back then reasoned like this: if the world were spinning from west to east, if you dropped a rock from a tower it should be left behind and fall substantially to the west,

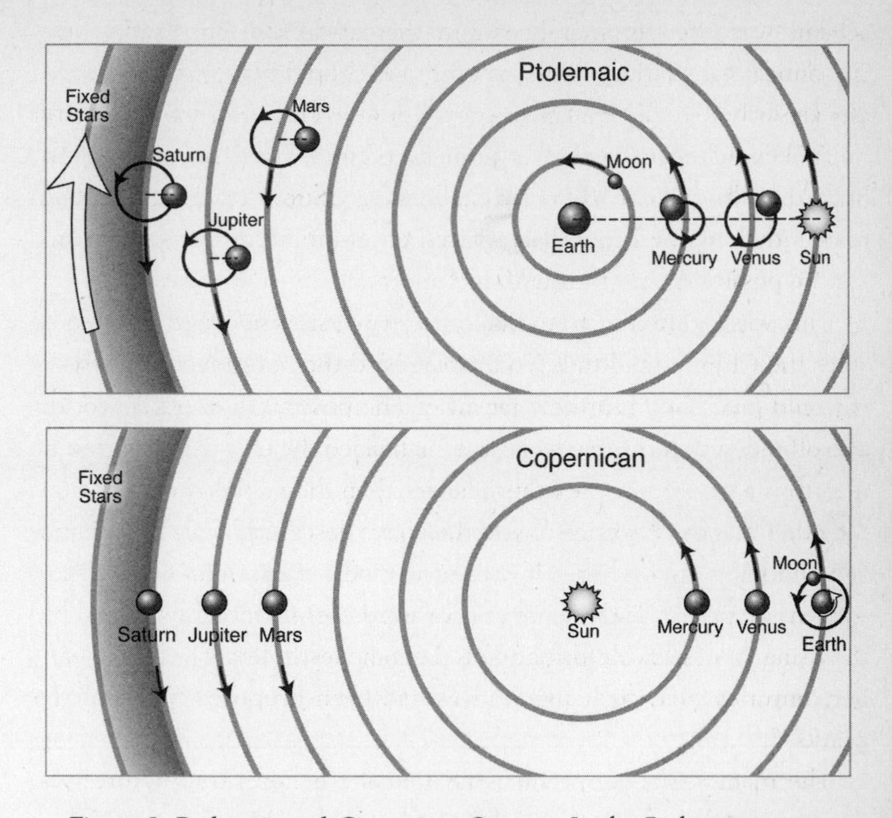

Figure 1. Ptolemaic and Copernican Systems. In the Ptolemaic system (top), the outermost sphere of the Fixed Stars rotates daily about the central earth (large unfilled arrow), with the various inner spheres also rotating daily but also moving more slowly with respect to each other. The epicycles of Mercury and Venus centered on a line connecting Earth and Sun, as shown. According to the Copernican system (bottom), the earth revolves about the sun like all the other planets, and only the earth rotates daily about its axis (small unfilled arrow).[20]

and yet it always falls at the foot of the tower. And if the earth were really swinging in a gargantuan orbit around the sun, then the fixed stars would appear in slightly different positions if you looked six months apart, from opposite sides of the earth's orbit ("parallax").[21] But no change was observed, and therefore either the earth does not in fact move, or else the stars are much farther away than the planets—which seemed utterly ridiculous, because why would God create such a useless void between the planets and the stars? The philosopher Jean Bodin wrote in 1628, "No one in his senses, or imbued with the slightest knowledge of physics will ever think that the earth, heavy and unwieldy from its own weight and mass, staggers up and down around its own center and that of the sun; for at the slightest jar of the earth, we would see cities and fortresses, towns and mountains thrown down."[22]

In 1572 a new star appeared in the sky—a star so bright it could be seen in the daytime. To the astounded medievals, new stars violated the fundamental principle of the unchangeability of the heavens. Every excuse for it was put forward, but none was convincing. It could not be some sort of meteorological phenomenon, as temporary phenomena were assumed to be, since observations by the great Danish astronomer Tycho Brahe showed that it moved with the stars and was definitely more distant than the moon. We know now that it was a supernova, and these are typically observed with the naked eye only once every few centuries. Another big problem was that the medieval crystalline spheres were cut through by the comet of 1577, which Brahe again showed was more distant than the moon. In 1604 another bright supernova was studied by Johannes Kepler. No nearby supernova has been visible in our galaxy since then; we are due for another any century now.

In 1609, Galileo Galilei in Venice heard about an optician in Holland who had invented a telescope, and from the description alone he was able to build one that could magnify about thirty times. He turned it toward the night sky and saw things that no human being had seen before. He discovered four moons revolving around Jupiter, disproving the belief that everything revolves around the earth. He saw spots on the sun, disproving the immaculate, unchanging nature of the universe beyond the moon. He measured the height of mountains on the moon by means of their changing shadows, showing that the moon was made of imperfect rock like the earth, not the supposed fifth element, quintessence. And less than seventy years after Copernicus's book, Galileo was able to observe the phases of Venus and show that they were as expected in a sun-centered system and completely incompatible with the predictions of Ptolemy's theory.[23]

The telescope was the first instrument to greatly amplify unaided human perception and it was initially greeted with skepticism, but with it Galileo definitively ruled out the Ptolemaic scheme, which had stood since ancient Alexandria. He had pulled down the medieval universe with data.[24] Galileo's book *Sidereus nuncius* (*Starry Messenger*) was published in 1610, and it rapidly made him an international celebrity. By 1611 the poet John Donne in England had understood how subversive the new cosmology was, and how the destruction of cosmic hierarchy threatened to break down human lines of authority. He wrote:[25]

And new philosophy calls all in doubt,
The element of fire is quite put out,[26]
The sun is lost, and th'earth, and no man's wit
Can well direct him where to look for it. . . .
'Tis all in pieces, all coherence gone,
All just supply, and all relation;
Prince, subject, father, son, are things forgot . . .

In 1616 the Church condemned the Copernican hypothesis, but Galileo kept challenging the Aristotelian legacy not only in astronomy but in mechanics and physics, and now he wrote not Latin monographs for scholars but engaging Italian dialogues for the public. By 1618, Kepler had discovered the laws governing the motions of the planets around the sun. He found that their orbits are ellipses, not circles as even Copernicus had assumed. The universe as Europeans had envisioned it for over a thousand years was crumbling. If earth is not the anchor and foundation of the universe and there are no crystal spheres holding the planets, people wondered, then what kept everything in its place? Why did the sun rise every day? Why didn't the planets and stars go careening off into space? How could there be any orderliness in the universe at all? But despite its unanswered questions, the "new philosophy," as Donne had called it, rolled through European culture.

What frightened the Church fathers perhaps more than the theological problems was the political threat to the Church hierarchy. Galileo ridiculed the Ptolemaic cosmology in his *Dialogue Concerning the Two Chief World Systems* (1632). It was a time of growing religious conservatism in Europe: the Pilgrims landed at Plymouth Rock in 1620 to avoid religious persecution, and in 1642 Cromwell's Puritans closed all the theaters in London. The Church in Rome responded to Galileo's *Dialogue* by placing its author under house arrest for life and forcing him to recant.[27]

The arrest and conviction of Galileo was a sobering event for scientists all over Europe. Following the lead of philosophers Francis Bacon and René Descartes,[28] scientists adopted—for their own protection—a policy of noninterference with religion: they would make no claims to authority over anything but the *material* world; they would defer to religion in all questions of *meaning, value,* and *spirit*. The Church, on the other side, needed to protect itself from endless battles over future scientific discoveries and the embarrassment or subversion of religious belief. It accepted this division of turf.

Galileo died less than a decade after his arrest. Within a few decades after that, a truce was in place and the spoils of this war over ultimate authority were clearly divided. This truce, which we will call the Cartesian Bargain, was never a written contract, but it could not have been more effectively enforced. As the Church and scientists both went on to develop rationales for their respective realms of authority, a kind of social schizophrenia entered the culture. The physical world and the world of values and meaning were for the first time in history seen as two separate realities. Many people who were impressed by the practical successes of science no longer took the authority of religion for granted. The Church began losing its power to shape people's reality when it became irrelevant to the physical world. Scientists often assume that science won in this deal, but all people pay a price for the distortion of reality. The Cartesian Bargain opened a chasm in the psyche that has led to the widespread assumption today that if there is a spiritual realm, it is separate from the physical universe.

The Cartesian Bargain was perhaps an inevitable response to a unique set of historical circumstances, but our culture has mistaken it for natural law. The "physical universe" is not at all what most people think, and as we shall see the meaningful universe can be understood to encompass both what people have considered the "physical" and the "spiritual." Today's scientific revolution is as powerful as the revolution wrought by Copernicus, Galileo, and Newton, but few people yet realize this. Historical shifts of such magnitude are rarely visible until later. The average person who was Copernicus's or even Newton's contemporary would not have fully appreciated the implications of overturning the medieval cosmology.

In the year 1642, Galileo died and Isaac Newton was born.[29] In England and Northern Europe scientists no longer feared medieval punishments, nor did they accept the authority of ancient books. Instead they required that nature be observed and objective data collected. Scientific societies were forming across Europe to discuss research, announce discoveries, and nurture their members' faith in the infinite possibilities of careful experimentation and observation.[30] The goal was to explain natural phenomena without resorting to spirits, evil influences, established authority, or God, but rather on the basis of principles or laws of nature.

One should not jump to the conclusion, however, that in Newton's time people in general now thought scientifically. The seventeenth-century goal of a complete explanation of the cosmos was a scientific one, but there was little reliable data, much unreliable reporting assumed to be data, and little experience in the design and conduct of rigorous experiments. In the popular mind the large gaps in what science understood were sometimes filled in with fashionable ideas from antiquity such as astrology and alchemy.

In this time of intellectual upheaval, Isaac Newton solved the tremendous mystery of how orderliness and regularity were preserved in the universe. Copernicus had suggested a notion of gravity a century before, but to Copernicus the "gravity" of the sun, moon, and earth was actually a "kindly enclyning" tendency of matter to come together and form Platonic spheres. Newton, however, was searching for a mathematical explanation of why the planets should move in regular orbits.

Supposedly Newton saw an apple fall from a tree, and at that moment he realized that the force that pulled the apple to the ground was the same force that held the moon in its orbit around the earth and held the earth in its orbit around the sun.[31] This insight had the potency of revelation: the same force that controlled ordinary objects on earth also worked on size scales that had been the stuff of myth. "Universal gravitation" held both heaven and earth together with a force proportional to the mass of the objects attracting each other. Newton's laws of motion explained the orbits of planets and comets, why objects did not fall off the spinning earth, and how the moon and sun cause the tides. The impact of Newton's laws was enormous, and the accuracy of their predictions so convincing that people extrapolated their applicability to the entire universe.

But psychologically, the universe that had once felt like a great cathedral filled with angels had vanished, and infinite reaches loomed. No place was special. There was no secure foothold in the universe, no anchor. The belief began, and it is even stronger today, that space extends indefinitely and is empty except for matter and forces. Gravity holds the planets in their orbits around the sun. The sun is a star, and other stars are scattered so incomprehensibly far away that they appear only as bright points. Heaven, if it exists, is in a separate spiritual realm that does not lie within the physical universe. God has no special place in the physical universe, and humans are not essential to the universe.[32]

God was banished to the heart. Physics claimed to define physical reality, yet it treated human beings like objects, and those objects were left wondering whether anything in the universe recognized them as more than that. Perhaps they were just a random occurrence on an average planet in a vast and uncaring scheme of things.

Counterbalancing the psychological discomfort of the Newtonian picture, however, was the unassailable fact that Newton's physics worked. The clean precision with which humans could now interact with the world through accurate prediction hugely empowered the human race. Although it was not until later in the Industrial Revolution that physics became the basis of technology, Newton's success was inspirational. Voltaire popularized Newton's physics and linked it with John Locke's liberal philosophy as a basis for the Enlightenment.[33] Science and technology made possible astonishing material improvements in life, and this created a new and different kind of security. The faith that "science can figure it out" became itself a new foothold and anchor in the universe.

But "universal gravitation," despite the name, was never applicable to the whole universe. Paradoxes arose when Newton tried to apply his theory beyond the solar system. For example, if the universe were finite in size, it would have a center, and gravity would make everything collapse to the center; therefore, the universe can't be finite.[34] But if the universe were infinite, then the night sky would be white because there would be a star along the line of sight in every direction; therefore, the universe can't be infinite.[35] As to the origin of it all, Newton never questioned the biblical story of creation: *however* the universe operated, God had created it that way. It was still ill-advised to question theology unnecessarily. Thus the Newtonian picture offered no serious explanation of how the vast and empty universe had gotten that way.

The story of origins continued to belong solely to religion until about two hundred years ago, when the great mathematical astronomer Pierre Simon Laplace proposed an astonishing—and essentially correct—theory of the origin of the solar system. The solar system began as a flat, rotating cloud of incandescent gas, he explained, and as the cloud gave off heat and shrank to form the sun, it left behind bits that congealed into the planets. If the sun and planets condensed out of a rotating gas cloud contracting under its own gravity, that naturally explains why all the planets go around the sun in roughly the same plane and in the same direction.[36] These were daring thoughts, inspired by the faith that scientific reasoning could legitimately reach even into the most remote question, the origin of the world. The Emperor Napoleon pointed out to Laplace that his theory did not mention God. "I have no need of that hypothesis," Laplace famously answered. But despite Laplace's confidence that science could explain everything without reference to God, neither he nor anyone had the slightest idea how to explain the creation of the matter that makes up the earth and sun, let alone the origin of space, time, and the universe itself. For those big questions there have been throughout human history until today only mythological stories no better supported than those of the Egyptians, Mesopotamians, or Hebrews.

The Newtonian picture left humans drifting in a kind of cosmic homelessness that persists to this day. Cosmic homelessness has been assumed by some existentialist philosophers of the nineteenth and twentieth centuries such as Nietzsche and Sartre to be the human condition.[37] The existential view of reality, while based on the Newtonian universe, is ostensibly about something completely different—the experience and meaning of being an individual human. The existentialists didn't try to understand the "ultimate nature of reality." They denied that that was possible, on the grounds that all systems of thought—whether philosophy, religion, or science—are inevitably inadequate, in conflict with each other, and can never present the whole truth; thus they can never be objective guides for life. We are each, in the existential view, confronted with endless decisions, forced to rely on our own personal interpretation of reality. Human beings are totally free to create the meaning of our individual lives, since "all knowledge is interpretation,"[38] but this freedom is also a terrifying responsibility, and each person faces it alone.

Perhaps the existentialists saw the universe as inherently meaningless because they understood it so poorly. Their view, while denying the possibility that science can guide human decisions, was unwittingly guided itself by old science. Existential terror is a response to a universe seen as infinite or at least incomprehensibly large, almost empty, and with no inherent purpose, so that to avoid total absurdity individuals must create meaning for themselves. This response has permeated our culture so thoroughly that many people assume that the existential stance is scientifically justified. But its fundamental assumption that the universe has no higher level of organization is wrong scientifically and an aberration historically. Human culture evolved over thousands of years in societies with self-confidence drawn from the belief that they were an integral part of their universe. Although Westerners today can hardly remember it, this self-confidence has historically been part of the human condition.

There may have been no way to avoid the four-century period of disconnection from the universe implicit in the Newtonian picture. But science has now run its course. Today we no longer have to extrapolate: astronomers can see to the edge of the visible universe, back almost to the beginning of time. With copious data on the early universe coming in from new instruments, cosmology has left the realm of metaphysics and become astrophysics. But simply learning that scientists have gone beyond the Newtonian picture cannot instantly erase the feelings of alienation that four centuries of conditioning have bred into our culture. The Newtonian picture will shape our thinking until it is replaced by a picture that is intellectually convincing and emotionally satisfying. This is the challenge of modern cosmology today.

To meet that challenge, our generation will have to absorb a complicated and counterintuitive cosmos as a model, but we have the tools to do it. If this new picture can inspire the writers, artists, and open-minded thinkers who are the real meaning-makers and visionaries of our time, it is possible that the painful centuries-long hiatus in human connection with the universe will end. The ancients always thought that humans reflect the larger cosmos, but this did not reveal the cosmos to them. What ancient peoples never understood is that it is not the *appearance* of the heavens—not the patterns of the stars or the distances to the planets—that is reflected in us or on Earth: it is the *fundamental principles,* which govern us as they govern everything. The emphasis on appearance leads to astrology; the emphasis on principles leads to astronomy. This book is about the principles that cosmology is discovering today that may have profound reflections in us and the life of our planet.

People have always needed to feel that they fit into something larger and more meaningful than everyday life, but in earlier cultures it was a lot easier to know what that "something larger" was supposed to be, because people basically agreed on it. In the Information Age, however, we have libraries full of creation stories—in addition to the Big Bang we have the two stories of creation that open the book of Genesis, and Hindu, Chinese, African, and hundreds of other stories reported by decades of anthropological and mythological research,[39] some still literally believed, some taken symbolically. Without a meaningful, *believable* story that explains the world we actually live in, people have no idea how to think about the big picture. And without a big picture, we are very small people. A human without a cosmology is like a pebble lying near the top of a great mountain, in contact with its little indentation in the dirt and the pebbles immediately surrounding it, but oblivious to its stupendous view.

Where We Stand Today

Imagine you have suddenly lost your memory. You see yourself in a mirror, but there is only this moment. You are unaware of any past, even the very moment before this one. You are solid, your cells are real, your heart is pumping, but how you got to the spot on which you stand, you have no idea.

Who are you? There is no answer.

You are not your family background, your personal history, the work you've done, your hopes for the future. These don't exist. You have no family, no associations. You're like a computer with hormones. You have lost your soul. You are listening to the latest music, buying the latest improved products, and believing the latest media interpretation of the world outside your room. This is all you know.

Now instead imagine yourself as a branching history and send your consciousness backward through the branches at lightning speed. Down past your parents and grandparents, down the countless generations before them, through your ancestors roaming from continent to continent, your primate ancestors, down through all the animals that preceded them, back through the earliest life, into a single cell, down into the complex chemicals that made it possible, down into the molten planet and the forming solar system, the birth of your carbon and oxygen and iron in exploding stars far across the galaxy, back through the universal expansion to the creation of your elementary particles in the Big Bang. This is not fantasy. This is science: you are all this. Who you are is the sum total of your history. How far back you take that history—how much of your own identity you claim—is up to you.

It feels unacceptable to many people even to think of having a cosmology based on science. They misinterpret freedom of thought as requiring a refusal to believe anything. They see fanciful origin stories as spicing up the culture. The problem is, however, that spices, even in the most artful mixture, cannot compensate for the fact that there is no food—no data, no evidence: such stories are not actually about anything beyond themselves. We are not arguing to throw away the spices but to start with some food and then only use those spices that improve the food at hand. Scientific reality is the food. Aspects of many origin stories can enrich our understanding of the scientific picture, but they cannot take its place. To make progress, we must make choices. Shouldn't we choose the ones supported by evidence?

Before the new cosmology can matter to most people, we need to perceive connections between the expanding universe and our own lives—not dogmatically but suggestively, leaving room for future interpretations but *letting science set a minimum standard*. Part II of this book begins to make that connection by explaining five fundamental ideas about the expanding universe and visually representing each with a symbol. Our hope is that once the idea is understood, the symbol will hold it together in the mind. The most powerful symbols we know are ones that represented mythic truths in long-lived cosmologies, some of which we have discussed. As we have shown, no long-lasting religious symbol has maintained a pure and unadulterated meaning, but has instead been recycled from one generation to the next and one culture to the next with ever new overlays of meaning. Thus our twenty-first-century overlay is as legitimate as any earlier one. Each of our reinterpreted symbols is accurate with regard to the specific feature of the universe it represents, yet taken alone as a "picture of the universe," it would be misleading—in a sense, a graven image—because it would rise in importance like a false god above the other *equally fundamental* features of the universe. However, when taken together, these symbols act as sophisticated lenses through which we can finally perceive a universe stranger than any fiction.

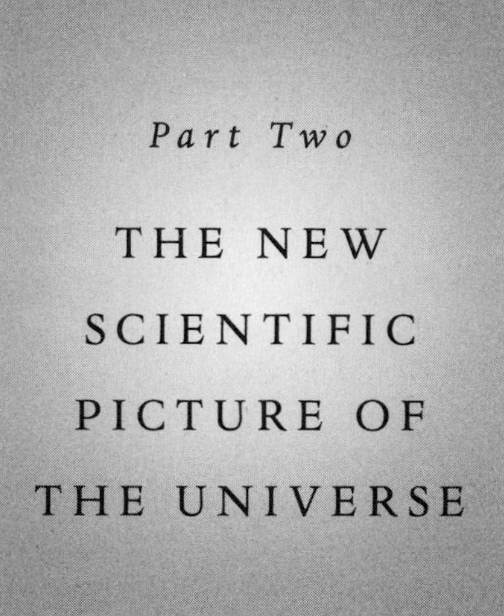
Part Two
THE NEW
SCIENTIFIC
PICTURE OF
THE UNIVERSE

FOUR

What Is the Universe Made Of?

THE COSMIC DENSITY PYRAMID

HUMAN BEINGS are made of the rarest material in the universe: stardust. Except for hydrogen, which makes up about a tenth of your weight, the rest of your body is stardust.[1]

What is stardust? It may bring to mind a shower of glitter, but a single fleck of glitter is made of many billions of atoms. Stardust is the atoms themselves. Hydrogen and helium, the two lightest kinds of atoms, came straight out of the Big Bang, while a little bit more helium and essentially all other atoms were created later inside stars. These atoms are the building blocks of everything in the universe that is visible, but most of the matter in the universe is neither atomic nor visible. It is not even made of the particles—the protons, neutrons, and electrons—that compose atoms. Most of the matter in the universe is an utterly strange substance called "cold dark matter," and its existence was only established late in the twentieth century. And stranger still, most of the density of the universe is not even matter—it's "dark energy," which is everywhere in space, causing space to expand ever faster. To understand what we are and what the universe is, we need to understand what the universe is made of, because those ingredients can behave only in certain ways, and how they interact tells us how the whole universe fits together.

The language of matter and energy makes many people uncomfortable about physics. Why do we, who live richly on the human level, need to know the boring details of what's going on below the surface? This book is about discovering our *real* connection to the cosmos—the most real and complete connection possible. What is happening on other size scales can seem utterly foreign to the reality we experience on the human size scale, but it nevertheless includes and affects us.

We fit into the universe in surprising and interesting ways on exotic size scales that most people never think about, and the size scale of atoms is one of them. Our minds are not free-floating spirits—they are embodied in atoms. Atoms are like letters. With just a few of them you can say infinite things, yet they have a strict grammar. Atoms are thus no more boring than the letters of the alphabet, which, in the proper order, bring us poetry, history, wisdom, and entertainment (and in the wrong order, bureaucratese, hate mail, libel, and spam). But even for the most literate person, the letters of an unfamiliar alphabet are meaningless, and similarly atoms and the other constituents of the universe are meaningless without some basic understanding. This chapter aims to provide that basic understanding, explaining how the universe is structured, where we fit in, what it's made of, and how discovering its structure was the key to understanding its composition. Then the chapter proposes a way to visualize the universe as a whole, based on what it's made of.

The Large-Scale Structure of the Universe

Just as ordinary matter on the small scale is built of atoms, the large-scale structure of the universe is built of galaxies. The two thousand or so stars you can see with the naked eye on the clearest, darkest night are all inside our own Galaxy, the Milky Way, and most are in our own small region of it. The Milky Way itself is rarely visible to the naked eye anymore except outside cities in places unpolluted by electric lights. Then it appears as a pale, narrow swath across the sky. It looks smooth like a milky cloud because it's made of so many faint stars. "Galaxy" comes from the Greek word for milk. Other cultures called it the royal road, the true Nile, the tree of life: the ancient metaphors were always of a long and narrow object because that's how the Milky Way looks to the naked eye. Individual bright stars dot the sky in front of it and near it, fewer in other parts of the sky.[2] But this naked-eye view gives no hint of its actual shape. The Milky Way is a flat disk with a bulge in the middle. It looks like a road or river because we're in it; from Earth we see most of the disk edge-on. Humans got the first sense of what the Milky Way's visible shape must be from photos of distant galaxies.[3] Our Galaxy, like many others, is a beautiful spiral, encrusted with gemlike star-forming regions that glow with the red of hydrogen or the green of oxygen spewed out from many dying stars. The naked eye and photographs give us a two-dimensional image of the Milky Way, but modern astronomical observations plus theory and supercomputer simulations have unveiled our home Galaxy not only in the three dimensions of space but the four dimensions of space and time.

There are about as many bright galaxies in the visible universe as there are people on Earth, and many more faint galaxies than people. Galaxies cluster like people on a beach, or high peaks in a mountain range. Most galaxies are in small or medium-size groups, like little villages in the country with each person in each village a galaxy—it's a rural universe. A small percentage of the galaxies are in clusters, which contain dozens of big galaxies and as many as thousands of smaller ones. The Milky Way is one of two big galaxies located in a small group called the Local Group with about thirty smaller galaxies. The other big galaxy is in the direction of the constellation Andromeda. The Local Group is in turn on the outskirts of a huge sheet of galaxies called the Local Supercluster, which includes the nearest cluster, the Virgo Cluster.

A supercluster is very different from a cluster. Clusters of galaxies are bound together by gravity. Superclusters are not bound and, because of the now-accelerating expansion of the universe, they never will be. On the scale of the visible universe there are about a million superclusters of galaxies. Between the superclusters are huge voids containing hardly any visible galaxies. Everything is moving: in addition to the uniform expansion of the universe, vast flows of galaxies have been observed heading away from voids and toward regions where galaxies are unusually abundant. One of the first such regions discovered was named "the Great Attractor."[4] All this is what astronomers call the "large-scale structure" of the universe, and much of it has been discovered only since the mid-1980s.

Below is a schematic picture showing our Cosmic Address in the large-scale structure of the universe. We are on the earth, eight light-minutes from the sun, which in turn is about halfway between the center and the outermost visible stars of the Milky Way. The Milky Way itself is in the region of our Local Supercluster called the Local Group. The whole Local Supercluster is at point zero on the radius of the cosmic horizon.

The large-scale structure helps us see how we fit into the universe. But why is the universe structured like this? Why are there galaxies at all? Why do they come only in special sizes and shapes, and why do they cluster the way they do? The key to answering all these questions lies in the dark matter that galaxies are made of, which can only behave in certain peculiar ways that are very different from the behavior of ordinary matter. The interaction of the cosmic ingredients, in the presence of dark energy, could only create this kind of structure. Furthermore, each cosmic ingredient has very different characteristics depending on whether we look at it on the vast size scale of its collective role in the cosmos or zoom in to the tiny size scale of its elementary particles.[5] We'll start with the most familiar ingredient and move to the exotic.

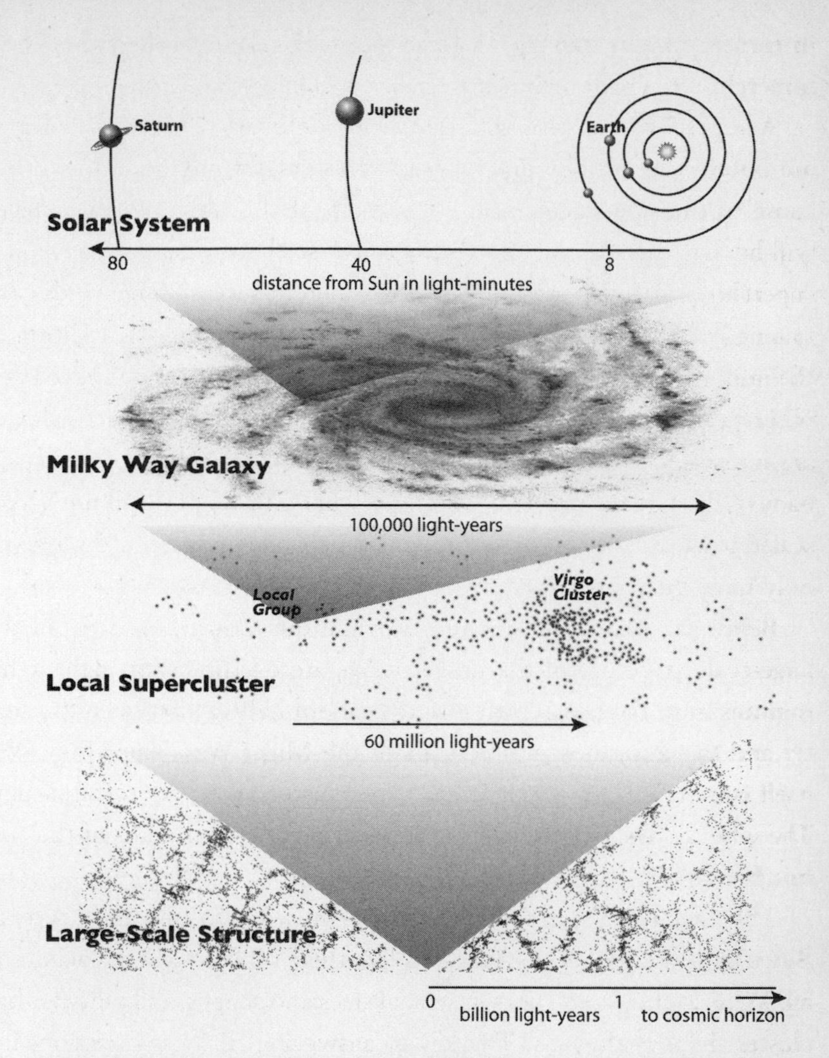

Figure 1. The Large-Scale Structure of the Universe. Our cosmic address is Earth, Solar System, Orion Arm, Milky Way Galaxy, Local Group, Local Supercluster. Every dot on the lower two panels represents a large galaxy. The Milky Way, our home galaxy, is at the bottom point of the lower two V diagrams. (The Large-Scale Structure data in the bottom panel is from the Sloan Digital Sky Survey.)

Cosmic Ingredient #1: "Ordinary" Matter

Each of us is an atomic pastiche: the iron atoms in our blood carrying oxygen at this moment to our cells came largely from exploding white dwarf stars, while the oxygen itself came mainly from exploding supernovas that ended the lives of massive stars, and most of the carbon in the carbon dioxide we exhale on every breath came from planetary nebulas, the death clouds of middle-size stars a little bigger than the sun. We are made of material created and ejected into the Galaxy by the violence of earlier stars, including some supernovas that exploded before the solar system formed four and a half billion years ago and some that happened only a few million years ago.[6] To understand how this happened—to appreciate the millions or billions of years it takes a star to produce a comparatively tiny number of heavy atoms, and the tremendous space journeys of those particles of stardust that have now come together to incarnate us—is a first step toward feeling some conscious contact with what we are made of and our cosmic history.

"Ordinary" matter is everything made of atoms. It also includes the parts of atoms—that is, protons (positively charged) and neutrons (electrically neutral), which make up the nucleus, plus electrons (negatively charged). The two simplest atoms, hydrogen and helium, have a unique cosmic role: they emerged from the Big Bang in the first few minutes and are the parents, in a mythic sense, of all the other atoms, because the others are all formed by fusing these primordial ones together. Hydrogen is the simplest and lightest atom. The common sort is just one proton and one electron.[7] The common kind of helium is two protons and two neutrons in the nucleus, plus two electrons. Everyone has seen helium balloons, which rise up because helium is so much lighter than the gases that make up air. Hydrogen is even lighter, but flammable. All other elements are made of atoms whose nuclei contain more particles and so are heavier than these two.

Anyone who ever sat in a chemistry classroom where a periodic chart of the elements hung on the wall will probably remember, possibly not fondly, that on that chart are more than a hundred "elements." An element is simply a lot of atoms of a single kind. If your homework is to memorize the names of all the elements, then their number seems exorbitant. But if you want a universe with infinite potential—with room for life, creativity, and growth on countless worlds over billions of years—you want a universe with lots of different kinds of atoms.

An atom is not a miniature solar system, with little electrons orbiting the nucleus the way the planets orbit the sun. This is a common but very misleading image. Planets are identifiable bodies, but electrons in atoms are not little bodies but *probability clouds* surrounding the nucleus; it is the *probability* of their being one place or another that is real. Quantum uncertainty makes it impossible in principle to trace the path of an individual electron in many circumstances.[8] On the atomic scale, particles can act like waves. There can never be half an electron wave in an atom, only a full crest and trough. As a result, electrons in atoms can only be in certain "quantized" energy levels and nowhere in between.[9] Each probability-cloud pattern that nature permits corresponds to a complete wave, and thus only certain shapes of probability clouds are possible, as illustrated in the figure below.

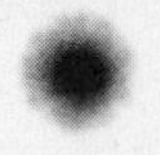

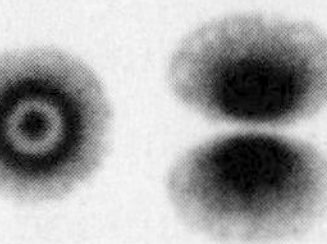

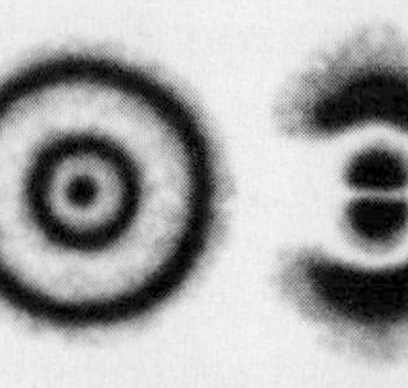

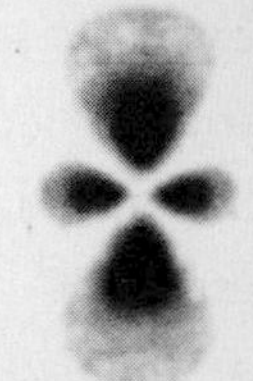

Lowest energy state | **Next-higher energy state** | **Third energy state**

Figure 2. Electron probability clouds for some of the lowest-energy states of a hydrogen atom. These figures represent slices through the center of the atom, and the darkest regions are where the probability of an electron being there is highest. The lowest energy state has only one possible pattern; higher energy states allow more possibilities.

If "existence" can mean anything on the scale of atoms, then these special shapes exist, and they each come in one size only, determined by the laws of physics. Unlike planetary orbits, chance cannot shrink or enlarge them. Atoms are precisely shaped and sized *regions of space* in which things are happening. These regions can fit together only in very specific, limited ways. These ways are further limited by a law of physics called the "exclusion principle," which says that no two electrons can be in the same state.[10] The end result of all these laws and constraints is that carbon, oxygen, silicon, and other chemical elements each have an immutable personality, which means, quite literally, a tendency to chemical aversion or attraction to other atoms. These atomic personalities in combination determine the kinds of molecules (that is, structures built of atoms held together by chemical bonds) that atoms can form, and also the kinds of light that the molecules absorb and reflect. These probability cloud shapes are therefore the basis for the entire physical world as we experience it with our five senses.

People rarely think about atoms because they are so small compared to the size scales within everyday consciousness, and with stars we have the same problem in the opposite direction. Most people think of them as shining dots in the night sky, not as gigantic thermonuclear reactors operating continually for millions or billions of years. But we and our planet are inextricably linked to the stars, because stars have changed hydrogen and helium into all the heavier elements we're made of. Stars begin as vast clouds of gas, mostly hydrogen and helium. Light as those individual atoms are, there are so many of them that together their gravity is immense and it gradually compresses the clouds until there is so much pressure and heat at the center that atomic nuclei fuse together. The energy released by this nuclear process—which is a continuous hydrogen bomb—ignites the cloud and a star is born. A tiny fraction of the hydrogen and helium in galaxies is in giant gas planets like Jupiter; some is still floating as clouds that glow only if they are heated and lit up by nearby stars; but the majority of all the visible matter in the universe is hydrogen and helium now shining in the form of stars.

Stars manufacture all the other elements—but not all in the same way. There are a few super-massive stars, many more middle-sized ones like our sun, and great numbers of faint stars about a third the mass of our sun. The faint small ones don't produce many heavy atoms, but the mid-to-large ones do, each in its own way producing different essential elements.

It is huge stars more than about eight times the mass of the sun that produce most of the oxygen and neon and all of the really heavy elements. After such mega-stars with their immense gravity crush and fuse the hydrogen nuclei in their cores, their outer layers swell and they become red giant stars. As such stars evolve further, their centers develop an onionlike structure, with hydrogen fusing in an outer layer, helium in a deeper layer, then carbon, neon, oxygen, with silicon and sulfur fusing in the core. Eventually the core becomes mostly iron, the most tightly bound nucleus, and then further fusion is impossible. But gravity never stops. The intense gravity of the outer layers of the star still pressing down on the iron core of the star causes the structure of the atomic nuclei themselves in the core to collapse, and the star ends up as one or the other of the two densest things in the universe: a neutron star or a black hole.[11] About 99 percent of the energy of the collapse is carried off by a brief burst of particles called neutrinos. The neutrinos can escape from the dense collapsing core of the star in only a few seconds since they have no electric charge and interact only weakly; they also have only very tiny masses.[12] It is the remaining 1 percent of the energy that produces the visible explosion and seeds the nearby universe with heavy atoms. The very heaviest elements like gold and uranium are apparently produced during the supernova explosion itself.[13]

Fifty times per second, a supernova occurs in some galaxy in the visible universe, spewing out into space enormous quantities of heavy elements that may travel millions of light-years before falling into the gravitational field of some newly forming solar system. Those heavy elements may join together to create in that new solar system a planet that billions of years later will pulse with life.

But not all the heavy atoms are made in such supernovas. Medium-sized stars more massive than our sun produced the medium-weight atoms like carbon and nitrogen that make up more than 20 percent of the weight of your body. Here's how: after millions of years of fusing hydrogen to make helium in their cores, those stars used up the hydrogen and started fusing helium to make carbon and oxygen in their cores, and they too (like the giant stars) swelled up and became red giant stars. Later the red giant evolved into an even larger "asymptotic red giant," and then blew off its outer layers to form a gas cloud called a planetary nebula.[14] This eventually dispersed into space, leaving behind a tiny remnant called a white dwarf, which typically has half the mass of the sun compressed to the size of the earth. A teaspoon of white dwarf would have a mass of several tons. The carbon atoms in our bodies mostly came from planetary nebulas. Planetary nebulas from stars that started out a little more massive than the sun also contain a lot of nitrogen, which is the main constituent of the air we breathe.

The brightest supernovas come from the deaths of white dwarfs rather than massive stars. Tycho Brahe's supernova of 1572, which so shocked medieval sky-watchers, was caused by the explosion of a white dwarf—and such explosions produce most of the iron in the universe. Quantum theory predicts that a white dwarf will explode if its mass becomes larger than a critical value about 1.4 times the mass of the sun.[15] Where does such a white dwarf get this extra mass? About half of all stars are orbiting around another star, forming a binary system. If one of these stars evolves into a white dwarf, and if the stars are close enough, then the great gravitational pull of the dense white dwarf captures gas from its companion, making the white dwarf more massive. When its mass exceeds the critical value, it explodes. These are the greatest thermonuclear explosions since the Big Bang itself—more than two million trillion trillion megatons apiece.[16]

Massive stars burn their candles at both ends. They are enormously bright, but they live only a few million years, while stars about ten times smaller like the sun live a thousand times longer. Massive stars were thus the first in the universe to explode, so the elements they produced were the first ones to find their way into newly forming stars.[17] Since when and where a star formed determines the mix of elements it starts with, only the earliest stars began as pure hydrogen and helium; later generations of stars like our sun contain mixtures of elements built up from many supernovas and planetary nebulas, and our planet is made of the same varieties of atoms. Stardust is thus part of our genealogy. Our bodies literally hold the entire history of the universe, witnessed and enacted by our atoms.

"Ordinary" atomic matter turns out to be extraordinary compared to what exists on the scale of the universe as a whole. All the stars, planets, gas, comets, dust, and galaxies that we see—all the forms of visible matter—make up only about half a percent of what's out there.[18] There is about ten times more atomic matter that is invisible than visible, and the invisible atoms are probably floating around in space in between the galaxies; they're mostly hydrogen and helium far from any stars whose light could illuminate them. Much of this material isn't whole atoms but just protons and electrons on their own. But even if you add together all the visible and all the invisible atomic matter, including the loose protons and electrons, the grand total is still only about 4 or 5 percent of all that's out there. We are sure of this, because astronomers have measured it in four completely different ways, and all methods agree.[19]

What is the other 95 or 96 percent of the density of the universe? Astronomers realized during the last two decades of the twentieth century that the density of the universe is almost entirely due to a combination of dark matter and dark energy, neither of which can be explained by the otherwise extremely successful "Standard Model" of particle physics. About 25 percent of the universe is cold dark matter,[20] which is not made of atoms or any of the elementary particles that compose atoms, but its immense gravity holds the spinning galaxies together. The other 70 percent seems to be even stranger than dark matter. It may be something Einstein called the "cosmological constant," and now it is often called "dark energy." Dark energy is a property of space itself, and it causes space to repel space, speeding up the expansion of the universe (see Chapter 5). Although dark "energy" is not "matter" in the usual sense, Einstein showed with his famous equation $E = mc^2$ that energy and matter are closely related and can be converted into each other. Thus dark energy, like matter, is a major portion of the *density* of the universe.[21] To understand what dark matter and dark energy are, we have to switch our mindset away from the tiny size-scale of particles and atoms to the very large size-scale of galaxies and even clusters of galaxies, since dark matter and dark energy shape the universe as a whole and were only discovered through their cosmic-scale effects. Then we will zoom in again to discuss what they may be on the small scale.

Cosmic Ingredient #2: Dark Matter

The very peculiar way that stars orbit the center of a galaxy is not at all the way that planets orbit the sun. This discrepancy was one of the clues that led to the discovery of dark matter and its momentous behind-the-scenes role in the universe. In our solar system the planets travel in elliptical but almost circular orbits around the sun, pulled by the sun's gravity. Since the strength of gravity decreases with distance, the farther away from the sun a planet is, the less it feels the sun's gravity and the slower it moves. The speeds of the planets vary this way because about 99.9 percent of the mass of the entire solar system is at the center of the system—in the sun—and therefore most of the gravity is at the center, since mass creates gravity. Jupiter is by far the most massive planet in our solar system, but its mass is only about a thousandth that of the sun—see Figure 3. All the other planets together add up to only 40 percent of the mass of Jupiter.

Mercury, the planet closest to the sun, orbits the sun in just eighty-eight days, while Earth takes one year and Jupiter twelve years. Since stars generally orbit the center of the galaxy, astronomers until the 1970s simply assumed that the same rules applied to stars as to planets. In a

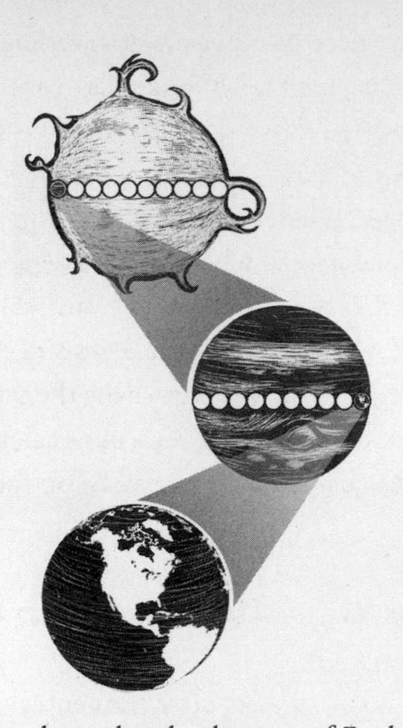

Figure 3. This figure shows that the diameter of Earth is a tenth the diameter of Jupiter, which is a tenth the diameter of the sun. Therefore the *volume* of Earth, or the amount of space it occupies, is only a thousandth the volume of Jupiter, which is a thousandth the volume of the sun.

typical large spiral galaxy like the Milky Way, most of the stars are in the central bulge and the inner part of the disk. The rest of the stars orbit in the outer disk, which gets sparser farther from the center. Astronomers reasoned that since the majority of the stars were near the center, most of the gravity of a galaxy should also be at the center, just like the solar system. It followed that the outer stars in the galaxy, like the outer planets in the solar system, should be traveling around the center of the galaxy a lot more slowly than the inner stars.

But in the early 1970s astronomer Vera Rubin and her colleagues made a discovery that shook astronomy: *the stars in the disk of a galaxy all orbit with about the same speed, no matter how far from the center they are.* The stars act like cars all traveling in circular tracks with their cruise controls set to the same high speed. Galaxies should fly apart, but clearly they don't. Why do high speeds mean a galaxy should fly apart? If you throw a rock up, it will fall down. If you throw it up much harder, it will go higher but the earth's gravity will eventually still pull it back down. But if you launch a rock—or a rocket—with what is called "escape velocity," it will exit the earth's gravitational field and never come back. Stars inside galaxies are moving with far *more* than escape velocity—if the gravity of the visible matter is all that needs to be escaped. But they are not escaping. The only explanation is that there must be much more gravity holding them in than the visible matter can be generating, and that means there must be more matter than can be seen.

At this point the concept of "dark matter," originally proposed in 1933 by the astronomer Fritz Zwicky[22] but ignored for forty years because it was so bizarre, was resuscitated. The term is unfortunately misleading, because dark matter is not dark: it's invisible. "Invisible," however, can have two different meanings. The common one is, *we* can't see it (it doesn't shine or else no light is reflecting off it toward us, for example because it is black). But dark matter is invisible in a much deeper way—it doesn't seem to interact in any way with light. Dark matter neither emits light as stars do, nor reflects it as gas clouds do, nor absorbs it as dust does. It does not emit or absorb X-rays, radio waves, or any other form of radiation that astronomers can detect. Dark matter is totally transparent. "Missing mass," another term journalists sometimes use, is also a misnomer. Dark matter is not missing, we just can't see it.

The only possible explanation of Vera Rubin's findings consistent with Newtonian gravity is that a vast cloud of invisible matter fills *and surrounds* each galaxy and each cluster of galaxies. A few scientists were so adamantly opposed to this conclusion that they were prepared to modify Newtonian gravity to avoid the idea of dark matter, but their attempts have not succeeded.[23] Unlike the solar system, the great preponderance of mass in the galaxy is not at the center, where most of the stars are; instead it is spread throughout the galaxy and far beyond the visible stars. Majestic as they appear, the spiral galaxies that we observe

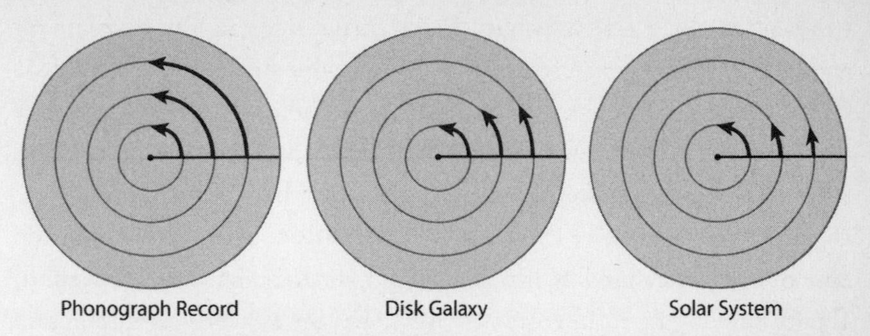

Figure 4. Three Different Kinds of Rotation. The length of each arrow represents the *speed* at different distances from the center (or equivalently, the distance covered in a fixed amount of time). On a solid rotating object like a phonograph record, the outer points move much faster than the inner ones in order to make a complete circle in the same amount of time. In the solar system, the opposite is true: the inner points (planets) move much faster than the outer ones because they're closest to the source of gravity, the sun. But in the disk of a galaxy, the stars all rotate at about the same speed.

through telescopes are only the glowing centers of these enormous "halos" of dark matter that extend to at least ten times the radius of the disk of visible stars. The dark matter is densest in the center of the galaxy and becomes thinner farther out. Even though the dark matter is very thin at great distances from the galactic center, the volume of space through which it is spread is so huge that the amount of dark matter at those distances is enough to maintain the cruise control of any visible star, gas cloud, or even satellite galaxy inside it.

The implications of dark matter are enormous. There is a huge logical jump from measuring the speeds and temperatures[24] of things we can actually see to the conclusion that a whole new class of matter not only exists but overwhelms everything we thought we knew. Astronomers as a whole are quite conservative; they were still skeptical about dark matter until its existence was shown by an entirely different method. Starting in 1986, astronomers have observed faint luminous arcs around the centers of rich clusters of galaxies. These arcs are optical illusions like a mirage: the light is real, but what it appears to be coming from is not. There's nothing arc-shaped out there. What's happening is something predicted by Einstein's theory of relativity: light is being bent by gravity. A little while ago we said that dark matter does not interact with light—but indirectly it can do so: since dark matter has mass, it causes space to curve, and that in turn bends the light.

Visualize a game of "monkey in the middle," where two children throw a ball back and forth over the head of a third child, who stands in the middle trying to intercept the ball. The galaxy cluster is the child in the middle; we on Earth are one catcher, and a distant galaxy *far behind* the cluster is sending out faint light. Some of that faint light coming straight in our direction would be intercepted by the cluster in the middle. But some light that would have gone off into space and missed Earth is pulled inward by the immense gravitation of the cluster in the middle and redirected on a new path straight toward us. In effect the distant galaxy's light has gone *around* the cluster.[25] Galaxy clusters are thus the greatest telescopes in the universe: acting as gravitational lenses, they collect and focus the light of far more distant galaxies, galaxies that otherwise would be difficult or impossible to see. This is called "gravitational lensing," and it turned out to be a completely independent method to measure the amount of matter in the galaxy cluster—by the size of the arc it creates as it bends passing light.[26] Since 1986 astronomers have observed many examples of gravitational lensing. Not just clusters but also individual galaxies bend light from background galaxies, so this is not a freak phenomenon. In every case, measurements of the total mass in the cluster (or galaxy) in the middle that are based on bending of light agree with results that are based on speeds and temperatures: a vast dark matter halo surrounds both galaxies and galaxy clusters.[27]

Many astronomers for a long time intensely disliked hypothesizing an outlandish amount of invisible matter to solve Vera Rubin's "cruise control problem," but the agreement of all these methods has finally convinced even most of the skeptics that almost all the mass holding galaxies and galaxy clusters together is dark matter. As Sherlock Holmes said, "When you have eliminated the impossible, whatever remains, however improbable, must be the truth."[28]

How much dark matter is there? What is it like? How does it behave? For several years these questions were answered in different ways by several different competing theories that managed to agree with all the reliable data, because the data were so rough and incomplete. But by 1998 all theories but one had been ruled out because they were inconsistent with new data from telescopes all over the world and in space, such as the Hubble Space Telescope. Astronomers call the one theory left standing "Dark Energy plus Cold Dark Matter," usually abbreviated LCDM,[29] but for simplicity's sake we will call it the "Double Dark" theory. The Double Dark theory predicted much of the avalanche of data that has descended since 1998, including what's been revealed by extremely detailed baby pictures of the universe etched into the heat radiation from the Big Bang[30] and extremely detailed maps of the location of galaxies in the universe.[31]

We have been speaking of the behavior of dark matter on the very large scale of galaxies and clusters of galaxies, but to talk about what the dark matter is made of, we now mentally zoom in to the very small size-scale of atoms and particles. As of this writing, no one yet knows what kind of particles dark matter is made of, though we know what it's not.[32] Joel and his colleague Heinz Pagels proposed in 1982 what remains the most popular candidate for the dark matter. It's called "supersymmetric weakly interacting massive particles," a clunky name chosen at least in part for its memorable acronym: WIMPs.[33] How do we unpack this complicated phrase, and why is it still in the running decades later?

The word "supersymmetric" means that the mysterious dark matter particle would fit into the unfinished but grand and potentially encompassing theory of physics called supersymmetry, which we discuss below. The word "massive" means that a single dark matter particle may weigh as much as the heaviest atoms, or maybe even more. But what does it mean to say that dark matter particles are only "weakly interacting?" The idea of a particle "weakly interacting" with others is central to the nature of the dark matter and explains why dark matter has been so hard to detect and identify, but alas, all anyone can ever say in words about these characteristics of elementary particles is metaphorical. "Forces" is the metaphor we use to describe how particles interact with each other. There are four known forces of nature: gravity, electromagnetism, the strong force (which holds protons together in the nucleus of the atom, overwhelming their mutual electric repulsion), and the weak force (responsible for certain kinds of radioactivity). Clearly if dark matter particles have mass they have gravitational interactions—that, after all, is how they hold galaxies and clusters together and bend light. But they don't have electromagnetic or strong interactions. The only interaction left is the weak kind. We don't know if dark matter interacts weakly, but supersymmetry predicts it must; and from a practical standpoint, if dark matter doesn't have weak interactions, we don't know how to detect it at all. So far, neutrinos—those tiny particles that carry off 99 percent of the energy when a gigantic star collapses into a black hole—are the only kind of particle that physicists have actually detected that has no electromagnetic or strong interactions, only weak and gravitational interactions. If supersymmetric WIMPs exist, they would be the second example.

Particles "interact" by exchanging other particles. In order for a weakly interacting particle to interact with another particle, it has to exchange a third particle that has at least eighty-five times the mass of a proton.[34] Otherwise the second particle will never notice it. Where does a little neutrino, whose own mass is millions of times smaller than a proton's, find so much mass or its equivalent in energy? It borrows it from the vacuum of space. If you could peer into space very, very closely, you would find quantum particles going in and out of the vacuum. The vacuum is like a bank, but the bigger the loan, the less time the neutrino can keep it. Thus the neutrino, if it can get the energy, can only have it for an incredibly short time. The large amount of energy that must be borrowed drastically cuts down the distance over which it can interact, making the weak force a very short-range force. The Bank of the Vacuum is constantly making and calling loans everywhere. But whatever the neutrino gets must be paid back *precisely*. The universe's accounting is perfect. Neutrinos so rarely interact that they stream from the sun and almost always go right through the entire earth without interacting with a single atom.[35] To detect dark matter particles is even harder: you've got to somehow get them to interact with your equipment, but unlike neutrinos, they don't stream toward us in huge numbers from the sun or in conveniently gargantuan bursts from supernovas. Thus experimenters have built small ultrasensitive detectors and are running them for a very long time in the hope of detecting a few WIMPs—and some are building much larger ultrasensitive detectors.[36]

One might wonder at this point why anyone would propose a candidate for the dark matter that is so strange that it doesn't fit into the known laws of physics and demands an entirely new theory like supersymmetry. The reason is that there is no room for any dark matter particle in our modern Standard Model of particle physics, so one must go beyond it. Moreover, our best established physics theories—Einstein's theory of general relativity (the modern theory of gravity, space, and time) and quantum theory (the modern theory of the very small)—are not entirely consistent with each other. Cosmology's recent discoveries have pushed beyond the limits of those theories and demand an encompassing theory. Supersymmetry is the only theory anyone has come up with that can encompass relativity and quantum theory. Physics and cosmology here dovetail, because the possibility that WIMPs are the dark matter is a natural outcome of a theory—supersymmetry—that physicists developed for completely different reasons.[37]

Supersymmetry is based on the idea of dividing everything into two classes. When we divide the world into us/them or similar dualities, we are usually simplifying—often oversimplifying—a far more complex reality. But in physics there sometimes are really only two possibilities, like positive and negative electric charges[38] or particles and antiparticles.[39] Supersymmetry says that for *every* kind of particle we know, there is a related particle, called a "superpartner."[40] In one fell swoop this theory doubles the number of particles in physics. But none of the predicted superpartner particles has yet been discovered. That may soon change, however, because the world's first accelerator powerful enough to make them, the Large Hadron Collider, is scheduled to start operating in Geneva in 2007. Why is all this important to the cosmic recipe? Because the lightest of these supersymmetric *partner* particles is expected to be stable. Therefore it may be the dark matter, which is so stable that it has been around since the Big Bang.[41]

Another possible candidate to be the dark matter is *axions*, hypothetical particles that were invented to solve a completely different problem in physics. If axions actually exist, they might have the right density to be the dark matter. A very sensitive search for axion dark matter is now underway at the Lawrence Livermore National Laboratory in California. It is of course possible that the dark matter is something completely different. Theorists have considered many other possibilities, but those with the unfortunate property that they have only gravitational interactions cannot be detected by any known method.[42] Fortunately the two possibilities that look most likely—supersymmetric WIMPs and axions—are both detectable. Stay tuned.

Cosmic Ingredient #3: Dark Energy

The most powerful ingredient in the universe is "dark energy," and yet until 1998 this was only a hypothetical possibility. A few theorists proposed that dark energy might exist along with cold dark matter, but most astronomers assumed that the expansion of the universe was gradually slowing down because of the gravity of everything in it. But in 1998 data from distant supernovas revealed that the expansion of the universe is not slowing down at all but instead accelerating.[43]

Einstein was the first to think of this possibility.[44] He had come up with the theory of general relativity in 1915, and when, two years later, he first applied it to the entire universe he discovered that his theory required the universe to be either expanding or contracting but not stable. Something must be wrong, he thought, since astronomers had at that time seen no shred of evidence for the universe either expanding or contracting, and in fact had never even considered it. Einstein went back over the assumptions that had led him to general relativity and realized he had assumed one thing that was not necessary, although not wrong, either.[45] There was room to add another term, and this term, which he called the "cosmological constant," represented a possible repulsive force that could balance the attractive force of gravity; with it he hoped that the theory could describe an unchanging universe. The Russian physicist Alexander Friedmann and the Belgian astrophysicist Georges Lemaître, however, took Einstein's original version of general relativity and, not fearing its implications, worked out the theory of an expanding universe. When a few years later Edwin Hubble *observed* that the universe is expanding, just as general relativity had originally predicted, Einstein was very disappointed in himself that he had not believed in his own theory. He thought that the cosmological constant was his greatest blunder,[46] and he abandoned it and accepted that the universe is expanding. However, as it turns out, Einstein was on the right track both in his original theory *and* his added idea. With modern theories of elementary particles, physicists have realized that instead of the cosmological constant being a *constant* cosmic repulsive force, it could have much more complex properties. This more complex version of Einstein's cosmological constant is what we now call "dark energy."

Dark energy causes space to repel space. Dark energy is in the nature of space itself. The more space there is—and increasing amounts of space are inevitable in an expanding universe—the more repulsion. The more repulsion, the faster space expands. The faster it expands, the more space, the more repulsion, and this can lead to an exponentially increasing expansion possibly forever. What seems to have happened, according to the Double Dark theory, is this: in the early stages of the universe there was relatively little dark energy because there was relatively little space—the universe hadn't had time to expand very much; however, there was the same amount of dark matter then that there is now. For about nine billion years, the gravitational attraction of the dark matter slowed the rate of expansion. But the expansion was still going on, and the dark matter thinned out. Since dark energy is a characteristic of space, it never thins out; instead its relative importance has increased. Now the repulsive effect of the dark energy has surpassed the gravitational attraction of dark matter as the dominant effect on large scales in the universe, and expansion is no longer slowing down but accelerating. The turning point was about four and a half billion years ago—coincidentally just when our solar system was forming.

One of the reasons that astronomers take the Double Dark theory so seriously is that it successfully predicted exactly the sort of large-scale structure and galaxy movements that astronomers actually observe today. But it also tells us that while the large-scale structure illustrated in our Cosmic Address (Figure 1) is how things look today, it will not be permanent. Galaxies and galaxy clusters *are* permanently bound together by gravity: they have become units and will travel together in the general expansion. Galaxy clusters were the largest objects to fall together and become gravitationally bound before the dark energy took over, but now that dark energy has become dominant, anything larger than a cluster—i.e., superclusters—will never collapse. There is too much space inside them, expanding too fast. Our Local Supercluster will never collapse but will be torn apart, ultimately faster than the speed of light. In the distant future, by the time the universe is about ten times its present age, all the superclusters will disperse, and the Milky Way's cosmic horizon—the farthest that intelligent beings can see in principle—will empty out. Only the galaxies in our Local Group and perhaps a few others bound to it will still be visible to our distant descendants.[47] Of course, this assumes the dark energy will keep acting as it does now, forever. That was what Einstein postulated in proposing a cosmological *constant*. But it might not. The dark energy could change in strength or possibly even change from repulsive to attractive, perhaps depending on how much matter there is. We don't know, because we don't understand what it is. The best data available at this writing indicate that it has *not* changed measurably as far back as we can look—but measurements of distant white-dwarf supernovas and other observations now underway and in the planning stages can allow astronomers to measure more accurately the properties of the dark energy in the past to see whether and how it has changed.

Visualizing the Universe as a Whole

This chapter could end here, but all you would have is a lot of information on the ingredients of the universe, and information alone is not a satisfying cosmology. As we have seen, cosmologies that have been meaningful for people's lives have been communicated with symbolic images and stories. Though we know far better than the ancients what the universe is made of, we have far less sense of what it might mean *for us*; we lack images and stories that can connect us to this new universe. A central goal of this book is to begin to provide some images by which we can visualize *our* universe—not random fragments of it, which is all that even the most stunning NASA astronomical photos give us, but its fundamental nature. Since most of the universe is invisible, the only way to visualize the whole thing is symbolically. Our ancient ancestors managed to do this: they saw their invisible gods through the gods' effects, as we see dark matter through its effects. They visualized their universe with the mind's eye; their images of the gods, the mythic animals, and even the sun were not photographic but symbolic. It was the well-founded fear that some people would nevertheless take such symbols all too literally that led the early Hebrews to ban graven images.[48] We want to be absolutely clear that the images in this book are not intended literally—they make no sense if taken literally—and the fact that each is so different from the others supports our argument that no single image could possibly be accurate alone. Taken together, these images begin to suggest a universe beyond all imagery but not beyond comprehension.

Our goal now is to visualize the combination of heavy atoms (very few, but more than a hundred varieties), light atoms (far more numerous, but only a couple of varieties), a huge amount of dark matter (maybe all one kind), and an overwhelming presence of dark energy (probably all one kind). The universe is made not only of these various things but of their relative amounts—the precise recipe counts. Our answer lies in the two images that follow. They represent the place of intelligent beings among all the ingredients in the universe.

The Pyramid of All Visible Matter is based on a picture everyone in the United States possesses: the pyramid topped by the all-seeing eye, which appears on the back of every dollar bill.[49] It is a pyramid, but its capstone is separated and floating above it, blazing with light, and dominated by an eye. To the Founding Fathers, who chose this symbol in 1782 for the reverse side of the Great Seal of the new United States, this pyramid represented a stable and enduring foundation upon which they were building a new reality. Its thirteen layers of bricks represented the thirteen colonies, crowned by the eye of Providence, which would, they hoped, protect their daring venture. This was a symbol that was used at the time by Freemasons, a secret society of which some of the founders were members. The Founding Fathers understood that an ancient symbol reinterpreted is almost always more powerful than a random new one, as long as the old one can be adapted appropriately to the purpose, and we agree. In this deeply layered symbol lies the possibility of representing all the visible matter in the universe—all the matter that people until the late twentieth century thought existed. One change we have made to assure accuracy is that, although on the dollar bill the capstone is two-dimensional, we have made the capstone three-dimensional so that it can represent matter in correct proportions. Volume in the pyramid is proportional to density in the universe.

The Pyramid of All Visible Matter is almost entirely made of the lightest atoms: hydrogen and helium, but there are so many of them in

Figure 5. The Pyramid of All Visible Matter.

the universe that they far outweigh all the heavy atoms. Hydrogen and helium are thus the massive bottom part of the pyramid. Above floats the glowing capstone, which represents stardust, the heavy atoms forged by stars. Not only we humans but all other living things, and not only Earth but all rocky planets throughout the universe, are made of the matter in this glowing capstone.

Now let's go another step. Within the capstone the fraction of stardust that is associated just with living things or the remains of living things is very tiny. Within that very tiny fraction, the matter associated specifically with *intelligent* life is vanishingly small—yet it is only *that* which looks at and grasps this pyramid. It is only *our* eyes. The very human-looking eye at the center of the capstone speaks powerfully for the minuscule bit of stardust associated with intelligent life. The eye is the only part not drawn to scale. Think of the eye inset on the capstone as being like an enlarged detail of a city center, inset on the corner of a large-scale map. The inset is way out of proportion to the size of things surrounding it, but everyone knows that an inset on a map is a zoom-in visualizing something important on a different size-scale. Here, in the same way, the eye on the capstone is a zoom-in to the presence of intelligent life. Without a zoom-in we could not even see the matter associated with intelligence on the pyramid, since it is so rare. The eye suggests mind or vision—unquantifiable things that transcend the issue of quantity. The eye of the glowing capstone of heavy atoms reminds us that intelligence bursts out only from tiny bits of stardust.

The base and capstone together form the Pyramid of All Visible Matter, and it stands on the solid ground of tradition and intuition, with a few plants for emphasis. But there is a hidden base beneath this stone pyramid that extends deep underground and all the way back in time to the beginning of the universe. This is shown in the second image, which we call the Cosmic Density Pyramid.

The white Pyramid of All Visible Matter is at the top, unchanged, but now it is revealed to be only the visible tip of a massive iceberg of invisible substances. Visible matter is only 0.5 percent of the total. The black part of the pyramid represents the invisible: that is, what we do not see but know is there. The top and smallest layer of invisibility is simply non-shining and unlit atomic matter—matter that is invisible in fact, but not in principle. It is about 4 percent of the total. Below it lies the far more massive layer of dark matter, which is invisible in principle. It represents about 25 percent of the total. And below that lies the energy that is accelerating the expansion of the universe, dark energy. It is 70 percent of the total density of the universe now.

You may have noticed that the amounts add up to a little less than 100 percent. Not included in the diagram are the small contribution from radiation (light of all wavelengths) and the uncertain contribution

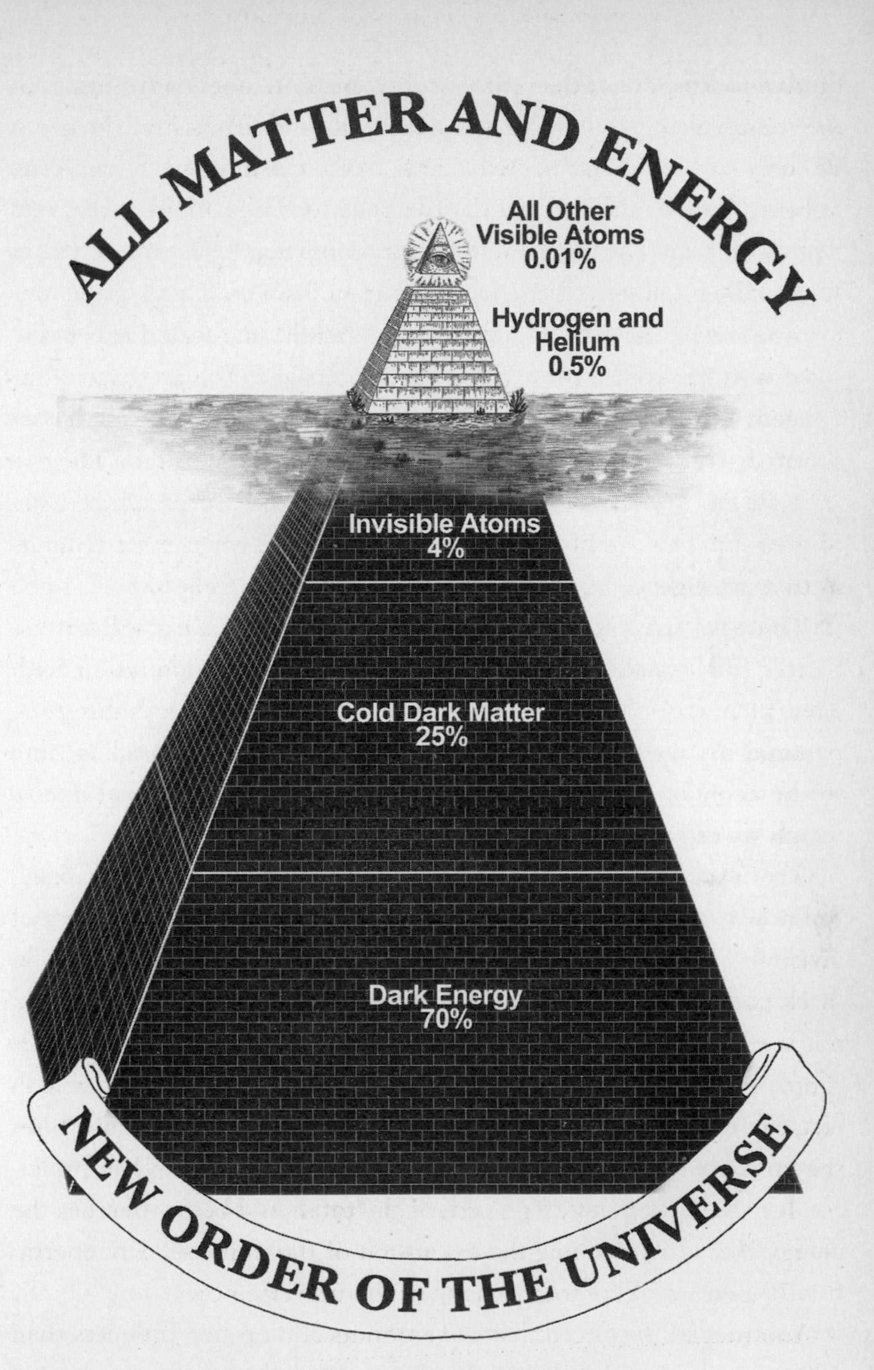

Figure 6. The Cosmic Density Pyramid.

from neutrinos. The heat radiation from the Big Bang represents only about 0.01 percent of the total cosmic density, and the light from all the galaxies that have ever shined is less than a tenth of that. The contribution of the neutrinos from the Big Bang to the cosmic density is between 0.01 and 2 percent; it is still uncertain because the masses of the three kinds of neutrinos are not yet known.[50] We can think of the neutrinos as being represented by the gap (of uncertain thickness) between the capstone and the white pyramid.

At first it seems amazing that there is so much out there that no one knew about until very recently. But then, once you understand that dark matter and dark energy exist, it seems just as surprising that the Cosmic Density Pyramid is so evenly divided between these different sorts of ingredients. One section could have been a billion times larger than the next, but it's not, and we don't know why—except that our Galaxy could never have formed if there were much more dark energy than there is. On the Great Seal of the United States below the pyramid, the Founding Fathers inserted a flowing ribbon, and on it they wrote in Latin: THE NEW ORDER OF THE AGES. That this was the new order was the overriding thought they wanted people to hold in their minds when viewing their pyramid, and the same is true here. The Cosmic Density Pyramid is the New Order of the Universe.

The Cosmic Density Pyramid serves at least four major purposes by helping us get some perspective on the universe.

1. It gives an understandable graphic of the latest scientific data on the relative proportions of what's in the universe, while simultaneously acknowledging what can't easily be pie-charted: the eye, mind, vision, intelligence—ourselves.

2. It emphasizes that there is a distinction *in reality* between the visible and the invisible; the distinction is not just a description of what we humans can see. Dark matter and dark energy are not simply unlit. By saying that they are invisible in principle, we mean that they have nothing to do with any form of radiation we know of that could carry

information—not only to human eyes but probably to the sensing organs of any living being in the universe.

3. The Cosmic Density Pyramid as a symbol has the potential to show not only what the universe is made of but how its composition has changed over time and will continue to change in the future. Dark energy in the early universe was insignificant compared to dark matter, but now it is almost three times more abundant. Since the amount of dark energy keeps growing compared to the amount of matter, in billions of years it will become exponentially larger than everything else. If we keep the total size of the symbolic pyramid constant, the dark energy is pushing upward, and in the distant future everything else including dark matter will be squeezed into the capstone, which if intelligent life continues will nevertheless retain its eye. The Cosmic Density Pyramid as shown above represents the composition of the universe today, but the symbol can be adjusted to show the *distinctive fingerprint of any era* in the history of the universe, past or future, by simply moving the dividing lines up and down to reflect the relative proportions characteristic of that era. In this way the symbol can communicate a lot of information by very simple changes.

4. The Cosmic Density Pyramid reveals that we are living in what might be called "the midpoint of time." That which in the symbolic pyramid is the rising tide of dark energy will in the real universe mean exponentially increasing amounts of space between our Galaxy and distant galaxies, making galaxies that are now visible disappear. The night sky will look much the same to the naked eye, since all the stars we see are inside our own Galaxy, and our Galaxy is bound together by gravity. But distant space, where with telescopes we can see hundreds of billions of galaxies, will look very empty to our distant descendants; few galaxies will be visible then from *any* point of view in the universe. Therefore, what the astonishingly evenly balanced Cosmic Density Pyramid of *today* represents is a special window of time that can only happen during a relatively brief epoch in the entire history of the uni-

verse: late enough that intelligent beings have evolved who have instruments to observe the distant galaxies, but not so late that the galaxies have begun to disappear. Without this period of overlap it might have been impossible for intelligent beings ever to get the chance to figure out the nature of the universe.

This is the mini-icon that stands for the Cosmic Density Pyramid. There is a point at the top—and we are the point. In all the mini-icons to come, our place will be shown by the point.

Discovering Our Place on the Cosmic Density Pyramid

When astronomers look into space with all their instruments, they see only the illuminated half-a-percent of what's out there. It is as though great fleets of ghost ships made of dark matter move through the ocean of dark energy, but in the blackness all we humans see are a few beacons lit at the tops of the tallest masts. The luminous/dark divide is a fundamental physical one. For the same reason we can't see dark matter, it can't "see" itself; it hardly interacts at all with other dark matter except gravitationally, and thus it can't create much complexity. Ordinary matter interacts with itself: particles interact to form atoms, atoms interact to form molecules, and under at least some circumstances molecules can form living cells and eventually evolve into higher life forms. But dark matter does none of this. When viewed in computer simulations that make dark matter visible, the dark matter behaves like nothing anyone has ever seen before. Often when we see a pattern in nature, for example the shape of a spiral galaxy, it reminds us of other things, perhaps the shape of a hurricane seen in a satellite photo; but there is nothing even vaguely reminiscent of the behavior of dark matter. Gravity swings clumps of it around in the presence of other clumps, but they can pass right through each other.

This chapter has talked a lot about quantities, but just for a moment we are going to talk about value, exploring the connection between the quantity and value of the various cosmic ingredients, and why the luminous matter is so much more important than its paltry amount might suggest. Children often think it would be great to be invisible because of everything you could do, but cosmically speaking, they are wrong: most of the universe is invisible, and with the same 14-billion-year opportunity our luminous kind of matter has had, it hasn't done nearly as much. Unlike dark matter, luminous matter has become vastly more interesting and complex than it was in the beginning, including the creation of life and intelligence. Our kind of matter does not take up much space or contribute much to the total density of the universe, but it contributes out of all proportion to the richness of the universe.

Our cosmic identity is a central theme of this book because consciousness is limited by identity, and identity is limited by ignorance. Most people tend to identify themselves with fairly narrow categories—a nationality, a race, a religion—which leads not only to conflicts but also to a stunting of imagination and potential. The wider our sense of identity, the more likely we will be able to experience our genuine connection to the universe. If a lost child who knew nothing of her background and had been raised by an indifferent family suddenly discovered that she was the direct descendant of an illustrious house traceable back many centuries, her sense of identity would expand momentously even before anything else changed. So many stories are based on this, from "The Ugly Duckling" to *The Winter's Tale*, that it must reflect a deep desire for transcending the everyday world and finding that one has a larger identity. The discovery of our own cosmic genealogies may have a similar expansive effect. We humans are luminous, stardust beings. When we see ourselves this way, we take the first step toward identifying with our place in a universe *that we have convincing evidence actually exists*. The idea of a larger identity is complicated and will be a major focus of the last chapter, but what we are made of is piece number one in the puzzle.

Imagine a mountain peak on the eastern horizon as the sun rises directly behind it. The mountain would still be dark, but as the first morning light streams over it, the peak would brighten and rays would glow from behind it much like our pyramid.

Full many a glorious morning have I seen
Flatter the mountain tops with sovereign eye.

—SHAKESPEARE, SONNET 33

For Shakespeare the sun itself is a sovereign eye, flattering the dark mountain with its radiance, just as the glowing top of our pyramid gives meaning to the dark, mute base. No matter how minuscule the relative share of cosmic density that is associated with intelligent life becomes in the future, the eye will always remain the focal point of this symbol. This is a value statement, but intelligent life will always be the most important thing to intelligent life. Nothing else in the universe is complex enough to hold our interest. The eye of the Cosmic Density Pyramid is truly sovereign to us.

When the Newtonian picture destroyed the comforting medieval universe and people stared out into endless space and shivered at how small they were, they felt for the first time the existential terror of cosmic insignificance. But even though the universe is overwhelmingly larger than those seventeenth-century people imagined, we humans are not insignificant, because we are citizens of the luminous and rare; the tremendous complexity of our minds lets us do what no amount of dark matter or dark energy can ever do. In astronomer Alan Dressler's words, "If we could but learn to look at the universe with eyes that are blind to power and size, but keen for subtlety and complexity, then our world would outshine a galaxy of stars."[51] Our Sovereign Eye—representing intelligent life wherever in the universe it may be—will outshine the rest of the Cosmic Density Pyramid as long as intelligent life goes on, no matter how the dividing lines among the ingredients of the universe change in the future.

There is a legend about Egypt's Great Pyramid that originally its capstone was plated with gold and on each of its four sides was painted a blue eye of Horus, the god embodied in the living Pharaoh. When the sun struck the pyramid, the reflected light could be seen for miles. At some point the priests removed the capstone and buried it secretly. It has never been found. But when found, the legend goes, it will be placed back upon the pyramid, and on that day will begin a "new order of the ages," a new awakening.[52] We moderns have lost our place in the universe for the past few centuries, but the scientific revolution in cosmology has revealed a new reality. *We* can replace the mythic golden capstone by awakening to its possible new meaning—that on the Cosmic Density Pyramid, we luminous, stardust beings *are* the Sovereign Eye that gleams from the capstone. We are no longer lost; we have discovered our place in the new order of the universe.

FIVE

What Is the Center of the Universe?

THE COSMIC SPHERES OF TIME

THE MILKY WAY *is the center of our visible universe.* How can our Galaxy be the center? Is this an astonishing coincidence or a sneaky manipulation to get a desired result? Neither—it's inevitable. Every galaxy is the center of its own spherical *visible* universe. The outer edge of this sphere is called the "cosmic horizon," and no light or information of any kind can have reached us from beyond it. There is no real edge at the cosmic horizon—any more than there is an edge of sea and sky at the horizon that you see from a ship; if you sail your ship toward the horizon, you simply see it move that much farther away. But in the universe, the view from our Galaxy is the only view we will ever have. The visible universe is *our* patch of the universe—it is unique to our Galaxy. Other galaxies' visible universes are the same size as ours and overlap with ours, but none can ever be exactly the same.

The reason why there is a horizon is that it takes light time to get here. Light travels at a constant speed in the vacuum of space, 186,000 miles per second, or 300,000 km/sec.[1] The distance light travels in a year is called a light-year. Any galaxy so far away from us that its light has not had time to reach us in the entire history of the universe lies outside our horizon, although it is inside the much larger universe created by the Big Bang.

When we look out into space, we look back in time. When we look across a room, we are also looking back in time, but only by the amount of time it takes light to reach us from there, which is so short that it is virtually instantaneous. When we look at the sun, however, we are looking back in time eight minutes. If the sun somehow imploded (it can't, don't worry about it), we on Earth would still see it placidly shining in the sky for eight minutes after it stopped existing. We see distant galaxies as they were billions of years ago when they emitted the light just reaching us now—but they have been moving away all that time with the expansion of the universe. Our cosmic horizon is the sphere surrounding us that represents the *present* location of the most distant things we can see—not their location when they emitted the light (which was much closer to where the matter that makes up our own galaxy was then).[2] The only way to know their present location is through theory—the Double Dark theory. Because of the fixed speed of light and the ongoing expansion of the universe, the size of our visible universe, which is a *spatial* quantity, is always changing depending on what we can see, which depends on *time*—specifically, the age of the universe. In modern cosmology space and time are more than intimate; they're inextricable.

"What, then, is time?" asked Saint Augustine in the fourth century. "If no one asks me, I know what it is. If I wish to explain it to him who asks me, I do not know."[3]

For centuries philosophers have debated whether time is something real or merely a human method of organizing our experience. In modern cosmology, time is as real as the cosmic horizon, as real as an electron, as real as the universe itself—but it's not what most people think it is. *The way we understand time determines how time affects us.* To understand time, we need to understand space. The goal in this book is always to find where we belong in the universe, and part of feeling that we belong is to be able to locate ourselves in time and space, but we need new concepts even to be able to think about what it means to "locate" anything. The center of our visible universe is not a place but a cosmic collaboration of space, time, light, and consciousness. Let's start with space.

A New View of Space

When astronomers say the universe is expanding, they mean that all distant galaxies are speeding away from our galaxy, and the farther away they are, the faster they are going. If the expansion were a movie and we ran the movie backward, all points would converge and we would see that the universe must have started in an extremely dense state—the Big Bang. When TV shows or movies try to illustrate the Big Bang, they generally start with a black screen—an empty "space"—and then there is an explosion *in the middle* of it and the fiery remnants spread out to fill the blackness. This is completely misleading. Before the Big Bang there was no preexisting space—at least, nothing like space as we know it. The Big Bang happened *everywhere*. It created the possibility of both "where" and "when." All wheres and whens are inside the Big Bang.

Time is a relatively sophisticated concept, and animals probably don't think about it; but the nature of space involves assumptions so deep biologically that we share them to a great extent with all animals that move and must therefore orient themselves in space and estimate distances. So don't be surprised if the new ideas about space seem absurd at first. That there was no space before the Big Bang is just one of the counterintuitive ideas of modern cosmology.

We have just said that expansion means distant galaxies are speeding away from us and from each other, but they are not speeding away *through* space, the way two spacecraft might fly away from each other; instead space itself is stretching between distant galaxies, and between those galaxies and us. Space is infinitely expandable. Space between distant galaxies and galaxy clusters expands at roughly the same rate throughout the universe, and thus astrophysicists say the universe is *uniformly* expanding. Galaxies are flying away from us faster, the farther they are from us—not, however, because space is expanding faster out where they are, but simply because there is more total expanding space between us and them. Therefore, in another counterintuitive result, every galaxy is not only the center of its visible universe, but it also appears to be the center from which the entire universe is expanding.[4]

Most people think of space as emptiness, or as a geographical location for things. But this is like thinking of skin as simply the geographical border of the body. Skin is a complex, living, sensual organ that happens to play the role of a border. Space, like skin, also plays the role of a border between things while having its own complex nature. According to relativity, space has an internal structure that is determined by the matter and energy at each point *in* it. In the solar system, almost all the contents of space are visible matter—the sun and the planets, moons, rings, asteroids, and comets. On the scale of the Galaxy and a little beyond, however, most of the contents is invisible dark matter. And on much larger scales, as we have seen, dark energy dominates, and this drastically changes the character of space.

Let's look at space by means of an analogy. Suppose you were a child who had spent your short life in New York City. You might think of the woods 150 miles away in the Catskill Mountains, of which you had only heard romantic stories, as the outside world. But then suppose you were somehow spirited into a trackless jungle, full of wild animals and terrifying sounds, and no friendly humans anywhere. You would realize that the real outer world was not just the civilized countryside of New York State but something you could not even have imagined.

Our New York City is Earth. Our romantic woods are analogous to what people call "outer space," a place very few of us have visited and which we mostly know about through pictures and stories. Outer space is generally thought to begin outside Earth's atmosphere where some artificial satellites orbit, and to continue beyond the moon and planets to the rest of the universe. But there is a huge distinction between the Catskills and the Amazon. If we want to think of "outer space" as starting just outside Earth's atmosphere, then we need to think of it as *stopping* at the edge of our "Local Group" of galaxies (the Milky Way, Andromeda, and about thirty little satellite galaxies). A conglomeration of matter like our Local Group, which is bound together by gravity, doesn't expand anymore but instead travels as a unit in the great expansion. Thus all space inside our Galaxy, and between our Galaxy and other members of the Local Group, is a special kind of space, because it's been tamed by gravity. Outside our Local Group lies the real outer space—wild space. On this huge scale, space is truly a jungle, where dark energy is tearing apart all large structures made of matter and accelerating the rate of expansion. Wild space is carrying hundreds of billions of galaxies away like petals on the Amazon River.

Meanwhile, if we use a microscope instead of a telescope and peer into space on the very small scale, something completely different is happening. Trillions of particles are blipping into and out of existence on time-scales so short that they fall between moments and exist only virtually. When people assume that space is nothing but the difference between here and there, they are making an accurate approximation for everyday life—but on both very large and very small size-scales they're very wrong. Space is the most dynamic thing there is.

A New View of Time

History is not background when it comes to the universe—the universe *is* its past to a great extent, as we will see. Light from the distant universe is stretched by crossing different amounts of expanding space before it arrives at our telescopes. Although its *speed* on arrival is unchanged, its *wavelength* (which determines its color) expands just as much as the universe has expanded since the light was emitted. This expansion of wavelength causes colors to become redder. This "redshift" is very precisely measurable and tells us how much the universe has expanded since the light left each galaxy, and (by plugging this information into the Double Dark theory) how far away that galaxy was when it emitted the light and how far away it is now.[5] All space is filled with such information moving every which way, readable by intelligent beings on any world. The universe is not only the way it looks today to us, but is also all the ways it looks to other worlds based on the information reaching them from different places at different times. Thus what is past depends on where you are; the past still exists, and it continues to shape the present and future. An object such as a chair possesses a history, but even if no one knows its history, the chair still exists. A play, on the other hand, does not exist except over time. It's not as though the curtain call at the end is the play and all the earlier scenes are its historical background. The play is the whole event, as well as the experience of the observers. The universe is more like a play than a chair.

Since the Industrial Revolution, most people no longer sell their handiwork; they sell their time. Our everyday lives are governed by the idea that time is divisible into small, equal, countable units; in short, time is money. If this is the fundamental understanding of time we have, it is a disastrous degradation, substituting hard little artificial bits for the majesty of nature's many rhythms. Unlike days and years, minutes and hours don't correspond to anything in nature; and since one is like another, they overshadow the drama of cosmic history with endless repetition. We need to burst out of this limiting notion of time, which tends to push us backward into the notion that time is cyclical like our watches. Time moves on cycles of many sizes, but it is also linear and possibly eternal; the universe is an orchestra of rhythms, and time is all of them creating a harmony that cannot be intuited without new concepts.

Time is measured by clocks.[6] Clocks, however, are not just machines manufactured by humans; a clock can be any regular motion in the universe. The first clocks discovered were probably night and day, the seasons, the phases of the moon, the beating of the heart, and the menstrual bleeding of women. These were all cyclical events and led ancient people to believe that time itself was cyclical. The ancient Egyptians invented the twenty-four-hour day[7] and the predecessor of our modern calendar.[8] They kept careful records for thousands of years, but every Pharaoh was the god Horus again. The patterns of the stars repeated. The cosmos was not going anywhere. Even today indigenous tribes may have no idea how old they or their customs are: 100 years, 1,000 years, and 10,000 years are indistinguishable if time is cyclical.

How Much Time Has There Been?

The Hebrew Bible broke with this tradition and related the story of the world in *linear*, non-repeating time, demarcated by a series of events that occurred only once, like Noah's flood, Abraham's pact with God, and Moses leading the Exodus. This had a huge impact on people's understanding of time. In the seventeenth century, Bishop James Ussher calculated, on the basis of the ages of the generations in Genesis starting with Adam, that the creation of heaven and earth occurred on October 23, 4004 B.C.E.[9] Clearly it mattered to him and many others to know how old the earth was, and he concluded that it was almost six thousand years old. In the eighteenth century, European geologists began to figure out how the earth wears down and new mountains form.[10] When they actually calculated how long these processes would take, they were stunned to see that geological change requires not thousands of years but hundreds of millions. Darwin also realized that evolution requires at least hundreds of millions of years. Most people had never before thought on such a vast time-scale.[11]

How could the earth be that old? Physicists initially didn't buy it. By the late nineteenth century the gloomy perspective of thermodynamics had most of them convinced that the sun was running down, that it couldn't possibly have been burning longer than maybe forty million years, and the earth would also soon cool and die.[12] But this was because no one yet understood atoms—they had no idea that an atom has a nucleus packed with energy, and that this nuclear energy is what keeps stars like the sun burning for billions of years and keeps the core of the earth molten and dynamic. The discovery of radioactivity led to this understanding, changed our physical picture of reality, and made it possible to figure out how old the earth and solar system really are.[13]

Radioactive decay is a natural clock based on quantum mechanics. In radioactive materials, what happens is that the *nucleus* of an atom spontaneously breaks its own internal bonds and spits part of itself right out of the atom with a blast of energy. Since an atom's nucleus determines what kind of atom it is, radioactive decay (as the spontaneous splitting is called) transmutes atoms into new elements (called decay products). This is what alchemists tried to do unsuccessfully for centuries. Think of the radioactive nucleus as a prison. An alpha particle (one piece that can split off) is trapped inside and knocks again and again on the wall. According to classical physics, it would have to jump over the wall to get out, but there's no way it can get that much energy. In quantum mechanics, on the other hand, it has a small probability of actually tunneling right through the wall, and suddenly it's out. The probability of getting through the wall is the same each time it knocks. Those still in prison just keep knocking, and sooner or later random others will also get out. *Statistically* the process is absolutely predictable: half of the atoms that make up the radioactive element will decay in a fixed period of time called the "half-life" of that element, and half of those that are left will decay in the next half-life, and so on, even though which *individual* atoms will be the ones to decay, and when, is absolutely unpredictable. The clock of radioactive decay is not cyclical and does not run smoothly like a mechanical clock, yet it is one of the fundamental timekeepers of the universe.

Radioactivity provided the first truly accurate clock that could measure ages on the scale of geological and even cosmic time.[14] By 1956 the age of the solar system was known, and over the last half-century this number has been repeatedly reconfirmed.[15] Even the hundreds of millions of years that shocked the early geologists was a puny underestimate. Earth is about 4.5 billion years old. All of recorded history—the past 5,000 years—represents only the last *millionth* of the age of the earth.

Why does it matter to know the true age of the earth? This book is about re-envisioning reality from a cosmic perspective in order to see how we fit in. Our species only evolved to its present form about 100,000 years ago, which means it took Earth almost the *entire* 4.5 billion years of its existence to produce our kind of intelligent life. Since the sun will become a red giant star in another six billion years, that marks the end of Earth as a habitat for our kind of biology. If so, we intelligent creatures have appeared at approximately the *midpoint of Earth's lifetime*. We are almost brand-new, and our descendants can look forward to a future of many billions of years. We help locate ourselves in the immensity of the age of the earth when we begin to appreciate the kind of processes, like our own evolution from a primordial cell, which can occur only over such a time-scale. Each of us only gets to live through a brief flash of that immense time, of course, but our consciousness is not limited to our flash, as anyone knows who's ever read a historical novel. Living in the consciousness of planetary-scale time deepens our identity and, most importantly, reveals the *opportunities of our time*, which, if we rise to them and embrace them, can provide us with the kind of cosmic meaning that humans have sought for thousands of years. There is no fundamental reason why our descendants can't be here for the second half of Earth's existence, or longer.

The Age of the Universe—Measuring Cosmic Time

The age of the earth is conceptually a much simpler idea than the age of the universe. We're all riding the earth together. This means we have simply ignored the fact that it's been spinning around its axis, rotating around the sun, spiraling around the Galaxy as our sun orbits the Galactic center roughly every 220 million years, and moving with the Galaxy itself in both the local flow of galaxies and the overall expansion of the universe. What does all this motion have to do with age? Age is measured by clocks, and in our relativistic universe motion affects the rate at which clocks run. When we talk about the age of the universe, we can no longer ignore this.

People often make three fundamental assumptions about time: that it is universal; that if clocks were absolutely accurate, they would all measure time passing at the same rate; and that time can be divided basically into past and future, with the past over and gone. These are useful approximations for getting along on Earth, but with regard to the universe as a whole, all three are wrong.

Weird effects described by Einstein's relativity happen to time as soon as people or things move *with respect to each other* at speeds comparable to the speed of light; one of those effects is that each one observes the other one's clock as running slow. There's nothing wrong with the clocks—*time itself* slows down for the fast traveler, compared to the rate at which it runs for those left behind.[16] The effect is real: if one twin could travel at a speed approaching that of light, from the viewpoint of Earth her time would slow down so much that she would come back younger than her twin sister.[17] Clocks also run slow when gravity is strong. For the universe as a whole, in which everything is moving with respect to everything else and some of the light that we see comes from near black holes where the gravity is very strong indeed, what clock can we possibly use to describe a meaningful age? What all cosmologists do is to imagine that there are clocks everywhere that started running at the Big Bang and that move with the expansion of the universe along with the nearby galaxies. Our creation story below will be measured against such clocks. This gives a definite cosmic time that all observers can agree on.[18]

The age of the universe can't be measured directly; but it can be calculated several different ways based on different sorts of observations. One method is to figure out how long it would have taken for the universe to expand to its present size. To do that, we need to know the expansion rate now and how much matter and dark energy there are; fortunately, we now know all three numbers fairly accurately. According to general relativity, the higher the expansion rate, the younger the universe. Independently of the expansion rate, there is another factor that depends only on how much matter and dark energy there is: for the same expansion rate, the universe is younger the more matter there is, and older the more dark energy. Consequently, until we knew what the universe was made of—that is, the contents of the Cosmic Density Pyramid—no one could tell for sure the age of the universe. There was even a period in the 1990s when the universe seemed to be younger than some of its stars.[19] At that time, no one knew whether the universe just contained matter or whether dark energy might actually exist. The age of the universe is the linchpin of our understanding, because interpretation of many aspects of the universe depends on its age, and those aspects all remained uncertain until we pinned the age down. In 1998, the discovery of convincing evidence for dark energy ended the uncertainty. Dark energy exists. All discrepancies have disappeared. The universe is about 14 billion years old. We finally understand how our cosmos fits together.

Today every measurement of the age of the universe, made in every way cosmologists have dreamed up, agrees.[20] The age will be refined, but it won't change. Like the age of the earth (4.5 billion years), we now know the age of the universe (about 14 billion years). We know how much time there has been, and we humans can locate ourselves in the immensity of it. Earth is almost incomprehensibly ancient, yet the universe was twice as ancient before Earth had even formed. The value of knowing the age of the universe is not just to toss around a big number but to start putting together a scientifically verifiable story, painting a picture on a scale no one could visualize before.

As recently as the seventeenth and eighteenth centuries, the general view was that the earth and universe were the same age—about 6,000 years—which was only five days older than the first humans, who according to Genesis were created on the sixth day. Even today on opinion polls almost half of U.S. respondents typically say that they believe that the earth and the entire universe are less than 10,000 years old.[21] But the evidence of the enormity of time fills space and is written in fossils, rocks, and meteorites here on Earth, and in the rocks that the astronauts brought back from the moon. The universe is so old that the people who wrote the Bible could not even have conceived of the number, let alone the reality of such a universe.

The Cosmic Spheres of Time

It is possible to visualize the modern idea of time. In the last chapter we borrowed the symbol of the Cosmic Density Pyramid from the dollar bill; here we will borrow our symbol of time from the Middle Ages. Below is a drawing of the universe as medieval people actually envisioned it. Medieval people felt themselves at the stable center of a great spherical universe. They were heirs to the Greek imagination, which burst out of the effectively two-dimensional flatland cosmos of Genesis to a three-dimensional universe, and they captured this insight for all time in the cosmology of the heavenly spheres. In their picture of the universe, God was outside and cradled the whole structure of nested crystal spheres. They were wrong astronomically that Earth is the center of the universe, but they were right psychologically: the universe must be viewed from the inside, from our center, where we really are, and not from some perspective on the periphery or even outside.

The classic image of the heavenly spheres remains useful. It expresses a truth not only about the universe but also about how people *experience* the universe. We do experience it surrounding us, and indeed we *can* accurately say that we are surrounded by nesting spheres, but in modern cosmology they are not hard crystal objects or orbits of celestial bod-

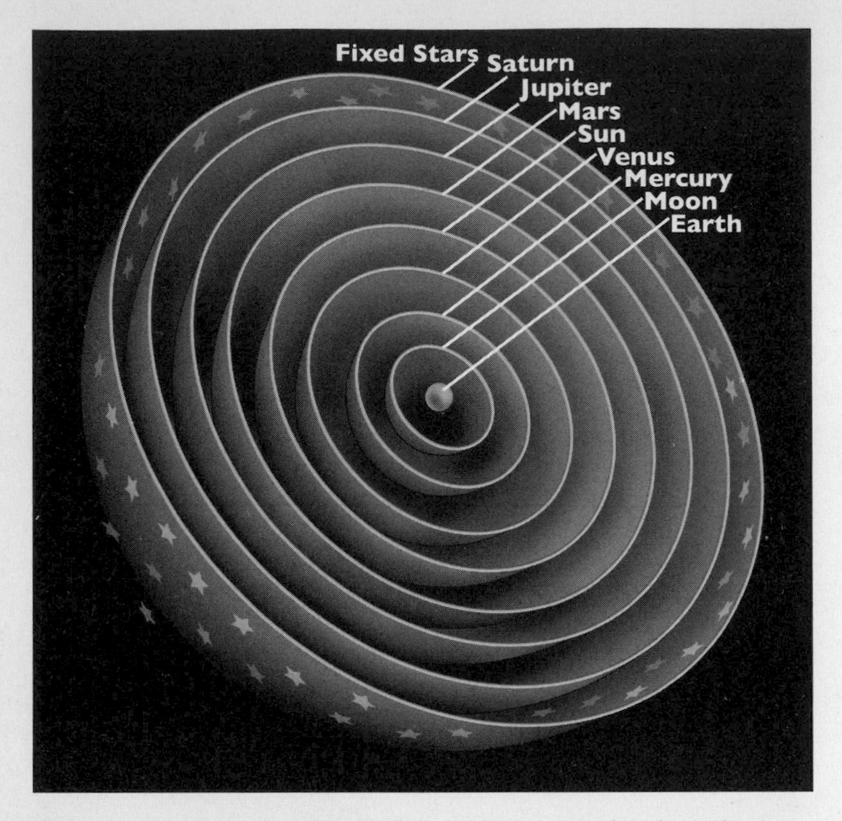

Figure 1. The Heavenly Spheres. They are drawn as medieval people imagined and described them. (They would have drawn them this way if they'd known how to draw in perspective. Instead they drew concentric circles, equivalent to the edges of this cross section.[22])

ies. They are what we will call "Cosmic Spheres of Time," and we truly are at the center in a sense never imagined in the Middle Ages. Since looking out into space is looking back in time, each concentric sphere in the figure below, moving outward from today, represents an earlier epoch in the evolution of the universe. The farther away from us a sphere is, the farther back in time are the objects that we observe in that sphere.

In the Cosmic Spheres of Time, our Galaxy today is the center of our visible universe. The innermost sphere around us contains the most re-

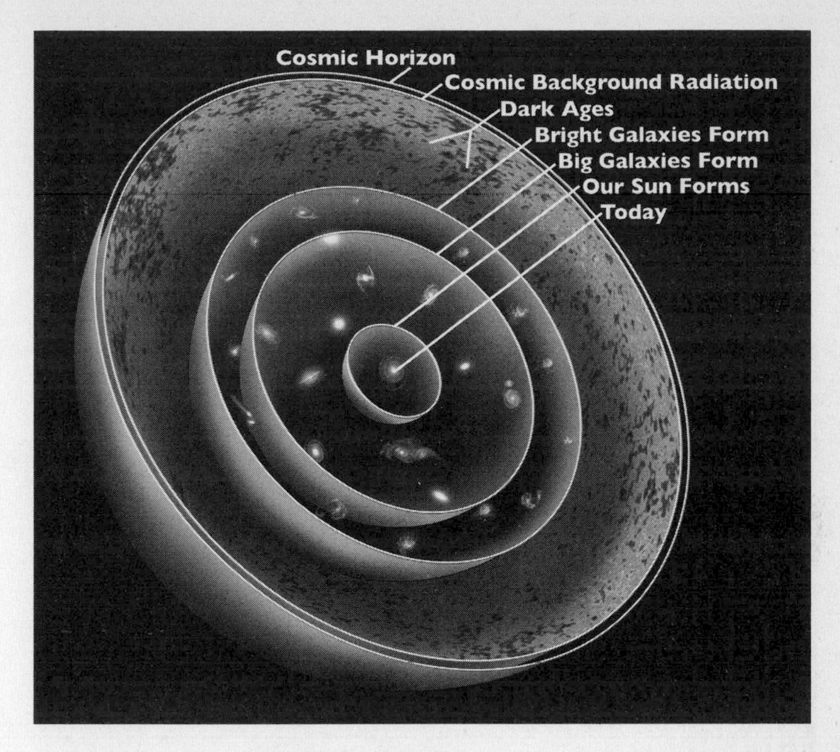

Figure 2. The Cosmic Spheres of Time—Our Visible Universe.[23]

cent past and the nearest galaxies. Light that started its journey toward us when the sun and the earth formed, about 4.5 billion years ago, is represented by the edge of the central sphere ("Our Sun Forms") in the figure. On the large scale, the universe looked much the same then as now, which is why the galaxies inside the second sphere ("Big Galaxies Form") look the same as those in the central sphere. But farther back in time lies the evidence that galaxies have evolved. Galaxies in the early universe—the next sphere out ("Bright Galaxies Form")—were irregular and small, but very bright because they were forming stars very rapidly.[24] Beyond that sphere there were no visible galaxies yet—between about ten million years and a few hundred million years after the beginning, it was literally the cosmic Dark Ages.[25]

The fourth sphere ("Cosmic Background Radiation") represents the time when the hot plasma-fog of the Big Bang explosion dissipated as electrons combined with nuclei to form atoms, and the universe became transparent to light. At this point, only 400,000 years after the beginning, the cosmic background radiation—the light of the Big Bang itself—was first able to escape through space and has filled the universe ever since. We see this radiation coming at us uniformly *from every direction*[26] (confirming that the Big Bang was everywhere). By about ten million years after the beginning, the cosmic background radiation was no longer visible light but had become invisible infrared radiation because its wavelengths had been stretched by expanding space. This was the beginning of the Dark Ages, because from a human point of view the universe was completely dark.

Today we receive the cosmic background radiation stretched even more, so that what started out like sunlight now reaches Earth as short-wavelength radio waves. When we observe the cosmic background radiation, we see within it images of the distant universe as a baby, long before it had galaxies or stars. Our observations have benefited from continually improving technology. Although the cosmic background radiation was first detected in 1965, no details could be seen in it until 1992 and sharp detail across the whole sky only in 2003.[27] Today's high-resolution pictures of this radiation show subtle patterns *ingrained in spacetime* clearly existing 400,000 years after the beginning. On the Cosmic Spheres of Time, the design reproduced on the inside of the sphere marked "Cosmic Background Radiation" is from the actual observations. These patterns are effectively the universe's genetic makeup. They have developed into galaxies, clusters, and superclusters, and they will continue to govern the evolution of the universe on the large scale forever. The origin of these patterns actually seems to predate the Big Bang, as we will discuss.

The baby pictures of the universe delivered to us by the cosmic background radiation are probably the earliest pictures we will ever get to see. The only way to see back even farther would be to learn how to "see" with neutrinos or gravity waves, rather than light, because long before the hot plasma-fog of the Big Bang became transparent to light, it was already transparent to neutrinos and gravity waves. With them we could see nearly back to the Big Bang, observing the universe not only in infancy but as an embryo in utero. However, neutrinos and gravity waves are much harder to detect than light.[28] Like light, their wavelengths will also have increased more, the closer to the Big Bang that they were emitted. But longer wavelengths mean decreased energy, which is why it becomes more and more difficult to see *any* radiation emitted close to the time of the Big Bang. The cosmic horizon is thus not a technological limitation to overcome but a limitation in principle. It creates the outer boundary of our visible universe and is represented by the outermost sphere, labeled "Cosmic Horizon."

The region outside the spheres is inaccessible, but if we could see galaxies that lie beyond our visible universe, would they look very different when their cosmic clocks read the same as ours—14 billion years after the Big Bang? Probably not. The galaxies for a long distance beyond our horizon should be much the same, although at some great distance we now think everything changes (see Chapter 7).

Everything beyond our visible universe is theory—but it's not guesswork. Scientific cosmology makes the strategic choice to assume that the laws of physics we have discovered in our laboratories and in the nearby universe apply everywhere and at all times. This strategy is not simply a guess; it's continually tested against observations and it has been repeatedly confirmed.[29] This gives us confidence to push into the unknown. No information of any kind can reach us *directly* from beyond the cosmic horizon, but that doesn't mean we can't find it out. We can reason it out from our theories. Theory is like a telescope or a microscope—it extends human abilities far beyond the power of our five senses, but is nevertheless limited, like any technology.

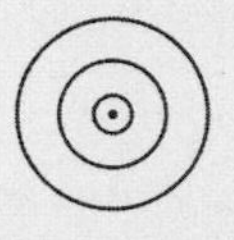

We are at the center of the Cosmic Spheres of Time, and this symbol ties together all these ideas and immerses us in them: that the universe is expanding; that the speed of light is the limiting speed for everything but space; that looking out into space is looking back in time; that the universe evolves and is very different from what it was in the beginning and will be in the future; and that human (or intelligent alien) consciousness is an essential element of what makes a visible universe. This mini-icon of the Cosmic Spheres (see page 137) reminds us of all these things. We humans feel ourselves central to anything we're actually participating in—that's the nature of our kind of consciousness. We embody the Sovereign Eye. We *must* have a central part. There's nothing wrong with that; it simply reflects what we are.

The Cosmic Spheres of Time as the Future

The Cosmic Spheres of Time are shown as the past, but we can alternatively draw a similar diagram of the future of the universe, where "Earth, today" is at the center and the nearest sphere is the near future, etc. It's also possible to represent the Cosmic Spheres of Time for not only the past but all of time, past and future, and this is what we do below in Figure 3, The Lightcone of Past and Future. However, in a drawing limited to a page of a book, putting in the future requires cutting something out. What we've cut out is the third dimension of the spheres, thus turning them into circles. This may sound confusing, but if you give it a few moments it can dramatically clarify many ideas. The basic visual idea is that the cones are actually circles (remember, representing spheres) changing size *over time.* The arrow of time is from bottom to top. The farther into the past the backward lightcone extends, the larger the circles. This is another way of representing the basic idea of the Cosmic Spheres of Time, where the farthest back in time, the Big Bang, is the largest sphere. But now we can see some very interesting connections.

Imagine that we stand at the center point, EARTH, TODAY. We stretch out our arms like the horizontal axis and extend them into space. Arms

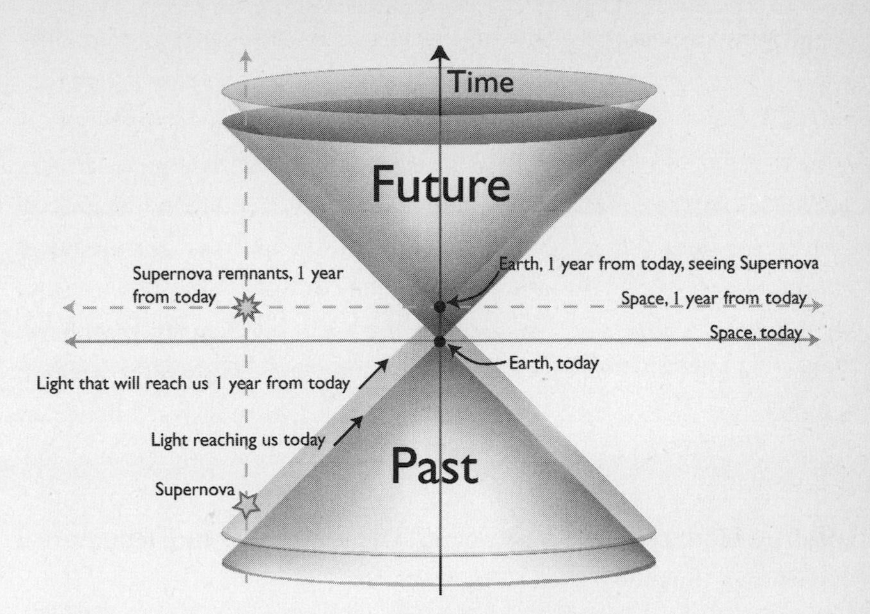

Figure 3. The Lightcone of Past and Future. This figure, two overlapping lightcones, represents our past and future today (the solid cones) and our past and future as they will exist one year from today (the ghostly cones). Space is plotted in the horizontal direction and time in the vertical direction. Think of each lightcone as circles that are changing in size over time. The cone labeled PAST is called the "backward lightcone." As information from the past of our visible universe pours into Today, it converges on the point labeled EARTH, TODAY. Thus the backward lightcone, labeled PAST, contains all events that could have had any causal effect on us as we are today. The cone labeled FUTURE represents the sphere of light expanding from where we are today. This cone and everything inside it is called the "forward lightcone." It contains every point in space and time that could be affected by what happens on EARTH, TODAY—and that is the meaning of our Future. Anything that happens in the region outside the lightcone cannot affect us yet, and whatever happens at EARTH, TODAY cannot affect it.[30]

outstretched, we slowly rise up our timeline by just standing still. Rising is simply a metaphor for time passing.

The forward lightcone, labeled FUTURE, represents the expanding sphere of light from a flash of light at EARTH, TODAY. Since nothing can travel faster than light, anything that happens here today can affect points in space and time only on and within this forward lightcone. The slope of the big diagonal lines that outline the cone represents the speed of light. Light travels on paths parallel to these diagonals, and anything slower is more up-and-down. Nothing can travel more sideways, because nothing can move faster than light. The ghostly lightcone represents where the lightcone of past and future will be one year from today. A supernova (lower left) that today is outside our backward lightcone, so that it has not affected us in any way and we don't even know about it, is destined to affect us next year as we drift up our timeline for one year to the point where the supernova first lies on the backward lightcone then. We'll see the supernova then as it was at the lower left, but at that moment—while we're seeing the initial explosion—the actual supernova, or more precisely its remnants, will be at the center-left, moving up its own timeline.

The lightcone shows the strange implications of the speed of light. Perhaps most surprisingly, it shows that much of the information we will have in the future is already on its way toward us, but right now it's not in our past or our future. It's outside our lightcone. Because of the limiting speed of light, time can't be divided into simply past and future. There is a third category: neither past nor future but "outside our lightcone"—perhaps "elsewhen."

The region outside the lightcone represents most of the universe, and it is not in our past or future because we're separated from it by too much "spacetime" for it to have affected us or for us to affect it. Einstein's concept of spacetime does not replace space and time—they are still meaningful on their own—but instead represents a deep interdependence between space and time. Time is dependent on the motion of the clock measuring it, and thus not all observers will agree on it. But because everyone sees light traveling at the same speed, relativity allows us to define a quantity called the spacetime separation that *all* observers must agree on. Within the concept of spacetime, the slippery individual quantities of space and time compensate for each other, revealing an underlying unity that is not relative but absolute—or "invariant."

There were occasions when Einstein regretted having named his theory "relativity" and mused that he should have called it the theory of invariance, since the invariance of spacetime is its central concept.

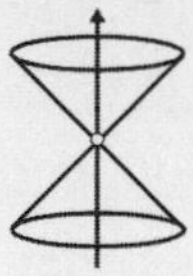

Whatever is not in or on our backward lightcone can have no effect on us today. But that doesn't mean it won't affect us in the future, or our children or our great-great-grandchildren. Light and other forms of information are already traveling toward Earth and will arrive in ten years, a hundred years, a million years from now. That information has been on its way for possibly billions of years. *Much of our future already exists—it just hasn't gotten here yet.*

The past exists, too—it's streaming away at the speed of light, and that's why we can't send it any information. But it's sending us plenty. We're reading the messages it wrote long ago that are arriving now. In some sense the past of the universe lies in our future, because more of the past will be found in the future as those ancient messages stream in, untouched and accurate after their long journeys. Imagine that a runner from ancient Greece were just arriving today, breathless, carrying the news of the defeat of the Persians, and we were the first to hear it from his lips. Runners are coming from all over the universe with news of their eras, in every form of radiation and the strangest languages anyone's ever heard; the cosmological revolution is human beings finally beginning to understand and put the whole story together.

"The Double Dark History of the Universe": Our Modern Creation Story

At last we have arrived at the beginning. This creation story is not only the truest story of our time, but the only one we know with any *chance* of being true. The opening characters are not Nun and Atum, or Adam and Eve, but neutrino and quark. Some parts of the story we are fairly sure we understand, and other parts we are less sure about. There are two different reasons for cosmological uncertainty: the first is simply that we don't understand the physics; the second is that we understand the physics but the situation is too complicated for us to be sure we can use the basic principles properly.

The period we are most confident about is from about one-thousandth of a second to about 400,000 years after the beginning. That period was unquestionably bizarre by earthly standards, but the universe was simpler then than it would ever be again—or than it was before (see Chapter 7). Cosmologists are uncertain about what occurred before the first ten-thousandth of a second, because the physics that was most relevant then is not yet fully understood. After that, the story becomes clear for quite a while, because the early universe was very smooth and governed by physics we have studied in detail in laboratories on Earth. But as time goes on, although the physical principles don't change, the complexity of the universe becomes so great that our uncertainty increases again. When we get to events such as the intricately interacting processes by which galaxies like our own Milky Way assembled and evolved—a subject of intense interest to the current generation of astronomers—we can't possibly reason it out from basic principles. What we do instead is to program into supercomputers the laws of physics and our best cosmological theories and let the universe evolve in cyberspace. Then we compare the result to actual astronomical observations to see if the cyberspace universe our theory produced is the kind of universe we actually live in. This is the first time in history that the technology exists to let us do this. Since the cyberspace universe does indeed match the observed universe *on the large scale*, we know that our theories are on the right track, because gravity and pure dark matter are relatively easy to understand. But inside galaxies what gets complicated is the ordinary matter. We still don't fully understand how crashing gas clouds form stars, or how the energy pouring out of a supernova or from the vicinity of a black hole affects the surrounding matter. We make nesting simulations, where in each successive one we zoom in to focus on a smaller size-scale, but there is a limit to how these extremely complicated simulations can be put together. The important thing is that we know what we don't know, and keep working to figure it out—with a lot of help from fabulous new telescopes and supercomputers.

We begin our story at one millionth of a second after the beginning: particles and antiparticles—quarks, electrons, neutrinos, dark-matter particles—exist free in a hot, incredibly smooth, non-turbulent, dense soup. For every billion antiquarks, there are mysteriously a billion and one quarks.[31] In a universal flash the quarks and antiquarks annihilate each other one-to-one, and of the two billion and one, only one survives. We and everything that we see are made out of the survivors. As the universe expands and the temperature drops, each surviving quark can no longer exist alone and binds with two more that have survived the same way. The three quarks create a single proton or neutron, some of which in a few minutes will begin coming together with other protons and neutrons to start forming nuclei of future atoms. But right now it's still too hot and dense for them to join together into nuclei. Even tiny neutrinos, which in the later universe will stream through the entire sun without interacting with a single atom, can hardly move without interacting. They are changing protons and neutrons into each other incessantly.

It is now one second after the Big Bang. The density has dropped so much and neutrinos have now lost so much energy that most will never interact with any other particle in the entire future of the universe. Since the neutrinos no longer interact, neutrons that the neutrinos were keeping active with their interchanges now start to decay with a half-life of only ten minutes. The temperature drops a little more, and now electrons annihilate with positrons (antielectrons).[32] Once again, there is one extra electron in about a billion pairs of electrons and positrons, and only that extra electron survives. The annihilated billions become energy, which fills the universe in the form of energetic photons, particles of light.

It is one hundred seconds after the Big Bang. Until now, the temperature has been too great to allow a proton and neutron to stick together to make a deuterium nucleus, or deuteron[33]—the frequent bombardment by the very abundant energetic photons breaks deuterons up as fast as they form, before they can fuse with anything. Meanwhile, the isolated neutrons are decaying. But as the universe expands the temperature falls further, and now deuterium can stick together. The great fusion of protons and neutrons into deuterons and then into helium nuclei sets the universe ablaze in a *universal* hydrogen bomb, which creates most of the helium there will ever be. The explosion is over in a few minutes as the universe continues to expand and cool. About three-fourths of the ordinary matter is now hydrogen nuclei (protons) and almost all the rest is helium nuclei (two protons and two neutrons).

It is now 50,000 years after the Big Bang. Streaming everywhere in the expanding universe, the photons that make up the heat radiation have been losing energy. The energy of the matter particles, however, is mostly locked into their mass (mc^2 again), which does not change with time. The matter, as will forever be the case, is mostly non-atomic dark matter. When the photons were full of energy, they were kings, dominating the density of the universe, and their energy (and that of the almost equally abundant and almost as energetic neutrinos) filled most of the Cosmic Density Pyramid for tens of thousands of years. But now the photons have lost much of their energy as their wavelengths have expanded with the expansion of the universe, and sluggishly moving dark matter, with its quiet gravitational power, has become the main component in the density of the universe. The universe has been expanding in almost perfect uniformity, but now regions that happen to have slightly more dark matter than average are expanding slightly more slowly than average—a process that will eventually lead to the formation of galaxies. But ordinary matter can't cluster gravitationally yet as the dark matter is doing, because gravity is still no match for the beating that electrons and nuclei are still getting from photons (a beating to which dark matter is immune since it doesn't interact with light).

It is now 400,000 years after the beginning. The protons and helium nuclei that burst from the universal hydrogen bomb in the first few minutes are now capturing electrons and becoming electrically neutral atoms. The hot plasma-fog of the Big Bang clears; space becomes transparent, and the photons escape—becoming the cosmic background radiation. Liberated from the beating they were getting from the photons, the hydrogen and helium atoms now start to follow the gravitational pull of the dark matter. The evolution of the universe has not happened smoothly and continuously; there have been and will be extremely exciting eras like the first few minutes packed with major events, and then long periods where nothing new happens.

It is 100 million years after the beginning. The universe is deep in the Dark Ages. The brilliance of the Big Bang fireball has long since redshifted into the invisible infrared, and no bright galaxies have as yet formed. We have to use theory—namely the Double Dark theory—to look back to this early time, because we cannot yet see what occurred then—although we hope to be able to do so soon as new radio telescopes on the ground and infrared telescopes in space begin operating.[34] The universe at 100 million years is still expanding pretty uniformly, but there are very subtle differences in density from place to place—density now of dark matter with primordial hydrogen-helium gas mixed in. Gravity is already revealing its fundamental nature: over time it will magnify any differences that already exist. As the denser regions expand more slowly, the less dense regions expand faster. Gravity continues to amplify the contrast between them, but then there is a discontinuity: when a region becomes about twice as dense as typical regions its size, it stops expanding and starts falling together, but not into a black hole because all particles are in motion every which way, so they don't fall straight to the center but swing around each other. Through a process named "violent relaxation" by its discoverer, the British astrophysicist Donald Lynden-Bell, the dark matter quickly reaches a stable configuration that is denser in the center and less dense farther away, and the average distance of the particles from the center is about half of what it was before collapse, as shown in Figure 4.[35]

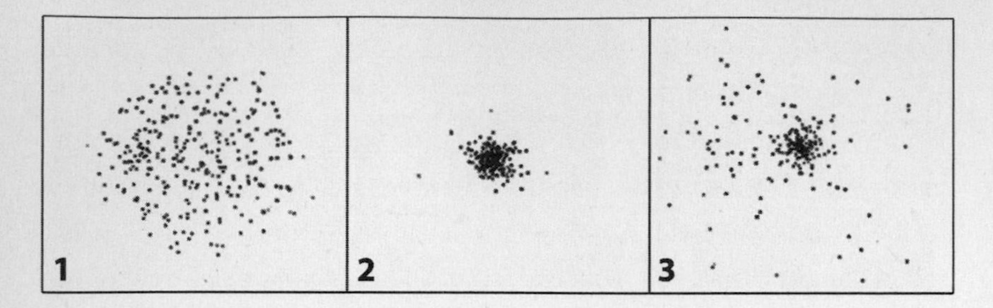

Figure 4. Gravitational collapse in the expanding universe. Panel (1) shows some representative dark matter particles in a region that has expanded as much as it can before gravity takes over and makes it collapse. As they fall together (2), the random locations and velocities of the dark matter particles prevent them from all ending up at the center. Instead, the particles quickly develop a stable distribution (3), denser in the center, less dense farther away. All three panels are to the same scale, showing that collapse doesn't much change the size of the dark matter cloud. By contrast, the rest of the universe keeps expanding.[36]

This picture shows the formation of a protogalaxy. Little ones form first, and then they are incorporated into more massive ones. At this time, 100 million years after the Big Bang, large protogalaxies are about a million times the mass of our sun (in contrast to the Milky Way, which is about a trillion times the mass of our sun). Such a protogalaxy is made mostly of non-atomic dark matter plus about 15 percent primordial atomic matter (three-fourths hydrogen and one-fourth helium with a negligible amount of heavier elements). Although the dark matter halos of these protogalaxies cannot collapse further because dark matter can't radiate energy, the hydrogen-helium gas can and does, and soon it will make the first star in each protogalaxy. As the gas atoms collide, their electrons are knocked into higher energy levels—and since this takes energy, the atoms rebound from their collisions more slowly than they came together. These atoms emit light as their electrons subsequently drop back down to lower energy levels, and this light is ultimately radiated away into space and lost to the forming protogalaxy. The gravity of the dark matter causes the gas to fall farther toward the center, becoming denser and denser. It radiates still more energy, and collapses further. Eventually the gas in the center becomes dense enough and hot enough to ignite as a star.

These are the first stars in the universe. Made of pure hydrogen and helium from the Big Bang, they are huge—typically about a hundred times the mass of our sun. Such massive stars race through their life cycles in a million years or so, creating inside themselves new kinds of atoms the universe has never seen before. Exploding in gigantic supernovas, they eject these heavy elements and leave behind massive black holes. Each protogalaxy may have only one such huge star, and when it explodes as a supernova, it blows much of the rest of the gas out of the protogalaxy. But the gravity of the protogalaxy is still there since its dark matter hasn't gone anywhere. Dark matter interacts only gravitationally. So eventually the gas cools and falls back into the protogalaxy. A little of it ends up falling all the way into the black hole left over from the first star, radiating brightly, as gas does when it is falling into a black hole. The returning gas has been enriched with heavy elements from the supernova, which makes it cool more easily, and the cooling gas condenses into new stars.

But what is happening outside the protogalaxies? Light is pouring out, both from the massive stars and the radiating black holes. It hits the gas that is floating around between the protogalaxies and knocks electrons right off the atoms, "ionizing" the hydrogen and helium between the galaxies. These fast electrons ionize more atoms. This transformation of almost all the atoms in space into ions will have enormous implications as the universe evolves, determining not only what kinds of radiation can pass through space to be later observed and pondered by intelligent beings, but also determining, in concert with the dark matter fluctuations, the size range of future galaxies.[37]

Meanwhile the expansion of the universe is slowing down because of the effects of gravity. The gravity of the dark matter is also causing larger and larger structures to merge, starting to form the galaxy groups and clusters where the primordial wrinkles were deeper than average—see Figure 5.

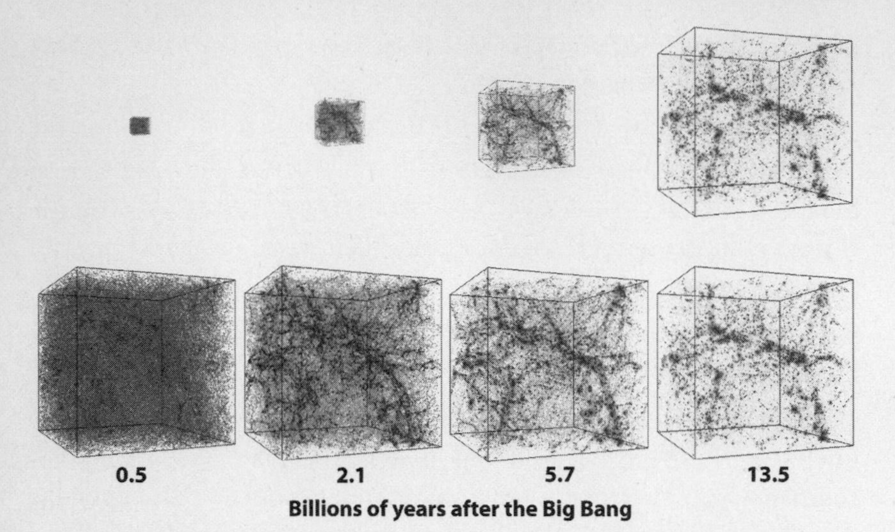

Figure 5. Expansion and Evolution in the Universe. Each panel shows the dark matter in a Double Dark supercomputer simulation of a cubic region of the universe that is 140 million light-years across *today*. The upper and lower panels show this region at different times after the Big Bang. The upper panels show how the cube has expanded with the universe, while the lower panels are shown all the same size in order to make it easier to see how the structure evolves. The smallest black dots represent dark matter halos of individual galaxies. Notice that most of these galaxies are part of a "cosmic web" of structure. The large filamentary structure from upper left to lower right evolves into several superclusters. There is a large void with few galaxies in the lower middle of each panel.[38]

It is nearly a billion years after the beginning. The Dark Ages have ended. Inside many large, violently relaxed halos of dark matter, gas is cooling and forming stars rapidly—especially when these young galaxies collide with each other. They are radiating light now that will travel for 13 billion years and be observed by twenty-first-century humans. These young galaxies lie in the sphere called "Bright Galaxies Form" in the Cosmic Spheres of Time. But if the dark energy continues to accelerate the expansion of the universe, that sphere, sadly, will be the first to disappear from our visible universe.

How will these odd and disheveled-looking early galaxies turn into gorgeous elliptical and spiral galaxies like our own Galaxy and those nearby? The universe is very crowded now at the close of the Dark Ages. Although it will expand by an additional factor of seven by the time humans come on the scene in 13 billion years, it already has as much matter as it ever will. The Double Dark theory predicts that in these congested conditions, galaxies are colliding and merging frequently.[39] As they come together, their dark matter halos interact gravitationally with each other, disturbing the paths of ordinary matter particles and stars, but the dark matter particles themselves pass near and even through each other like ghosts. The stars inside each galaxy are so very far apart compared to their sizes that they almost never collide. When galaxies merge, what crashes dramatically is the gas. First the thinner, outer gas of the two galaxies starts heating up, while the denser gas flows inward toward the centers of the galaxies, forming new stars of many sizes. The most massive stars explode first in core-collapse supernovas. This has been happening since stars first formed in the early universe, and each generation of stars is enriched with the heavy atoms of earlier supernovas. The richer the supply of atoms in a newly forming planetary system, the more material there will be for the evolution of planets and ultimately of life.

It is now two billion years after the beginning. In the centers of the now-maturing galaxies lurk black holes as massive as a billion suns, but their spheres of invisibility (the region from which no light escapes) are only a few light-hours across, comparable in size to our solar system. Two spiral galaxies are passing each other very closely. Gravity is pulling them together, but their motion makes them sideswipe each other, setting off great bursts of star formation in each galaxy. The central bulges of the two galaxies then merge quickly, but the outer parts are flung out in two tails a hundred thousand light-years long, making the merging galaxies look like two mice cuddling nose to nose. In the center of the newly forming galaxy, the massive black holes whip around each other, but they cannot merge without giving up tremendous amounts of energy. To do this the black holes kick nearby stars out of the galaxy, and with each kick, the black-hole pair, already bound by gravity and circling ever closer together, recoils in the opposite direction, eventually reaching more stars, kicking them out, recoiling again, until at last the black holes have lost enough energy to coalesce. The new black hole is extremely massive now, since gas has been falling in all along, feeding it. As the compressed and heated gas reaches the sphere of invisibility of the black hole and is about to fall in, the gas radiates into space primeval energy that has been locked in its mass (mc^2). This beaming black hole is a quasar, a beacon across the universe, shining from its tiny region more brilliantly than ten or even a hundred galaxies of a hundred billion stars each.[40] By four billion years after the beginning, quasars will become rare. There will never again be as many as in this galactic bumper-car era.

Imagine how much energy it would take to heat your house all winter with all the windows and doors open twenty-four hours. Now imagine how much energy is required to heat all the gas for many thousands of light-years around. But it's happening. The energy is coming from the gas falling into the black hole and from the new stars that are forming and exploding in supernovas as a result of the galaxy merger. Hot gas is blowing out of the galaxies in vast winds reaching speeds of many hundreds of kilometers per second.[41] After the galaxies merge, the new galaxy, given another billion years, may form a new disk of gas and stars around the merged center, with a rounded bulge or bar of stars in the middle. In this case the resulting galaxy will look very much like the Milky Way.[42]

It is nine billion years after the beginning. Gravity has been fighting against expansion since the Big Bang, and in denser regions it has won and created stable structures like solar systems and galaxies, which are bound together gravitationally and travel as a unit in the great expansion. But on larger scales where the universe is less dense, expansion is winning over gravity. Now the balance of power is shifting from dark matter to dark energy, and the steady expansion of the universe is beginning to accelerate. Any cluster of matter that has not already begun to collapse together because of gravity, never will. The great superclusters of galaxies, the largest structures in the universe, have too much still-expanding space inside them for gravity to overcome. They begin to disperse. Meanwhile, in the suburbs of one spiral galaxy, a star and its planetary system are forming—our own.

It is 14 billion years after the beginning. We have reached the present era. It is the midpoint of time for our solar system. Our sun and its planets are almost five billion years old. Like other middle-aged stars, the sun has been growing steadily hotter. It has about six billion years to go until it becomes a red giant and expands to swallow the planets Mercury and Venus and burn Earth to a crisp. But in a deep sense, this is also the midpoint of cosmic history itself: at this stage, just as intelligent life on Earth is acquiring the ability to see to the distant reaches of the universe, accelerating expansion is carrying the most distant galaxies away ever faster, and they are disappearing over the cosmic horizon. In short, our visible universe is emptying out: our distant descendants, no matter how advanced their telescopes may be, will never be able to see as many galaxies as we can see now. In this sense, our era is the midpoint of cosmic time.

The history of the universe is in every one of us. Every particle in our bodies has a multibillion-year past, every cell and every bodily organ has a multimillion-year past, and many of our ways of thinking have multithousand-year pasts. Each of us is a kind of nerve center where these various cosmic histories intersect. Time is one key to appreciating what we are. Ancient creation stories reflected the universe as people saw it with the naked eye through the lens of their cosmology. But humans have now gone far beyond in discovering what exists, how it works, and new ways to think about it. We can measure time not only by seasons and years but by radioactive decay, by the life cycles of stars, and by the expansion of the universe. Our theories, such as relativity, illuminate the whole cosmos conceptually, letting us think productively on an awesome range of scales and phenomena.

But thinking is not enough. Like the ancients, modern people want to *connect* to the universe, but the strangeness of a creation story without human or animal personalities, one that is hard to visualize because things happen in it following unfamiliar laws—this kind of creation story may seem at odds with our intuitive sense of what makes a satisfying explanation. But when we open our consciousness to the universe, we have to remember that we are at the center of the spheres of time, and only close to home do things appear "normal"; the farther away in spacetime, the more strange. Our animal/human creation stories may have their place as metaphors for our lives—but no longer for the universe beyond the solar system. They're misleading metaphors for photons, neutrinos, quarks, and dark matter; they connect us to the wrong universe. We need to connect to *our* universe. If all knowledge were intuitive, there would be no need for science.

Relishing the counterintuitive can be liberating. The only human that has to be in the story is you. That's what it means to participate, but you can do this in your imagination. It is one thing to appreciate intellectually that time may be as this chapter described it, but quite another to suspend disbelief. Suspension of disbelief is the chance we give to any play or movie to touch us, independently of whether it's true. Suspension of disbelief is the chance we give *ourselves* to participate in the drama and the comedy. Let's give ourselves and our universe at least that chance. Let's visit the cosmic sphere that contains our neighborhood—the center of our visible universe.

Contemplating the Cosmic Spheres of Time

Imagine that you can leave your body behind and send your consciousness out on a journey through space. When your consciousness travels alone, it feels no cold, no need of oxygen, none of the effects that happen to ordinary matter when it accelerates to speeds close to the speed of light, as you will do. Your consciousness simply sees the changing scene around it and understands.

So now let your consciousness rise out of the room, out of the building, and off the surface of the earth. You are accelerating constantly, so that although you begin to rise slowly, you move always faster. Rather than looking outward into space, you turn your focus toward home and watch what you are leaving behind. In perfect peace you see your hometown far below you, then mountains, plains, and coastlines. You can see the curvature of Earth now, and a moment later the shape of your entire continent. Faster and faster you rise until the whole planet comes into view then shrinks into the distance. You speed out of the solar system. The sun becomes just another star.

Movement becomes a mere idea. There is no visible way to detect it, only the tricky consciousness of time passing. A few stars give the illusion of creeping but do not actually move. Or do they? Now they seem to creep, but with excruciating slowness, like the minute hand of a clock. With increasing speed they definitely begin to move. In fact the constellations are changing. Now a billion stars are rolling into your range of vision. Still accelerating, you see our spiral arm of the Milky Way, the Orion Arm, wrap around you like a cape of diamonds.

Out of the disk of the Galaxy you shoot, approaching the speed of light, streaking away on the perpendicular. The entire Galaxy sprawls before you, rigid and magnificent like a photograph. You watch the shining disk recede as you move without resistance through its massive halo of invisible dark matter, so powerful in its gravitation that it holds the Galaxy together. The dark matter thins on its outskirts. The brilliant galaxy in Andromeda appears and races the Milky Way into the distance, until the two are a pair of jewels, set in blackness flecked with a few dozen small satellite galaxies.

Now you look around and there are galaxies everywhere, hundreds of them. They grow denser as you penetrate far into the great galaxy cluster in Virgo. Yet those galaxies are still so far from you that even in the very center of their brilliance, there is no warmth at all. The pleasure of utter wonder that has filled you on this journey becomes tainted by discomfort and overwhelmed by foreboding. Your consciousness experiences total disconnection from everything you have ever known, and the feeling is one of terrifying loneliness. A force millions of years deep within the biology of your mind sweeps out and cradles you in its arms, pulling you back, back, and you make a huge loop and speed toward your own Galaxy once more, still facing the Galaxy.

Light that could barely keep up with you on your outward journey now rushes toward you, inundating you with information. Streaming now toward the Milky Way, you see half a million years pass every second. The Galaxy, which had appeared as still as a photograph on your outward journey, now revolves before you like a whirling pinwheel of fireworks as supernovas go off constantly. As they circle the galactic center, a hundred billion stars move up and down like stately horses on a carousel, some crossing the entire thickness of the disk with each oscillation, others tracing smaller zigzags. Enormous density waves slowly circle the Galaxy like barely moving traffic jams. These are the spiral arms of the Galaxy. Hydrogen gas, circling as fast as the stars, crashes into gas slowed by the density waves so violently that new stars burst forth, lighting the immense clouds and swirls of gas around them, creating nebulas that look like multicolored flames many light-years across. Exploding spectacularly into supernovas, thousands of the new stars quickly disappear, their short lives lighting up the spiral arms.

You can imagine the gravitational pull of the immense ball of stars at the core of the Galaxy and the black hole pulsating within it, the gravity of every star, every speck of dust, every bit of dark matter in the Milky Way. Far out on the disk, about halfway to the visible edge, travels an inconspicuous star that pulls you more powerfully than all the rest of the Galaxy and all the universe beyond. It is our sun, moving on its individual journey around the Galaxy.

Suddenly as if through an invisible tunnel, you are sucked across the Galaxy toward a point near the sun, a very small point. Time decelerates. When the flaming nebulas have frozen like a photograph, when a supernova occurs somewhere in the Galaxy only once in a lifetime, you have returned to the pace of Earth. Earth rises before you, blue and sparkling in the sun. It is radiant with life, familiarly shaped land masses, and ever-shifting clouds, the curvature of its horizon softened by the eggshell-thin gauze of an atmosphere, and you understand why every culture has great myths of homecoming. Earth is the center of the universe for you, but the universe is the Cosmic Spheres of Time. Decelerating, slowly, slowly descend through the crystal-clear atmosphere, returning to your home from the sky at the speed by now of a horse and buggy, floating down to the humming world, your neighborhood on this vibrant planet. Reenter your body, and truly open your eyes.

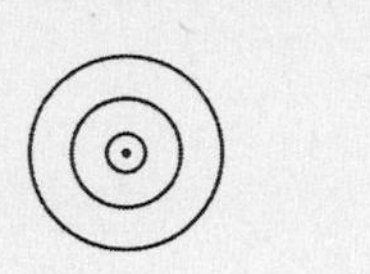

SIX

What Size Is the Universe?

THE COSMIC UROBOROS

THE SIZE OF A *human being is at the center of all the possible sizes in the universe.* This amazing assertion challenges not only the centuries-old philosophical assumption that humans are insignificantly small compared to the vastness of the universe but also the logical assumption that there is no such thing as a central size. Both assumptions are false, but we have to reconsider the key words of the assertion—center, possible, size, and universe—to reveal the prejudices built into them that constrict and distort our picture of reality. In the modern universe there is a largest and a smallest size, and therefore a middle size. The size of a thing is not arbitrary but crucial to its nature, which is why scale models can never really work. Only by understanding size and its role in determining which laws of physics matter on different size-scales can we can get an accurate perspective on anything outside the narrow realm of human experience. This chapter will develop a new symbolic picture of the universe that portrays us in our true place among everything else that exists.

There are wildly different-sized objects in the universe. Betelgeuse is the bright star at the upper left corner of the familiar winter constellation Orion. It is a red giant so monstrously larger than our sun that it could fill the orbit of Mars. Earth is a mere pebble beside it. But compared to our home Galaxy, the Milky Way, Betelgeuse is just a spark in a raging forest fire. Clearly, these sizes are relative, and we need a language in which to discuss them without falling back on vague words like "huge," "tiny," or the most misleading of all, "infinite." (Just because a size is too big to grasp does not mean it is literally boundless.) The concept of "size scale" is a mental framework with which to define a chunk of reality. It describes our range of conscious focus at any given moment. *A size scale is not a physical entity but a setting of the intellectual zoom lens.* The size scale of an object is a region large enough to include the entire object but not so large that the object becomes insignificant.

Even if you don't like math, you can still easily compare $10 and $10,000. Comparison and manipulation of numbers are such different activities that they are performed in opposite hemispheres of the brain.[1] You think differently about buying something if it costs $10 than if it costs $10,000. In this case, the numbers are not something you have to manipulate; instead they define a certain general category. Price alone does not tell you everything. Is $1,000 a lot? Well, it is for a dinner but not for a car. So as with prices, the number that defines a size doesn't tell you whether something is "big" or "small." It only helps you compare it to something else. Unlike prices, which are usually expressed to the dollar or even penny, the numbers we use are mostly not intended to be precise but instead to suggest general ranges of size. Size scale is an approximate concept, but for the universe that is all we need.

The basic length unit we use is the centimeter, which is a little less than half an inch. There are 100 centimeters in a meter. The height of most people is between 1 and 2 meters (that is, between about 3 feet, 3 inches and 6 feet, 7 inches). This chapter is not going to pay any attention to size *differences* this small—one meter is the same as two meters for our purposes. Since the number 100 is written exponentially as 10^2, the height of people is in the 10^2 cm range. The raised exponent refers to the number of zeros after the 1. Mountains, the height of which is measured in kilometers (thousands of meters) are in the 10^5 cm range (100,000 cm). In the direction of decreasing size, a typical cell in your body is in the 10^{-3} cm range (0.001 cm). Negative exponents tell you in which place after the decimal point the first non-zero digit occurs. The ratio between 10^2 and 10^3, or between 10^{-6} and 10^{-5}, is a factor of 10, and this is called one "order of magnitude." If we want to describe the difference between the size of a human being (10^2 cm) and the size of a cell (10^{-3} cm), we can say that they differ by 5 orders of magnitude. A human differs in height from a mountain by only 3 orders of magnitude (10^2 compared with 10^5 cm). Thinking exponentially, a human being is far closer in height to Mount Everest than to a single cell, even though "common sense" leads many people to assume the opposite because they have no experience of how small a cell really is. Orders of magnitude may sound complicated at first, but as we use them they will begin to seem an essential part of language.

There Are a Largest and a Smallest Size

The powers of 10 go on infinitely in both directions in pure mathematics, but not in the physical world. The smallest size exists because of the interplay between general relativity and quantum mechanics. We have already mentioned that relativity redefines space and time as spacetime; now we come to something it says about space and gravity. General relativity tells us that there can't be more than a certain amount of mass squeezed into a region of any given size. If more mass is packed in than the region can hold, gravity there becomes so intense that the region itself—the space—collapses to no size at all. This is a black hole. Nothing can escape from inside a black hole, not even light; hence the term. Any object compressed enough will hit this limit and suddenly become a black hole.[2]

Meanwhile, quantum mechanics sets the minimum size limit, but in a very peculiar way. Electrons, protons, and other particles have extremely small masses and are always whizzing about. They are very hard to pinpoint. The "size" of a particle is actually the size of the region in which you can confidently locate it. The smaller the region in which the particle is confined, the more energy it takes to find it, and more energy is equivalent to larger mass. There turns out to be a special, very small size where the *maximum* mass that relativity allows to be crammed in without the region collapsing into a black hole is also the *minimum* mass that quantum mechanics allows to be confined in so tiny a region. That size, about 10^{-33} cm, is called the Planck length, and it's the smallest possible size. We have no way to talk or even think about anything smaller in our current understanding of physics.[3]

The largest size we can see is about 10^{28} cm, which is the distance to our cosmic horizon. From the Planck length to the cosmic horizon is about 60 orders of magnitude. The number 10^{60} is extremely big, but it's not infinite. It's comprehensible. With it, we can begin to define our cosmic context.

Size Matters: The Key to Cosmic Perspective

The ancient Egyptian god Nun, the great unknowable and indescribable source of all the other gods, was sometimes portrayed associated with a serpent or even as a serpent. There is something about the image of a serpent that has led many cultures to associate it symbolically with the creation of the world and the unity of all things, especially when the serpent is represented as swallowing its own tail. In ordinary speech the word "serpent" is sometimes used interchangeably with "snake," but a snake is an animal, while a serpent is the symbolic, mythic, sometimes dreamlike representation of that animal. Snakes do not actually swallow their tails, but serpents can do anything humans can imagine. Adapting an idea of Sheldon Glashow, the 1979 Nobel laureate in physics, we

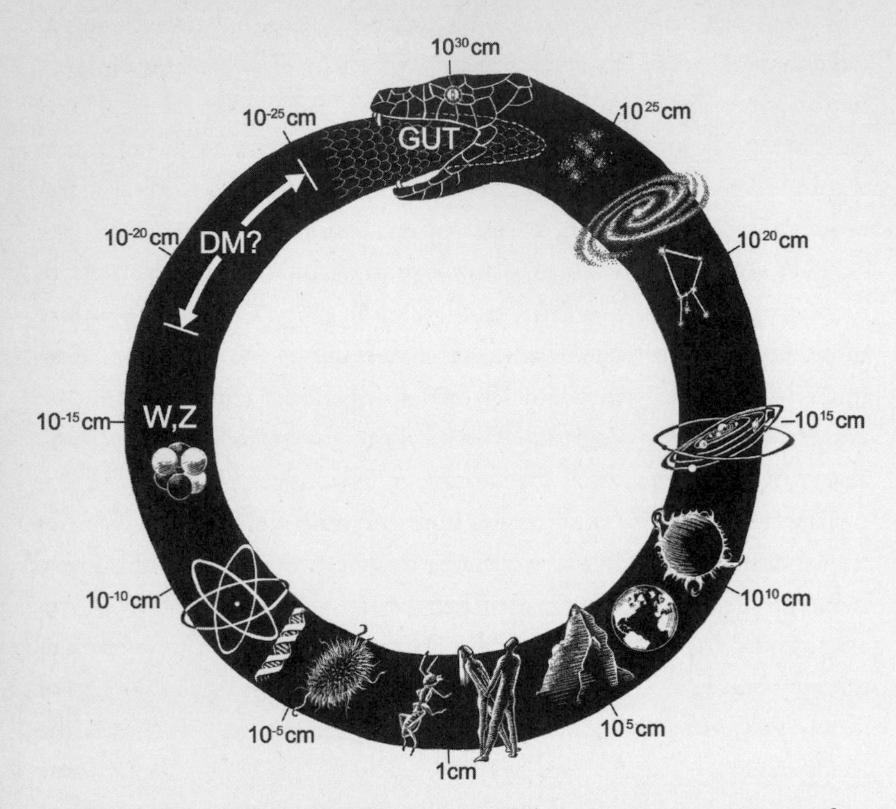

Figure 1. The Cosmic Uroboros represents the universe as a continuity of vastly different size-scales. As Figure 3 of Chapter 4 shows, the diameter of the earth is about two orders of magnitude (10^{-2}) smaller than that of the sun. About sixty orders of magnitude separate the very smallest from the very largest size. Traveling clockwise around the serpent from head to tail, we move from the maximum scale we can see, the size of the cosmic horizon (10^{28} cm), down to that of a supercluster of galaxies, down to a single galaxy, to the distance from Earth to the Great Nebula in Orion, to the solar system, to the sun, the earth, a mountain, humans, an ant, a single-celled creature such as the *E. coli* bacterium, a strand of DNA, an atom, a nucleus, the scale of the weak interactions (carried by the W and Z particles), and approaching the tail the extremely small size scales on which physicists hope to find massive dark matter (DM) particles, and on even smaller scales a Grand Unified Theory (GUT). The tip of the tail represents the smallest possible scale, the Planck length. Human beings are just about at the center.

turn to the multithousand-year-old symbol of the serpent swallowing its tail and give it a modern interpretation.[4] "Uroboros" is the ancient Greek word for a serpent swallowing its tail. We will call the symbol in Figure 1 the "Cosmic Uroboros." The tip of the cosmic serpent's tail represents the smallest possible size scale, the Planck length, and its head represents the largest size scale, the size of the cosmic horizon.

Let's get oriented on the Cosmic Uroboros. Most of the time we humans are conscious only of things from about the size of ants to the size of mountains. This range of sizes corresponds to the bottom of the Cosmic Uroboros—if it were a clock face, it would fall approximately between 5:00 and 6:30, just about the middle. This is humanity's native region of the universe, our true homeland. This is the "reality" in which common sense works and normal physical intuition is reliable. It's not a geographical location: it's a point of view. We will name this range of size scales "Midgard," a name for Earth borrowed from the Norse creation myth, the Edda, in which the world of human beings was seen as midway between the land of the giants and the land of the gods.[5] For much the same reason, the ancient Romans named their sea the Mediterranean, literally "middle of the earth." We have chosen the name Midgard for our human-scale homeland in the modern universe not because it is between heaven and hell or any other spiritual dualities, but because it is midway between the largest and smallest sizes. This turns out to be the only size that conscious beings like us could be. Smaller creatures would not have enough atoms to be sufficiently complex, while larger ones would suffer from slow communication—which would mean that they would effectively be communities rather than individuals, like groups of communicating people, or supercomputers made up of many smaller processors.

Different physical forces control events on different size-scales. Electrical and magnetic (electromagnetic) forces control what happens from atoms up to mountains, even though gravity also plays a role. But around the size scale of mountains, gravity starts to gain the upper hand. The maximum size of mountains is determined by a competition between electromagnetism and gravity. The electromagnetic force is the glue of the chemical bonds that hold together the atoms that mountains are made of, and the strength of the glue is the same everywhere, regardless of the size of the planet. But the strength of the gravitational force grows with the increasing mass of the planet or of the mountain. When the mountain becomes big enough, its gravity overcomes the electromagnetic forces that hold mountains together, and the roots of the mountain flow or break, causing earthquakes. The smaller the mass of the planet, the weaker the gravity pulling the mountain down. Consequently, mountains can be much higher on smaller planets like Mars than they are on Earth.[6] Since the strength of gravity continues to grow with mass, once we reach that part of the Cosmic Uroboros where gravity controls, all larger scales are also controlled by it and all other forces become less important.

Moving counterclockwise from Midgard up into the larger size-scales means adjusting our conscious focus, zooming out to encompass vaster regions, where gravity has counteracted the headlong expansion of the universe by collecting matter in those regions that in the early universe happened to be slightly denser than average. Gravity eventually stopped the cosmic expansion in those regions, and gravity has ever afterward shaped and held everything in the region together in a beautiful, dynamic, yet stable structure—a galaxy, in which stars and planets formed and evolution has had time to work its wonders. The largest structures astronomers see are the great sheets of galaxies known as superclusters. In the old Newtonian view, there was no known object larger than a star, and stars were randomly distributed forever. But in the new cosmology not only are there galaxies, each containing hundreds of billions of stars, but there are superclusters of tens of thousands of galaxies, which astronomers have been mapping since the mid-1980s. That, however, appears to be the end of the line. We see no structures larger than superclusters. On scales much bigger than superclusters, the universe becomes increasingly smooth. If each supercluster were a dot, the visible universe would look much the way Newton expected. He was right about the universe being essentially uniform, but on the wrong scale: he thought the *stars* were scattered more or less evenly, but instead it's the superclusters.

Moving clockwise now on the Cosmic Uroboros, zooming way inward past Midgard to the very small, we reach the size scales of subatomic particles. This is the region controlled by what are called the strong and weak interactions. These forces are active only on scales smaller than atoms.[7] Gravity is of no importance at all on these scales. In fact, gravity's power fades out at the small end of Midgard. It can't hurt a mouse. You can drop a mouse down a thousand-yard mine shaft and at the bottom, as long as the ground is soft, it will walk away.[8] Gravity plays virtually no role in the life of bacteria, which are at about 7:00 on the Cosmic Uroboros. From there until about 12:00, gravity is completely irrelevant.

But then a strange thing happens. As we continue along the Cosmic Uroboros to the very tip of the tail, gravity becomes extremely powerful again. The reason is that gravity's strength increases as objects get closer to each other, and at the tip of the tail distances between particles are almost unimaginably small. The Cosmic Serpent swallowing its tail represents the possibility that gravity links the largest and the smallest sizes and thereby unifies the universe. This actually happens in superstring theory, a mathematically beautiful idea that is our best hope for a theory that could unify quantum theory and relativity. In string theory, sizes smaller than the Planck length get remapped into sizes larger than the Planck length.[9]

The latest breakthrough in particle physics was the realization in the 1960s and 1970s that the strong and weak forces are closely related to the electromagnetic force. In the very successful Standard Model of particle physics based on this, elementary particles are treated as if they are points with certain properties. But the Standard Model cannot be the final word on the subject, since it cannot explain why, for example, electrons and other elementary particles have the masses and other properties that they do. So physicists have been trying for several decades to go beyond the Standard Model. The very speculative but promising physics of string theory suggests that not just electrons but all elementary particles might just be the ways a single kind of tiny looped string can vibrate, and in that case an electron would be just a *way* a string vibrates. An identical string vibrating in a different way would be a different particle. Just as only certain shapes of electron clouds are allowed in atoms, only certain sorts of vibration (and thus of particles) are possible. An electron is a special sort of vibration: it is the lowest-mass vibration having the property of electric charge. String theory has striking mathematical elegance: it might even be true, and it's so powerful that it might eventually allow physicists to understand the reason for quantities like the masses of elementary particles. However, string theory works only if you assume a world with ten dimensions—one time and nine space dimensions. No one has figured out what string theory implies for the world of one time and three space dimensions that we actually experience—not only with our senses but with our most sensitive scientific instruments. Consequently, this beautiful theory has not made a single testable prediction yet (except possibly for the existence of supersymmetric particles like the WIMP dark matter particle), so we don't yet know how to evaluate its claim that particles are "really" vibrating superstrings.

There is a second meaning to tail-swallowing that may seem strange at first. Swallowing may have existed *before* the serpent. At the beginning of the Big Bang, if our present understanding of the laws of physics is right, there was nothing but the head of the Cosmic Uroboros with the tip of the tail in its mouth. There was little of the body because there was little difference between the smallest scale and the largest scale. The smallest scale is fixed by the constants of nature, and the largest scale, the size of the cosmic horizon, was only a little larger than that because the universe was so young and had not yet had time to expand. The body filled in later as the universe expanded and evolved. Thus tail-swallowing may express a fundamental aspect of the evolution of an expanding universe.

The Cosmic Uroboros represents not only a way to structure the universe but also a dream that has been an underlying personal motivation for many scientists. "What I'm really interested in," Einstein said, "is whether God could have made the world in a different way; that is, whether the necessity of logical simplicity leaves any freedom at all."[10] This question is still open. The universe could possibly have been organized in many ways and just happened to end up the way we find it. But it is also possible that there was only one way everything could have worked together. The dream of physicists is to find the theory that answers such questions and ties everything together—a "theory of everything." The Cosmic Uroboros swallowing its tail thus symbolizes the dream of a theory of everything, which will tie together our understanding of the universe. Through this dream, physicists are expressing a desire perhaps even more ancient than the uroboros symbol: to feel coherent and at home in the wholeness—to experience reality as One.

Even if there is no success in that quest for years to come, the Cosmic Uroboros can help us right now to appreciate our extraordinary place in Midgard. The centrality of Midgard on the Cosmic Uroboros has nothing to do with the units we choose to measure length. Whether measured in centimeters or light-years, Midgard would always fall in the middle. Midgard, as we have said, is not a special location in space—it is a special size-scale, and it is *everywhere* in the universe.

As a serpent, the Cosmic Uroboros is much more than a circle, because every point on it is unique. There are a head and a tail, and therefore every point in between has a relative position. There are a beginning and an end, even though they overlap and are interdependent and inseparable. On a circle, all points are identical. On the Cosmic Uroboros every point has its own meaning. The uroboros has been used to represent the continuity of whatever universe a tribe or people perceived themselves to be living in. Something about the serpent swallowing its tail has resonated in the human imagination for thousands of years. We humans are not yet able to explain the perennial attraction of this symbol, and it may be deeper than our conscious understanding. The serpent's exceedingly simple and flexible body has been endlessly twisted and artistically embellished. It has been seen as both goddesslike and evil, fascinating and repulsive, finite and infinite, yin and yang.[11] None of this rich history would have been implicit in a circle. The uroboros symbol as we interpret it here is capable of representing the modern universe at least as completely as it represented the universes our ancestors imagined. The Cosmic Uroboros resurrects an ancient symbol whose possibilities are by no means exhausted.

Why Scale Models Never Work

Galileo was probably the first to understand the physics of size scales and to realize that size is not arbitrary but crucial to the nature of each thing, determining both its shape and its function. In his 1638 book *Discourses Concerning Two New Sciences* (in which he invented the fields we would call today "mechanics" and "strength of materials"), he explained with simple arithmetic why large animals cannot look like small ones.

At the top of Figure 2 is the delicate bone of an animal, and below it is the bone as it would have to be if the animal were three times taller or longer. The reason the second bone would be so much bulkier, Galileo figured out, is that the strength of bones is determined by the area of their cross-section. Area is a two-dimensional quantity (length × width). If you make the bone three times as long, always keeping it the same shape, the area of its cross section goes up nine times, but the weight is three-dimensional and goes up 3 × 3 × 3, or 27 times. The animal's new weight would crush its bones. This is why an elephant does not look like a large gazelle.

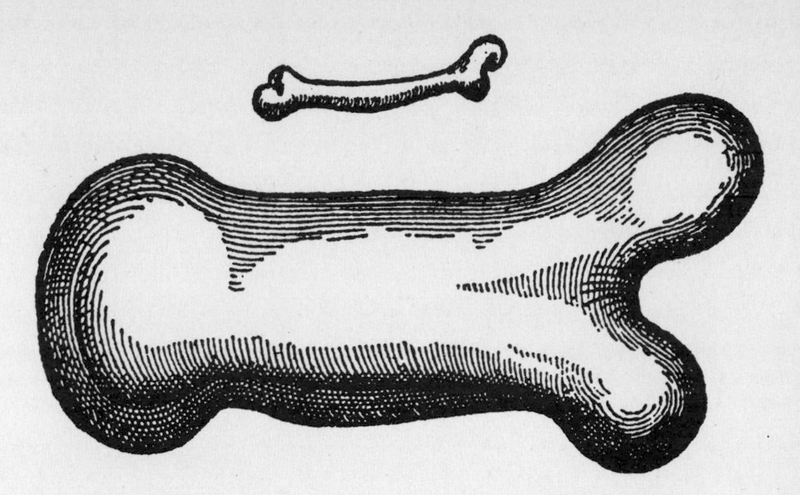

Figure 2. Galileo's bones: illustrating scale-model failure.[12]

With this simple calculation Galileo ruled out entire classes of what prescientific people had always considered natural possibilities—like elves and giants. He did it by thinking through what would happen if you scaled up only three times. But from the smallest scale to the cosmic horizon is 10^{60}! There is absolutely no way that unaided human intuition can predict or even imagine how things work on such distant scales. *Physical laws that apply at one scale do not cease to be true at other scales: they merely cease to matter.* Conversely, laws whose effects were nonexistent at one scale become overwhelmingly important on another, and reasoning that is valid on one scale may not work on another.[13] Atoms are the same size everywhere in the universe, a size that is set by fundamental quantities such as the masses of the elementary particles and the strengths of their interactions. The properties of materials are very different on small size-scales—closer to the dimensions of atoms—than they are on larger scales. The smallest living creatures have to be large enough to contain a huge number of atoms in order to have sufficient chemical complexity to do the things that living things do—such as using energy to process material and to reproduce.

Scaling ideas are very broadly useful, from understanding the possible forms of life on other worlds to understanding politics on our own, since what works for a family may not work for a town, and what works for a country may not work for a world. We'll return to this later, but for now we are concerned with pointing out two classic problems that result from assuming that reasoning that works on the human scale must also work everywhere else.

Mental Muddle #1: Scale Confusion

There is no single physical phenomenon that occurs on all size scales. There are no galaxies the size of atoms. Different kinds of things happen on different size-scales, and to talk about something in the context of a size scale in which it cannot occur creates a mental muddle we call *Scale Confusion*. Scale Confusion is applying laws and understandings appropriate to one size scale to phenomena on another scale where those laws and understandings don't apply.

Consider water. Water comes in the forms of ice, liquid, and steam, but how little of each of these things can you have? Can you have a single molecule of ice, or liquid, or steam? No. A water molecule is just H_2O—two hydrogen atoms and an oxygen atom. Ice, liquid, and steam are all meaningless concepts on the scale of a single water molecule. It takes millions of water molecules to make the smallest snowflake. "A molecule of ice" would be an oxymoron because it involves Scale Confusion. If ice, liquid, and steam are actually qualitatively different, and they don't exist in a single molecule of water, where do they come from? Can they be latent in the water molecule? The way a water molecule interacts with others of its kind may be characteristic of the molecule. But entirely new phenomena, such as "phase transitions" between liquid water and ice, become possible only on a larger scale involving vast numbers of molecules. *Complexity itself generates new kinds of behavior every few powers of ten, all around the* Cosmic Uroboros. This is analogous to the change in perspective between a child of five and an adult of twenty-five, and between adults of twenty-five and seventy-five—a maturing perspective does not mean more of the same: perspective eventually changes qualitatively as it changes quantitatively.

As another example, the second law of thermodynamics describes the tendency toward increasing disorder, or "entropy." If you leave a bottle of perfume open, the perfume will evaporate. The molecules, so nicely ordered in their compact container, will float off randomly in the room, increasing the entropy. They will never come together by chance and reunite inside the bottle—entropy doesn't spontaneously decrease. We can't easily recognize increasing disorder in very simple systems. If someone runs a movie of two billiard balls hitting each other and bouncing away, there is no way to tell whether the movie is being run forward or backward. But if you see a movie in which fifteen scattered balls suddenly come together to arrange themselves in a perfect triangle and the cue ball comes flying off, you can be certain the movie is running backward. Although entropy is irrelevant with two or even three balls, with sixteen balls entropy is already a clearly observable property of the system. Entropy is called an "emergent property" because it only emerges when a system becomes sufficiently complex.

The idea of emergent properties is becoming well known, and many examples have been described in different fields.[14] Consciousness may itself be an emergent phenomenon. Roger W. Sperry, the 1981 Nobel Prize winner in medicine, who discovered the functions of the left and right hemispheres of the brain, wrote that contrary to the traditional thinking in consciousness research, consciousness was in his view not reducible to physical and chemical processes. Beliefs and values are what control conscious behavior, not the underlying brain processes that have traditionally been thought to be fundamental. Consciousness, he concluded, is an "emergent property" that does not exist at a lower level of complexity.[15]

Although we can't say what consciousness is, we may be able to say what it is not. In the late nineteenth and early twentieth centuries in Russia there was a philosophical movement called "Cosmism," which was especially popular among scientists.[16] The Cosmists believed the universe was filled with intelligent aliens, and in preparation for humans perfecting themselves and establishing utopias in space, Cosmists like Konstantin Tsiolkovsky laid the technical foundations for much of rocketry, orbiting space stations, and space colonization. But their movement went beyond science. They believed that consciousness exists in every particle in the universe, including every star and galaxy, and that humans and intelligent aliens have the highest concentration of this consciousness and therefore the greatest responsibility to further the peaceful development of still higher consciousness. The Cosmists had many ideas that are still worth developing, but we don't believe that this view of consciousness is one of them. Consciousness can only exist at those high peaks of concentrated stardust where life and intelligence can evolve. It is a rare jewel in the cosmos.

There are many individual examples of emergent phenomena, but considering them one by one is like reading notes for an unwritten book on scraps of paper thrown into a shoebox. The Cosmic Uroboros provides a *structure* that makes sense of the notes, fills in the gaps, and tells a story. Along the Cosmic Serpent new properties emerge with each sufficiently large change of scale, and new laws dominate at new levels of complexity. The whole is incomparably greater than the sum of its parts.

Here is perhaps the most complicated (and controversial) instance of Scale Confusion: the question "Does God exist?" The Scale Confusion inherent here (we will ignore the difficulty of defining God) is that "existence" is actually a property of the middle of the Cosmic Uroboros. It is not a word that means the same thing at different size-scales. On a small scale, do electrons exist? The electron is a very useful concept—a particle with specific properties that we can talk about—but there is no solid thing, only a "probability cloud." In other words, the probability of its being somewhere is what's real. It doesn't make sense to say electrons exist in the commonly understood meaning of the word. But it makes even less sense to say they don't exist, because when you flick the switch, electricity flows and lights turn on. The confusion is due to the fact that when speaking of elementary particles, we lack any intuitive sense of their strange state, and "existence" is at best a metaphor. Niels Bohr, one of the founders of quantum theory, made this point, saying, "When it comes to atoms, language can only be used as in poetry. The poet, too, is not nearly so concerned with describing facts as with creating images."[17]

In the same way, very large-scale things can only metaphorically be said to "exist." For example, look in the direction of the constellation Virgo (a pattern of nearby stars) but look through it—look much, much farther away. There lies something astronomers have labeled the "Virgo cluster of galaxies." Certainly it is simpler to ask whether a galaxy cluster exists than to ask whether God exists, so let's ask the question: do galaxy clusters exist? Since we see each galaxy in a cluster at a different time in its history, due to the finite speed of the light coming to us from it, the cluster as we see it is a construction of our thought. Even the constellations Orion or the Big Dipper, which are made entirely of nearby stars inside our own Galaxy, don't really exist, since if we looked at those stars from a different vantage point, the pattern would be entirely different. The appearance of the Big Dipper is not even stable from the vantage point of the earth but is slowly changing as the stars that make it up move in their separate ways around the center of our Galaxy. The truth is that as we move further and further away from the human scale toward either larger or smaller scales along the Cosmic Uroboros, the concept of "existence" becomes increasingly metaphorical. "Existence" is a clear property only in the middle of the Cosmic Uroboros, our solid, reassuring, comforting homeland of Midgard. Unless God is by definition confined well within these limits, God can't be said to exist—or not to exist.

Mental Muddle #2: Scale Chauvinism

We propose the name Scale Chauvinism for the natural assumption that the way things look on some particular size scale is fundamental, and everything else can more profitably be viewed from this fundamental point of view. The most common chauvinism, of course, is chauvinism of the human scale. As Protagoras said, according to Plato's *Theaetetus,* "Man is the measure of all things." But human-scale chauvinism isn't the only possible kind. Richard Dawkins has written a fascinating book called *The Selfish Gene* in which he argues that living creatures such as human beings are actually DNA's method of propagating itself.[18] Humans exist for the sake of the DNA, which is, in Dawkins's terms, "God's Utility Function"—that is, in this metaphor from economics, DNA is the good that God tries to maximize. In Dawkins's view, the molecular level is fundamental. In a different mindset, James Lovelock argues that Planet Earth may be a self-regulating organism that he calls Gaia, after the ancient Greek goddess of the earth.[19] According to Lovelock's "Gaia Hypothesis," what individual plants and other organisms are doing is not understandable if considered only from the point of view of survival of the organism itself. The behavior of plant and animal populations is best understood as being part of Gaia's constant adjusting of its temperature and atmospheric constituents to maintain homeostasis, or stability. Lovelock regards the global scale as fundamental, with smaller size-scales serving the goal of the health of Gaia herself, life itself. Both these theories are wonderfully eye-opening—*if* they are understood as potentially useful approaches to the many faces of the universe. Many size scales hold critically important perspectives on reality. But none is more fundamental than the others. They all have a place on the Cosmic Uroboros.

A particularly dangerous form of Scale Chauvinism is the kind of thinking called "radical reductionism." This is not the same as "reductionism," which in science is the perfectly reasonable idea that explanations of large-scale phenomena must be *consistent with* scientific knowledge about smaller scales.[20] "Radical reductionism"—which is usually what people mean when they criticize "reductionism"—is the argument that larger scales can be *explained* by knowing what is going on at smaller scales. This is the "nothing but" argument: politics is nothing but psychology, psychology is nothing but biology, biology is nothing but chemistry, and chemistry is nothing but physics. To the radical reductionist, the Cosmic Serpent does not swallow its tail—there is no serpent. All that "really exists" is the tip of the tail. Everything always "boils down" to physics. But it is difficult to find even a single example where such thinking has led to deeper insight, let alone scientifically useful predictions.[21] Larger scales don't boil down to smaller ones. To the contrary, scientific laws and organizational principles that were irrelevant on small scales come into play on larger scales. Subatomic, human, galactic—the universe unfolds fully on all scales, and all are fundamental even though human consciousness can usually only focus on one scale at a time. The essence of Scale Chauvinism is failing to use the zoom lens. The essence of Scale Confusion is arbitrarily sliding the zoom lens and not realizing that this has consequences. The key to seeing the universe in clear focus is to learn to operate the zoom lens and to respect the uniqueness of every size scale.

Are We Insignificant?

Many people today contemplate the stars and the vast distances in between and conclude how insignificantly small we are compared to the universe. This view has contributed to a sense of alienation and sometimes even despair that have for more than three centuries been a reaction to humanity's demotion from the pinnacle of God's creation to a tiny speck floating in endless space. But now we understand something we didn't know before.

There is no thing and no force in the universe that is significant on all size scales. Gravity is certainly a significant force in the universe. On the grand scales of galaxies and clusters of galaxies, the motion of all matter is controlled by gravity. In the headlong expansion of the universe, the only thing that pulls matter together and maintains it in rare oases of stability is gravity. Only gravity's absolute stability could hold together a solar system and give evolution the time to create, by the interplay of pure chance and natural selection, levels of complexity like intelligent life. Gravity may even be the key to the Cosmic Serpent swallowing its tail. How could anything be more significant than gravity? But gravity plays no role whatsoever in the attraction between magnets or people, or, as we said earlier, in the lives of very small creatures. Similarly, the "strong force" is overwhelmingly powerful inside the atomic nucleus, but it falls off quickly with distance and outside the nucleus it is quite insignificant. If the forces of nature themselves are insignificant on some scales, humans are doubtless also insignificant on some scales, although not on our own. *Everything in the universe is significant on some scales, insignificant on others.* All human knowledge could be stored in something the size of a computer chip. If it were, would its smallness make it insignificant? Living on the Cosmic Uroboros, absolute size has nothing to do with significance.

The Meaning of Midgard

Intelligent creatures in the universe have to be midsized. There is a kind of Goldilocks Principle: Creatures much smaller than we are could not have sufficient complexity for our kind of intelligence, because they would not be made of a large enough number of atoms. But intelligent creatures could not be much larger than we are, either, because the speed of nerve impulses—and ultimately the speed of light—becomes a serious internal limitation. We are just the right size. You might expect that a galaxy-scale intelligence would think at a fabulously deep level. But in fact the number of thoughts that could have traveled back and forth across the vast reaches of our Galaxy in its roughly 10-billion-year lifetime is perhaps the number an average person has every few minutes. The speed of light seems dizzyingly fast to us, but on the scale of the visible universe it is excruciatingly slow and would prevent the parts of any large intelligence from communicating with each other in a reasonable amount of time compared to the age of the universe. Thus the cosmos can't have a central brain or government. Thinking must be decentralized to make any progress, given the limit of the speed of light.

Real thinking is the job of our size scale—beings more or less our size, bigger than an ant, smaller than a mountain, beings of Midgard. We humans exist on the only size scale where great complexity on the one hand and immunity from relativistic effects (like the speed of light) on the other are both possible. Our consciousness is as natural a blossoming on this special scale as a star is on its size scale or an electron on its own.

Not only do intelligent creatures have to be approximately the size we are, but the universe had to be more or less the size and age it is to have produced us. Atoms formed as the universe expanded and cooled. Galaxies formed as gravity resisted the expansion of the universe in regions that were denser than average. The first generation of stars could not have had planetary systems supporting life. Those stars contained only the elements that emerged straight from the Big Bang. All the heavier elements came into existence by being manufactured over time inside stars. The most massive stars burst into supernovas at the ends of their brief lives, dispersing heavy elements into space. For us to have come into existence, some of those elements had to find their way across vast, empty space into star-forming regions of galaxies and be sucked in by gravity to become later-generation stars like our sun, together with their new planetary systems. This process and the subsequent evolution of intelligent life on our planet required many billions of years, during all of which time the universe was inexorably expanding. Thus we could not have evolved until the universe was about as big as it is now.[22]

The Cosmic Uroboros shows that no being occupies an isolated niche. Stars are not merely stars; some of them are the centers of planetary systems and sustain whatever life there may be. Dark matter is not merely a lot of individual invisible particles; collectively it shapes the galaxies and holds them together as they spin, and it shepherds thousands of galaxies into the largest-scale structures in the universe, the superclusters. Electrons, bacteria, humans, and galaxies are phenomena that occur on different size-scales along the Cosmic Serpent, but all have effects on other scales than their own, sometimes even across the Cosmic Uroboros.[23] The way to think about the universe is not as actually being a certain way, but as one way from the perspective of one scale, and another way from the perspective of another. In the end all scales are unified by the universe itself, and thus the serpent swallows its tail. Our Galaxy is at the center of the Cosmic Spheres of Time because every galaxy is, but we humans are at the center of the Cosmic Uroboros by the interplay of the complexity of our brains and the age of the universe. In yet another sense that could never have been foreseen before modern cosmology, we truly are at the center of the universe.

In seeing themselves as central to the universe, our ancestors were right, but like Newton, they too were right on the wrong scale. They saw their own little population as central, when in fact all living beings of about our size are central. Their error was to define themselves far too narrowly, as most people still do. In Shakespeare's *Hamlet*, when the traveling players visit the court of Denmark, Hamlet asks them "to hold, as 'twere, the mirror up to nature; to show virtue her own feature, scorn her own image, and the very age and body of the time, his form and pressure."[24] In a sense this book too is trying to hold a mirror up to nature, to use art and image and symbolism, as every cosmology has done, to reflect what cannot be directly described, and in so doing to show the very age and body of the time that we mean something in the cosmic context. Modern cosmology presents a new perspective that can help us not only to appreciate the awesome completeness of the universe but also to find what it means to each of us to be the human part of it.

I do not see a delegation
For the four-footed.
I see no seat for the eagles.
We forget and we consider
Ourselves superior.

But we are after all
A mere part of the Creation.
And we stand somewhere between
The mountain and the ant.
Somewhere and only there
As part and parcel
Of the Creation.

—CHIEF OREN LYONS, ONONDAGA NATION, IROQUOIS CONFEDERACY[25]

CONTEMPLATING THE COSMIC UROBOROS

The Cosmic Uroboros has no beginning, because it is all here all the time. But to speak about it with words, we must start somewhere. So let's begin with that spot at the bottom that represents our own world, the world of things that are measured in meters or miles, the realm of Midgard.

In Midgard you—personally—are midway between the size of a living cell and the size of Earth. Think of a single cell on the tip of your finger. That cell is as tiny compared to you as you are compared to Planet Earth. A single atom in that cell is as tiny compared to you as you are compared to the sun.

Now imagine that you curl yourself up into a ball and you become that atom in the cell on your finger. What does the world look like? Your electron cloud touches the electron clouds of the atoms all around you. It is a cozy world.

But now imagine that you, much more tightly curled up, are the nucleus of that same atom. You look outward, but it is six miles to the next nucleus of your kind, and there is little comfort in knowing that three miles away its electron cloud is touching yours.

Imagine now that you are a star. It is an even lonelier world. You are sitting here in California, and your closest neighbor is in Australia. You are the only two people on Earth. Even if you are a star in a globular cluster, that tightest of all star clusters, your closest neighbor is still a thousand miles away.

Imagine now that you are a galaxy. Things become almost cozy again. Other galaxies are not far away. In this room, your nearest neighboring galaxy is sitting only twenty feet from you. If you are in a rich cluster of galaxies, your near-

est neighbor is only a few feet away, and you feel like a person at a cocktail party. But conversation is virtually impossible. For you to think a single thought takes many thousands of years, because you can't think faster than the speed of light. It takes 100,000 years for one thought to cross your mind, and many times that to formulate an idea. You have only had time for a few galactic ideas in the ten billion or so years that you have been forming. Thinking is the privilege of creatures who live in Midgard.

Now imagine that you are a supercluster of galaxies. You are touching the next supercluster, and it touches the next, like people holding hands and encircling large voids. But your consciousness is wavering and flickering because unlike a galaxy, you, the supercluster, are not really bound together by gravity. Your parts are expanding away from each other. In time you will drift apart like clouds in a blue sky. And yet whatever you are will last many billions of years.

The universe looks different and works differently on different size-scales, but you can't tell that from looking around, because any size scale you focus on appears to be reality itself.

Now let your attention slide down the Cosmic Uroboros back to Midgard. This journey through other size scales was a gift of thought, something possessed only by the luckiest citizens of Midgard. All the channels of imagination that stream throughout the universe start in Midgard, and they all return here. Midgard is the Eden of the universe.

SEVEN

Where Do We Come From?

THE COSMIC LAS VEGAS

In the ancient cosmology of the flat earth, the Primeval Waters were believed to have existed before "heaven and earth" and to surround them forever. In modern cosmology, there is similarly a mysterious eternal something that may have been everywhere before the Big Bang and may surround our universe forever: it is the state of being called "eternal inflation." Unlike the ancients, however, who could only mythologize about the untouchable realm beyond heaven and earth, we have an astoundingly imaginative but mathematically rigorous theory. A few theoretical cosmologists have thought a lot about what eternal inflation must be like—that is, if it exists. Their work could be wrong; no one has figured out yet how to test it. It hovers at the border between physics and metaphysics. If you put the emphasis on its current untestability, then the theory of eternal inflation is "metaphysics." If you put the emphasis on the fact that mathematical intuition has in the past led to theories that were later tested and confirmed,[1] then eternal inflation, like string theory, is an "untested physical theory." Neither view is wrong, but the second view may be more conducive to productive scientific research.

Cosmologists' methodology is simple: come up with the best theory you can, and "run with it." Assume it's true and try to find out all the consequences. Live mentally in that kind of universe, sometimes for many years. (If anyone figures out how to test the theory, there will suddenly be many more cosmologists who start to run with it, since it's much more satisfying to work on a theory that has some chance of being shown either right or wrong.) If a future discovery rules out eternal inflation, it will have to do so within the framework of an even grander theory. But as we said at the beginning of this book, a theory is like a house: you can rarely find its problems and limitations—or its promising secret passageways—unless you're willing to move in with all your furniture. We are unlikely to find a grander theory unless we have the courage to move into this one.

With the theory of eternal inflation we humans can now think scientifically on a size scale enormously larger than our universe, and so we need an origin story on that new scale. But the story can only be satisfying if it has a chance of being true. Unlike traditional stories, which people have the luxury to tell from start to finish, it's not possible to tell the story of the origin of the universe chronologically, because the beginning is shrouded in mystery and our only path into it starts from current knowledge. Current knowledge is fairly secure from about one one-thousandth of a second after the Big Bang forward in time, so in this chapter we'll start at the Big Bang and turn our attention backward into the mists before it. This is a two-step process. The first step will take us to the very edge of scientific data. The second will plunge us into purely theoretical realms beyond data, and here those of us who are not physicists will find, like every people that ever went before us, that mythology is the best language for giving meaning to these realms.

To the Edge of Scientific Data: Cosmic Inflation

There is a standard theory today that accounts for the instant just before the Big Bang and explains where the Big Bang came from. "Cosmic inflation" is the name of the theory. This is not the same as eternal inflation, although they are connected, as we will explain. Cosmic inflation makes testable predictions, and so far it has passed all tests. Although there is always the chance a future test could shoot it down, cosmic inflation looks very promising.

Without cosmic inflation, Big Bang theory explains only a few early stages of the history of our universe—nothing about what happened before (what set up the initial conditions for the Big Bang?) or after (how, if everything came flying uniformly out of the Big Bang, did it end up forming complicated structures like huge galaxy clusters and long filaments of galaxies?). If the Big Bang had been uniformly smooth, the only thing that gravity would have done afterward is slow down the overall expansion of the universe, but the distribution of matter would have stayed just as smooth. For many years, astrophysicists played around with theories of other forces such as "cosmic strings" that could have shaped matter after the Big Bang, but no evidence ever turned up for such theories. By the late 1990s, data on the cosmic background radiation definitively ruled out cosmic strings as the cause of structure in the universe. But gravity alone could not have created the complex structures of galaxies—unless it had something to work with from the beginning.

The way gravity operates is that it magnifies differences that are already there. If one region in the expanding universe is ever so slightly denser than average, then that region will expand slightly more slowly and eventually be relatively denser than its surroundings, while regions with less than average density will expand faster and become increasingly less dense than average. Since there are more- and less-dense regions on large scales in the universe today, there had to have been differences in density from the beginning—and this was a clue that something happened before the Big Bang that needs to be explained. We live in an unusually dense region of the universe. Out on our Galaxy's disk where the sun orbits, the density is about a million times the cosmic average. In the solar system it is about a billion billion times the cosmic average. On Earth, the densest planet in the solar system, the density is a trillion times higher still! The average density of the earth is almost four times greater than that of the sun. What could have caused these differences in density? The old Big Bang theory was silent on this question.

The theory of cosmic inflation was proposed in the early 1980s by Alan Guth to solve a different problem.[2] But cosmic inflation ended up explaining these density differences, and much more. Keep in mind that the goal is to explain how the Big Bang could have had all kinds of irregularities built in from the start, which gravity could then have amplified over billions of years to create very large irregularities in the universe, such as galaxies like our Milky Way. The theory of cosmic inflation says that for an extremely small fraction of a second before the Big Bang—less time than it would take light to cross the nucleus of an atom—the region that was going to become our universe expanded *exponentially:* that is, in every instant it expanded to twice the size it was in the instant before.

Being scientifically accurate is important but very difficult with ordinary language. Stay with us for a moment and we will get there. This is what we meant when we warned earlier that mythic language is not an option—it's a necessity. We need it, and very soon, because we're getting to that level that the ancients were trying to penetrate. Slogging through technical details is not the best way for most people to understand. Here is a story that illustrates how blindingly fast exponential growth can be.[3]

A sultan was thrilled when his grand vizier invented chess. Full of gratitude, the sultan asked him to choose his reward. "You may give me one grain of wheat on the first square of the chessboard," said the grand vizier, "two grains on the next square, four on the next, eight on the next, and so on. That would be enough." "Such a modest gift for so great an act?" the sultan exclaimed. "You shall have it today!" But when the sultan tried to prepare the chessboard, he discovered that the amount of wheat needed grew faster and faster. By the sixty-fourth square, he would need about a trillion metric tons—a thousand years' worth of the modern world's production of wheat![4]

In this case what was inflating was the quantity of wheat. But in cosmic inflation, what exactly is inflating? You can think of it in various ways, and the more ways, the closer you get to a real understanding. We explained earlier how dark energy causes space to repel space. The basic cause postulated for cosmic inflation is that a much higher density of dark energy existed for a very brief time and subsequently became all the matter and energy in the universe. While this dark energy was in control of the region that would become our visible universe, the size of this region increased tremendously while the density of dark energy in it remained nearly constant. All this new positive energy filling space was balanced by negative gravitational energy as the universe inflated.[5] Alan Guth likes to say that cosmic inflation is the "ultimate free lunch," and that it put the bang in the Big Bang. Once physicists realized that this possibility is built into the standard quantum theory of elementary particles, the theory of cosmic inflation was born. During inflation, space is repelling space, so that the more space there is, the more repulsion—that explains its explosive, exponential growth.

The defining characteristic of inflation is its exponential rate of growth. All inflation, whether of space, money, planetary resource use, or the human population, doubles in every fixed interval of time. Inflation always starts relatively slowly: if you graph it, the curve of growth over time is almost flat in the beginning, but it rises more and more steeply until it is shooting almost straight up. The only mathematical difference between cosmic inflation and inflation of wheat is the amount of time between doublings. The growth of cancer works the same way, although different types have different rates. A single abnormal cell begins to divide and doubles, say, every hundred days. A cell is the size scale of about 10^{-3} cm. Two cells, four cells—it can take seven or eight years until a tumor crosses into the macro world where it can be detected. But by the time it is detected, doubling at that same rate it can be fatal fairly quickly. If tumors maximized their own survival, they would annoy but never kill their hosts, and their hosts would learn to live with them. But they don't. The cancer kills the organism and thus itself; the sultan runs out of wheat; the population overruns the planet and self-destructs. In any finite environment, whether involving wheat, money, energy, or population, inflation will quickly hit material limits—but inflation of space can go on endlessly, according to general relativity. Cosmic inflation far exceeded the speed of light, because although that universal speed limit applies within each little region of space, the regions themselves can expand apart much faster than light.

How does hypothesizing a wild moment of cosmic inflation before the Big Bang answer our original question—that is, how does it explain the creation of density irregularities, which gravity could then have amplified over billions of years into the galaxies that astronomers observe today? The answer requires a brief detour away from the early history of our universe into quantum theory.

A Brief Detour Through the Quantum World

In the quantum realm—which is the region on the Cosmic Uroboros from about 7:30 clockwise to the tip of the tail—a particle and its antimatter twin can burst out of nothingness, come back together, and disappear in a flash. If you could look very, very closely at apparently empty space, you would see zillions of particles popping out of the vacuum and disappearing. Such particles don't even come into existence unless they're "observed," but even if no one observes them we nevertheless know that many must be there.[6] Individual events involving particles cannot be predicted, but the *probability* of an event can be predicted, because the probability obeys deterministic laws.[7] For many decades quantum events have been observed and measured experimentally in tabletop experiments and in gigantic machines called particle accelerators. There is no way to explain *why* individual quantum events happen—they have no cause—but *that* they happen is absolutely standard physics. Much of the world's economy is now dependent on electronic products such as computers and cell phones, which exploit quantum theory.

How can things be so *predictable* at the macro level here in Midgard when at the subatomic scales the fundamental nature of reality is *random*? This is an example of how the relevant laws of physics can change drastically on different size-scales along the Cosmic Uroboros. But still, how can these utterly different modes of reality coexist? The answer is that they're not really different modes of reality but different modes of seeing reality. Newtonian mechanics describes the *average* for any given phenomenon, but quantum mechanics describes the bell curve. An average is limited to the norm; a bell curve permits even the wildest possibilities to occur on the scale of elementary particles—just not very often. In the quantum world, as Nobel Prize–winning physicist Murray Gell-Mann summarized it, "whatever is not forbidden is compulsory."[8]

Einstein never believed in quantum reality, even though he made major contributions to the theory. He always thought the randomness was only superficial and reflected physicists' lack of understanding. Someday, he expected, causality would be discovered below the apparent randomness, and he famously declared that God does not play dice.[9] But it turns out that the kind of underlying simplicity he sought is something we wouldn't really want. Life itself depends on quantum reality. In the same quantum theory that gives us infinite variety, there is clarity and stability. Every electron is absolutely identical. And every electron in an atom or molecule can furthermore be only in certain configurations and not in between. This is the secret of life's own longevity: DNA can encode vast amounts of information and preserve and pass it on through countless generations with hardly a mistake because on the quantum level, only certain arrangements of atoms are possible.[10] Classical mechanics, which assumed that matter is continuous, could never explain such perfection.

Cosmic Inflation Drew Blueprints for Galaxies

Now we can return to the question of how cosmic inflation could have created the tiny irregularities in the infant universe that later became galaxies. In the brief instant of cosmic inflation, the part of the universe that we can now see inflated from the quantum size all the way up to the size of a newborn baby. In other words, at the beginning of cosmic inflation, the part of the universe we can now see was so small that it would sit near the tip of the tail on the Cosmic Uroboros, but by the end of cosmic inflation, roughly 10^{-32} seconds later, it had exploded fully halfway around the Cosmic Uroboros to become as large as something on the human, macro scale. As the universe was inflating during that 10^{-32} seconds, the countless random quantum events inside it got expanded by twenty, thirty, forty orders of magnitude or more, depending on whether they occurred at the end of the inflationary process or earlier.[11] These quantum events left their tracks in spacetime, and the tracks are extraordinarily subtle, like the extra height of a single bacterium on a soccer ball.[12] Then the Big Bang occurred, marking the end of cosmic inflation. After the Big Bang, as the universe continued to expand, but incomparably more slowly, the tracks (or irregularities or wrinkles) expanded proportionally. At the sluggish, relatively stable rate of expansion since the Big Bang, it has taken the universe almost 14 billion years to grow in size by the same factor as it grew in that instant of cosmic inflation, that is, to cover the nearly thirty orders of magnitude of the right half of the Cosmic Uroboros, from the size of a baby up to the size of the cosmic horizon.

How big was the original region that inflated during cosmic inflation and became our universe? So small it would sit near the tip of the tail of the Cosmic Uroboros, the Planck length. As we have explained, physics does not permit us to talk meaningfully about anything smaller than the Planck length, about 10^{-33} cm. Smaller than that size-scale, space would be constantly collapsing into black holes that evaporate just as rapidly, in a "Planck time," about 10^{-43} second. This would make space and time granular and endlessly fluctuating on this tiny scale. The region that inflated into our universe was probably close to the theoretical limit, maybe only a hundred times the Planck size. In this book we coin the word "sparkpoint" to refer to such a region—the size that can inflate into a universe.[13]

So now we have a universe of spacetime being born from the Big Bang about the size of a baby, subtly wrinkled on every size scale. What are the wrinkles? Wrinkles are differences in the density of energy and matter that correspond to a warping of space. Space is not, as Newton assumed, an emptiness in which matter attracts matter from a distance; instead, space bends and curves, and these curves guide matter to move along them. Einstein's theory implies that this is what gravity really is: *space tells matter and energy how to move, and matter and energy tell space how to curve.*[14] This cosmic collaborative process that we call gravity—this subtle, endless dance between spacetime and its contents of matter and energy—is what gradually built up the long filaments, sheets, and clusters of galaxies that astronomers observe throughout the universe.

If the theory of cosmic inflation is right, it has two major implications for any future scientifically based, meaningful cosmology: first, given the wrinkled spacetime that emerged from the Big Bang, gravity alone created the galaxies, and therefore *no outside force imposed order on chaos* as in the old myths, and even in Plato's cosmology. Second, the wrinkles became the blueprint for the large-scale structure of what would only later become our universe, and therefore *the blueprint for our universe existed before the Big Bang.* But this "blueprint" was not planned by any intelligence; it was a random slice of random quantum events, immortalized and enormously expanded. The blueprint required no Demiurge or Craftsman, to use Plato's terms, to implement it. (This, however, says nothing about larger or subtler concepts of God.)

Is there anything in our normal world reminiscent of cosmic inflation, so that we can get some kind of intuitive handle on it? At least in one respect there is, and it's called a "phase transition." A phase transition is what happens, for example, when water being cooled reaches a certain temperature, stops getting colder, and begins to turn into ice. The Big Bang was a kind of phase transition. The dark energy that had driven inflation changed into ordinary energy at the end of inflation, and this was the source of the matter and energy that burst from the Big Bang and fill our universe today.

But wait, this raises a huge question: How can energy turn into particles? The answer is $E = mc^2$. This equation of Einstein's says that energy is equivalent to mass, although a little bit of mass makes an enormous amount of energy: you have to multiply the mass (m) by the speed of light (c) squared to find out how much energy is in it.[15] Physicists observe energy turning into elementary particles all the time at particle accelerators.

What is the actual mechanism by which the inflationary energy became the energy and matter that we observe? We don't know. There are not only many ways, there are many categories of ways in which it could have happened. On this particular question cosmologists are in the position of those ancient Greek thinkers who realized that not all theories of the universe could be right but had no way of figuring out which one was. We're working on it by trying to find differences between the sorts of universes that would have been left by the various alternative ways that inflation might have converted to expansion. Meanwhile, there is one aspect of this question on which there is a consensus: the conversion of inflationary energy into the energy and matter of radiation and particles tremendously increased the entropy of the universe.

Roughly speaking, entropy is a measure of disorder. Entropy is a quantity which, like energy, is not observable directly but can be reliably calculated. The Second Law of Thermodynamics says that the entropy of a closed system (for example, a tightly sealed thermos bottle) never decreases. The impending "heat death" of the universe was popularized in the early twentieth century, but the situation looks completely different from the twenty-first-century perspective. After the discovery of the cosmic background radiation, scientists realized that almost all the entropy of the universe is in the radiation from the Big Bang. All the starlight and everything else that has happened since then has added only a minuscule additional amount to the entropy.[16] The universe will be forming new stars ever more slowly as the supply of gas in galaxies diminishes, and ultimately, after many *hundreds* of billions of years, even long-lived small stars will start to run out of fuel. That, rather than "heat death," is what will cause the universe to run down.[17]

This chapter has gone into some detail on cosmic inflation because there is impressive evidence for it. All of its major predictions that have been tested so far have turned out to be right.[18] But cosmic inflation is only the first step back from the Big Bang. Where did cosmic inflation come from? We are in search of an origin story as potentially satisfying to us as the stories of the ancients were to them. To find the source of cosmic inflation, our only method is to extrapolate the relevant equations backward in time, beyond where we have any known way now of finding data. Extrapolation can be risky, but unless we push our equations as far as possible, we will never find out where they might fail—and what unexpected possibilities might appear. Mathematical extrapolation can be almost magical in its power and is not something to dismiss as pure guesswork.

We are now ready to take the plunge into the purely theoretical realm that is perhaps science's grandest myth and a modern version of the ancient idea of the Primeval Waters.

Eternal Inflation

What caused cosmic inflation? It turns out to be more fruitful to ask instead, why did cosmic inflation end? Because when we extrapolate the equations back to find the origin of cosmic inflation, the most likely possibility is that inflation is a state of existence that is eternal: once it starts it goes on forever—although it does stop in tiny pockets, which become big bangs that evolve into universes, one of which is our own. That unstoppable state of existence is called eternal inflation; its realm, which according to the theory surrounds our universe, is sometimes called the "superuniverse" or "multiverse" or "meta-universe."

Eternal inflation is a purely quantum realm. The size of the regions over which anything can interact is so tiny that only quantum laws apply. This means that, in terms of the Cosmic Uroboros, no such region is larger than the tail end of the Cosmic Uroboros. It can't get bigger because every sparkpoint (every region on that tiniest size-scale) is inflating away from every other so fast that no two can ever interact to form even something the size of an atom. Our universe is a rare aberration inside the meta-universe; it is one of those tiny pockets where inflation stopped—just a bubble of spacetime. The theory of eternal inflation, largely worked out by cosmologist Andrei Linde, says that inflation stopped only in the minuscule part of the universe visible to us (that is, inside our cosmic horizon) and also a larger unknown region beyond our cosmic horizon that corresponds to our bubble (that is, the part of the Big Bang we can't see). The size of our bubble must be many times the size of the visible universe.[19] Outside our bubble, which is infinitesimally small compared with eternal inflation, inflation continues forever.[20] We said earlier that the Big Bang created the possibil-

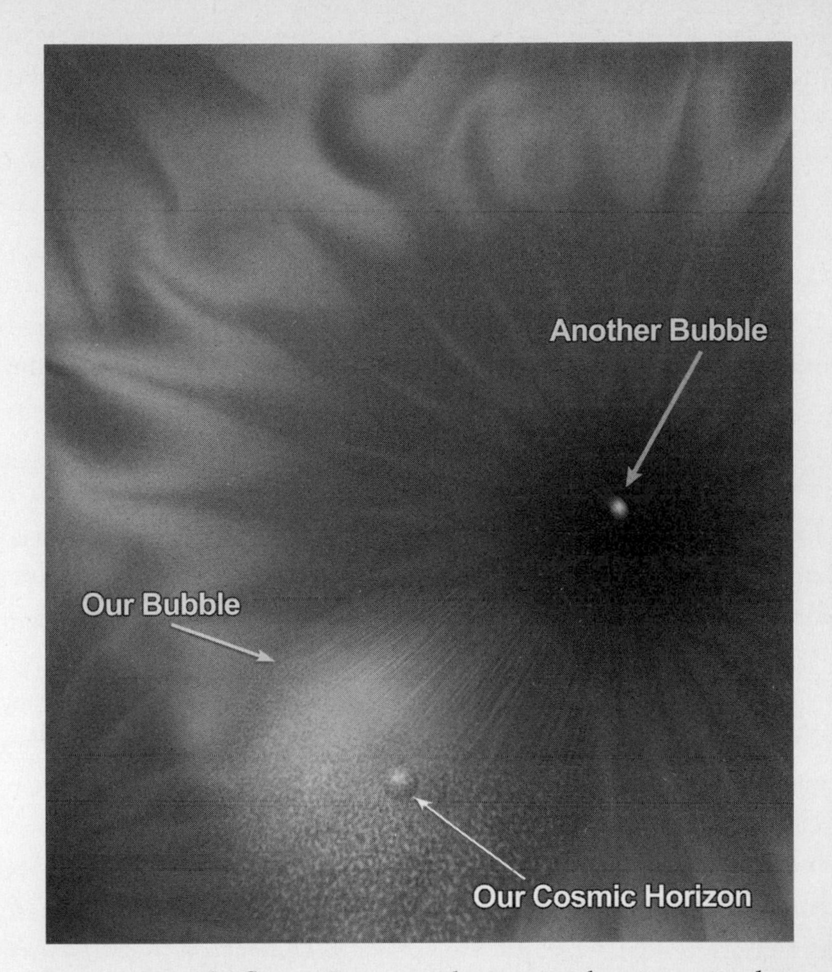

Figure 1. Eternal Inflation. Our cosmic horizon is only a tiny region deep inside Our Bubble, which represents our Big Bang. Other bubbles are constantly forming in eternal inflation, but they are expanding away from our bubble much faster than the speed of light, so we can never detect them.

ity of "where," and all "wheres" are inside the Big Bang. But this doesn't mean that the Big Bang is all there is. It just means that in eternal inflation, "where" is no longer a very meaningful word.

We find ourselves now in the situation where we can begin to describe eternity scientifically in substantial detail, but the picture is far from intuitive, and even far from being a picture. We have reached the limits of ordinary language and are faced with the choice of mathematics or myth. Therefore, we here shift from theory to myth—but feasible myth, not arbitrary myth. By the nature of language, one word follows after another; by the nature of thinking, a story must run in time. Thus we must tell it that way, even though in eternal inflation nothing runs in time. In eternal inflation, time doesn't mean anything on the large scale because it has no direction. What determines the difference between past and future in our ordinary experience is causality: past events cause present events, which in turn cause future events. Since everything in eternal inflation is flying away from almost everything else faster than the speed of light, which prevents even tiny regions from being in causal contact for more than an instant, there is no way to distinguish one time from another: everything is changing, yet overall nothing is changing. Eternal inflation is the ultimate steady state universe![21] Thus "eternal" means not only that it never ends but also that there is no direction to time, and

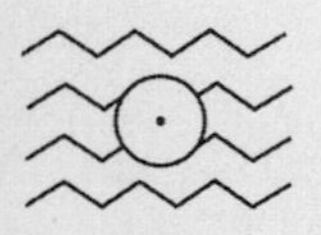

every time is the same as every other. On this mini-icon of our Big Bang in the midst of eternal inflation, the point at the center of the Big Bang is us—our cosmic horizon, our Cosmic Spheres of Time—and the zigzag waves that represent eternal inflation are an ancient Egyptian symbol for the Primeval Waters.

Eternal inflation is probably the hardest concept to visualize that we discuss in this book. Our suggestion is to keep in mind that the more metaphors through which we approach non-human-scale realities, large and small, the easier it becomes to accept that no metaphor can be entirely accurate. Below we present two: the Cosmic Las Vegas and the Kabbalistic Creation Myth. By absorbing both metaphorical ways of thinking about eternal inflation—and others that people will think up[22]—we may come closer to a sense of the truth that lies beyond them all. There is some danger in this approach, since sometimes multiple metaphors just confuse things and might lead a reader to wonder why we can't just stick to the clearest one and develop it. We will take this risk, however, in the hopes that at least one will work for you and that, ideally, together they will communicate a sense of a realm beyond all metaphors.

Eternal Inflation: The Cosmic Las Vegas

In our beginning there was—and almost everywhere else there still is—nothing but creativity: infinite potential, dense and hot, maybe 10^{30} degrees, so hot it makes no difference if it's Fahrenheit, Kelvin, or Celsius. Creativity was wildly experimenting with every possibility that quantum uncertainty permits. There were billions of events per second expanding from every sparkpoint for all eternity, unlimited by the speed of light or lack of space.

Imagine a cosmic Las Vegas, its real estate inflating forever, lights flashing, money rolling out of slot machines, coins multiplying blindingly fast, everything hot and dense. There are no gamblers, no minds with intentions, but in every tiny region small enough that light can cross it before it expands away faster than the speed of light, there is the equivalent of a coin flipping. Every flip is a quantum fluctuation. In the Grand Casino of Eternal Inflation, the rules are as follows: every time a coin comes up tails, it shrinks to half its size; every time a coin comes up heads, instantly there are two of them, they're both twice as big, and the bigger they are, the faster they flip. Since the consequences of heads versus tails are highly unbalanced, the probabilities favor heads, i.e., inflation, and most coins grow enormously and reproduce ever faster—although quantum physics sets an upper limit on the sizes of the coins. There are microscopic holes in the floor. The probability that a particular coin will come up tails again and again, so relentlessly that it gets tiny enough to fall through a microscopic hole, is extremely small but not zero. Eternity is long enough for very low probability events to happen, and once in a while a coin will shrink enough. At the instant it passes through the floor, it exits eternity, and eternity's rules no longer apply to it. Time begins. It becomes a universe.

Figure 2. Eternal Inflation as the Cosmic Las Vegas.

In a chain of events as inevitable sooner or later as a losing streak to a gambler, one sparkpoint got tails every time. Each flip was a random event. A single heads could have pulled the sparkpoint back and vastly increased the probability of another speedup until it merged forever in the cauldron of eternal inflation. But that did not happen. Tails continued, again and again. Suddenly the sparkpoint passed an invisible threshold built into the laws of physics and disappeared through the floor. For this vanishingly small capsule of eternal creativity, there was no turning back. Quantum events had taken it, a little like Alice, through an invisible looking-glass.

One sparkpoint that fell out of eternity became our universe.

Our sparkpoint had not lost its creative dynamism, any more than a child changes her character upon leaving home. It was still inflating and emanating quantum fluctuations. But the moment it exited eternity and entered time—the moment it left the Cosmic Las Vegas—its inflating became destined to die out, like all who live in time. Down the hill of potential energy our sparkpoint now rolled, unable to regain eternal potential, compelled to express its finite potential immediately. It had only the instant of actually passing through the floor, the instant we call cosmic inflation, to create the blueprint for a cosmos. And it did so. The region that would become our present cosmic horizon inflated from the size of the sparkpoint to that of a newborn baby in perhaps 10^{-33} second. In the process it spawned all the quantum impulses that will continue to reverberate for hundreds of billions of years, appearing as wrinkles in space that developed into all the big structures in the universe from galaxies to superclusters. The face of eternity as it looked in the last instants of cosmic inflation is imprinted forever on our universe. The lines of that face are permanent wrinkles in spacetime.

People often think of eternity as being "the same thing forever," but it's actually much more complicated. Eternity on the quantum scale is dynamic beyond all human imagination, spawning universes that can burst out into time. Eternal inflation is endlessly creative and lavishly profligate. Coins fall out very rarely, compared to the number that keep inflating, yet every sparkpoint in eternal inflation has the possibility, sooner or later, of becoming a universe. In its details, every resulting universe will be unique because the quantum fluctuations during each one's instant of passing through the floor will be completely different.

There may be a blizzard of universes in which our bubble is a single snowflake, which crystallized at the outset into a unique pattern of matter and energy and set off on a journey to realize itself as a universe. But unlike a blizzard of real snow, these universes don't pile up or ever land. Each one is isolated forever from all other universes by more and more eternal inflation between them. No one knows if the laws of physics are the same in other bubbles, nor do we yet have any way of finding out. We may be further than ever from answering the question that Einstein said was the only one that really interested him: "Did God have a choice?"[23] That is, could the universe have been put together in other ways, or is there only one way it could work? Einstein assumed this question could be solved by studying the nature of our universe, but if eternal inflation is right, the answer—if there is one—might lie in other universes.

How could the laws of physics be different in the different bubbles that form in eternal inflation? In quantum field theory, and also in the hoped-for encompassing superstring theory, what we think of as the laws of physics are the outcome of an underlying highly symmetric theory plus the details of how its symmetries break—and they could perhaps break in several, possibly very many, different ways. What then determines the laws of physics in our bubble? We don't know.

If the laws are different in different bubbles, the one thing we know is that in our bubble, they have to allow creatures like ourselves to exist. This is called an "anthropic" argument.[24] It's not the same as the "argument from design" invoked by nineteenth-century theologians such as William Paley, who tried to prove the existence of God the Designer on the basis of how wonderfully animals fit into their environment—an observation that was subsequently better explained by Darwinian natural selection.[25] Paley was trying to explain the *outcome* of the natural laws—the visible results—but the anthropic cosmological argument tries to explain the laws themselves. Many scientists are intrigued by the fact that if the values of the constants of nature, such as the strength of gravitational attraction or electrical repulsion, were even slightly different, then the entire universe would have to be very different, and in most cases creatures like ourselves would probably be impossible. The utility of such arguments remains controversial among scientists.[26] It seems a bit excessive to postulate a vast, possibly infinite, number of other universes just to help explain (or to avoid having to explain) why ours is the way it is. But that might be the only explanation available. Whatever the answer, we need to appreciate that pondering why our universe is set up so perfectly to last long and be infinitely creative is a wonderful problem to have. Consider the alternative.

Very, very deep inside our bubble, probably as minuscule compared to the bubble as a child's sandbox is to the visible universe, is our cosmic horizon. Our bubble represents the rarest of phenomena: the evolution of a long-lived universe. In eternal inflation, nothing persists. When all possibilities exist, none is realized. Our universe is vanishingly small compared to the meta-universe, but in it, events reverberate. It takes time to play out the great possibilities, time to grow and become something. The great miracle of our universe is that something is happening. Galaxies are evolving. Life is evolving. We are not just eternal potential—we are part of a great story.

Eternal Inflation: The Kabbalistic Creation Myth

The Cosmic Las Vegas metaphor illustrates the random, quantum nature of eternal inflation, but it gives no sense of what the theory of eternal inflation could *mean* if it is true. To approach this, we return to the medieval religious movement called Kabbalah. The Kabbalists were a secret sect of Jews concentrated mainly in Provence (now in southeast France), Spain, and later in the town of Safed near the Sea of Galilee. In their cosmology, the universe began in a point containing the blueprint for the future world, and that point expanded to become the universe. We are not Kabbalists, nor are we trying to promote Kabbalah. We do not think that Kabbalists were prescient and somehow knew mystically what science is now discovering. Although they believed the universe began in a point and expanded, they almost surely believed that it stopped expanding when it had become the medieval universe of nested spheres. But they devised a brilliant metaphorical vision of how an expanding universe could have begun, and they took it so deeply to heart that it became part of a great centering cosmology that affected every aspect of their lives. For them every good act derived cosmic value from the way the universe was created. Understanding not only how they made this connection conceptually but also how it uplifted them in a very dark period could be of great value to us.

The Kabbalists taught that creation had happened in ten stages called "Sephirot," which formed the metaphorical rungs of the spiritual ladder by which God had descended into human life and which humans could learn to use to ascend toward God. The ten were symbolized in many ways, including as a tree of life with its roots in the heavens, as a human figure standing upright, or as concentric spheres. Below we have drawn them as a tree of life with its roots in the heavens.

The upper three are drawn large because they are the cosmological Sephirot; they represent the earliest stages in the transition from nothing-but-God, (not) shown outside the tree and called "Without End," to something-else-existing. These three aspects of God are the most abstract and most remote from daily life. The following seven are the divine characteristics progressively closer to everyday life, such as love, judgment, balance, splendor, down to Shekinah, the presence of God that talks with and comforts people. For our cosmological purposes, however, the first three are the most relevant.

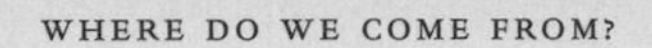

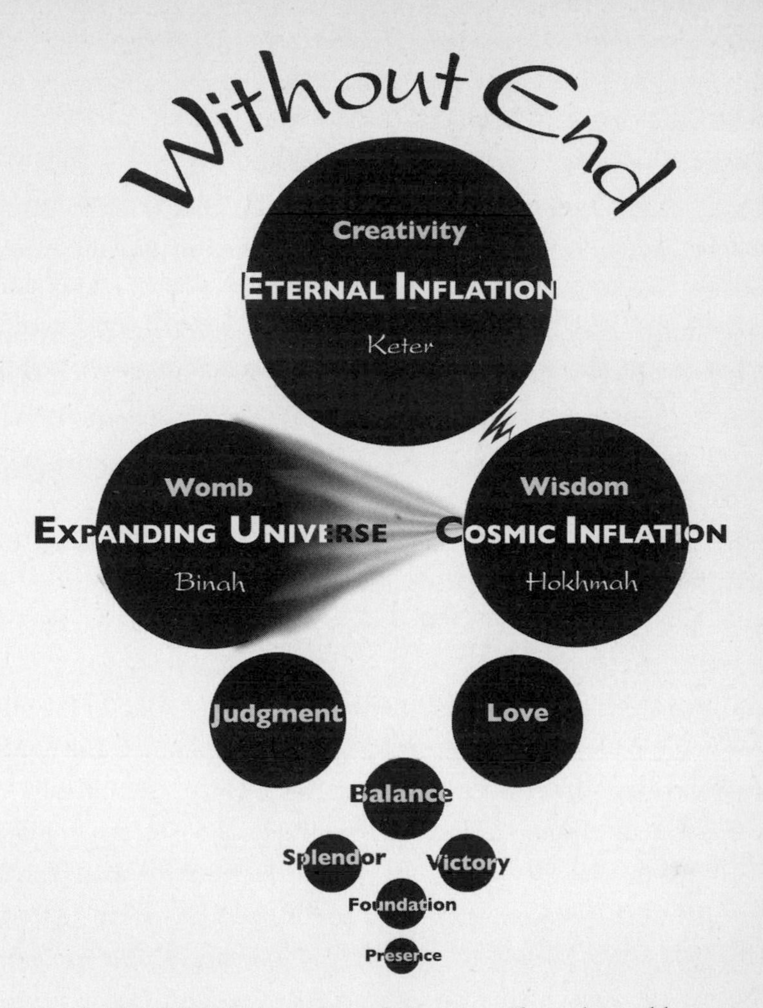

Figure 3. The Kabbalistic Order of Creation. The unknowable monotheistic God is called "Without End" (*Ein Sof*) and is represented by the white around the entire drawing. Without End emanates the ten Sephirot, or "aspects" of God identified here.

Keter (which in Hebrew means "crown") symbolizes the first tingling of God's desire to create and to enter into a relationship with a future creation. Keter is the infinite *potential* for creation, God's "getting ready" to create.

From the potential bursts *Hokhmah*, a Hebrew word for "wisdom," in the sense of a flash of insight. Hokhmah is the point containing the blueprint for the future universe. When it flashes into existence, it is the first thing separate from God.

Then comes *Binah*, both "womb" and "understanding" in Hebrew, in the sense of a mother's understanding of her child. Binah is the womb in which Hokhmah grows; Binah transforms the blueprint into the universe.

These concepts correspond closely to the ideas of eternal inflation, cosmic inflation, and the expanding universe, although these theories being developed by modern cosmologists are responding to completely different questions. In the figure we have taken the sequence eternal inflation, cosmic inflation, expanding universe and superimposed it over the first three Sephirot, Keter, Hokhmah, Binah.

Like Keter, eternal inflation is the awesome and infinite source of all possibilities, yet without the realization of any. It is, from a human point of view, pure potential. In its infinite reach, a single sparkpoint emerges: the sparkpoint is Hokhmah.

Like Hokhmah, cosmic inflation contains the blueprint for the future universe, a unique and complete package of information whose implications will reverberate forever. Like Hokhmah, cosmic inflation is the exit from eternity, the beginning of time, the instant with no instant before it.

Like Binah, the expanding universe has turned the primal plan into the reality of spacetime and galaxies. Binah was seen as female in Kabbalah. It is a bit startling to identify the expanding universe with a womb, yet there is something reassuring about positing motherly understanding as a cosmic quality. Hokhmah was traditionally seen as male, and Keter was beyond gender, nicely balancing things out. The idea that any aspect of the universe as a whole is either male or female is of course not to be taken literally; it is only useful to the extent it helps us experience connection with the universe.

On this mini-icon, the point at the center of Binah is us—our visible universe, our Cosmic Spheres of Time—just as it is on the mini-icon for eternal inflation above.

If terms such as Keter, Hokhmah, and Binah did not already exist, bearing metaphorical significance, someone would have to coin new ones, because there are no English words we know of that describe these conceptually pivotal moments in the transition between eternal inflation and the universe as we know it. But coining arbitrary new words might be as unsuccessful as Esperanto. Science communicates a special part of its meaning best by couching it in mythological terms that have worked for centuries—if such terms are accurate. And here they are, because in the Kabbalistic understanding creative potential arises first, and from it the *plan* that will shape and govern the world, all *before* the world comes into existence.[27] There is probably no name more appropriate for the Big Bang, as we understand it scientifically today, than "Hokhmah-Binah."

But this is not the end of the story. In sixteenth-century Safed, the Kabbalist Isaac Luria bound together the search for the true creation story with the search for life's meaning. Luria asked this question: If before the beginning God was everywhere, then how could there have been room for the creation to expand? Rather than thinking of an initial expansion from a point, Luria turned the idea inside out and thought of God initially *withdrawing* from a point—going into "self-exile"—creating empty space for the universe, which would then be inside Himself. And in this way Luria made a connection between an abstract process of creation and the meaning of his time in history. According to Luria, it is the absence of God from the space that makes evil possible within it. But God is not completely absent: as He withdrew, His divine light poured into the divine vessels that were the foundations of the world. The divine light was so powerful that the vessels shattered. Not even God's own creation could withstand His light. And consequently this world God made is cracked and broken and in need of repair. But every shard of the vessels carries a spark of divine light, and these sparks lie scattered in unlikely places, often buried under suffering and injustice. The purpose of life for the Kabbalistic community was to find the sparks and raise them up, and in this way to help God repair the world. *Tzimtzum* was the name for God's act of withdrawing into self-exile.

Shevirah was the shattering of the vessels into countless pieces, each with its spark of divine light. *Tikkun* was the cosmic purpose of repairing the world by finding and bringing together the divine sparks.

Parallels with scientific cosmology continue. In the early moments of the universe, according to our best scientific theories, spacetime did in fact shatter. One of the most powerful ideas of modern physics is that the symmetries underlying the observed universe are broken, and many of them are thought to have broken during the very early evolution of the universe.[28] Not only did the original symmetries break: cracks permeate spacetime on all large size-scales. This permanent pattern of cracks or wrinkles (metaphorical words) in the universe gives spacetime the irregular character that has resulted in matter clumping in some places and not others. If spacetime had been perfectly smooth, there would be no galaxies, no stars, no planets, no life, just a thin soup of particles. Life could only have evolved in an "imperfect" universe.

There are, of course, fundamental conceptual differences between Kabbalistic and scientific cosmology. Most importantly, modern science parts company with the Kabbalistic view that everything from Keter down to daily life was imbued with moral content. *There are no "defects," either moral or aesthetic, on size scales other than those of human life,* because the very idea of being "defective" is a purely human opinion. The wrinkles in spacetime are as perfect as anything can be.

Furthermore, Luria thought the present universe was unchanging.[29] He assumed, as did other Jews, that in the beginning God was everywhere. Putting these two assumptions together, God filled all space, and the total amount of space was unchanging; therefore, some explanation was required as to how God found the room to create anything at all. Luria's idea of God withdrawing from a point (*tzimtzum*) was that explanation. In eternal inflation, however, everything is flying apart, space is eternally expanding, and thus finding room for new universes is never a problem. Immersed in eternal inflation, our cosmic bubble can expand forever and it will still never come into contact with any other universe.

How did Kabbalistic cosmology enrich people's everyday lives? The Lurianic cosmology of God's self-exile might, in some happier era, have been just a colorful theory, but it became a powerful mythology because it overlay the painful, all-too-personal reality of exile. For Jews in the century or so after their expulsion from Spain in 1492 and from other European countries, the idea that God had to go into exile in order to create gave cosmic significance to their own traumatic and seemingly endless history of expulsions and exiles.[30] Mythology, as we have said, is different from theory because mythology makes "me" part of the story. The enrichment of everyday life came from this juxtaposition: persecution became uplifting rather than demoralizing when Jews were able to understand their ordeals as reflecting the exile that God Himself chose to go into in order to become creative and begin His collaboration with humanity.[31] Tending the garden or kissing the children good night—ethical, useful actions, no matter how mundane—became cosmically valuable acts of raising the sparks of divine light and thereby helping to repair the pain and injustice of their world.[32]

Kabbalah was a medieval movement. By the time of the European Enlightenment, educated Jews read Descartes and Newton, and they dismissed the Sephirot as being as silly as the medieval disputes about angels, and Kabbalah was almost forgotten until the early twentieth century.[33] But today we can see that the Kabbalistic metaphors suggest a reality closer to our modern astrophysical view than Newton's unchanging empty space does. And this picture of reality once helped ordinary people discover a cosmic purpose in their lives. *There is no deeper source of meaning for human beings than to experience our own lives as reflecting the nature and origin of our universe.*

Neither Kabbalistic cosmology nor eternal inflation are "true" in any ultimate sense, but by seeing each from the perspective of the other, we begin to get a feeling of what to demand from any cosmology intended to function for human society in the twenty-first century. Just as light cannot be fully described accurately as either a stream of particles or a wave but only as something beyond either metaphor yet with characteristics of each, the universe cannot be adequately described as either something scientifically observed or something spiritually experienced. A future centering cosmology must speak on both levels. The emerging scientific cosmology and Kabbalah can be understood as alternative metaphor systems whose juxtaposition reveals something larger than either can express alone.[34]

Both the ancient world and the Kabbalists were right about what lies beyond the known universe: we are surrounded by something mysterious and other. But unlike the ancient watery chaos, eternal inflation is far from inert. *So awesome is the energy of eternal inflation that the smallest possible amount of it, so small that it fell below eternity's minimum threshold and disappeared through the floor—that subminimal bit became our entire universe.* And the mysterious no longer surrounds just the earth and its little patch of visible heavens: eternal inflation surrounds the entire universe that was created by the Big Bang. Science has pushed back the borders of mystery and expanded the known universe immeasurably.

Creation Stories of the Future

Creation stories around the globe mainly fall into three categories: *the world is unchanging* (although if created, it went through changes in a distant, irretrievable past); *the world is cyclical* (it changes in the short term, but in the long term the cycle itself is unchanging); or *the world is always changing and time goes in one direction.* In science the debate in the mid-twentieth century involved one theory from each category. The "Steady State" proponents argued that the universe is basically unchanging, but this theory was undermined by the discovery that very distant galaxies are not like nearby ones, and disproved by the discovery of the cosmic background radiation from the Big Bang. Early Big Bang proponents thought that the Big Bang could be either cyclical or one-way. The cyclical Big Bang predicted that the universe will eventually slow down, stop expanding, and contract in a Big Crunch, conceivably followed by another Big Bang. This symmetrically attractive model was weakened by the discovery that there is not enough matter (and thus gravitation) to cause a re-collapse, and then weakened further by the discovery that the expansion of the universe is actually accelerating. At that point, the one-way Big Bang appeared to be the only theory left standing. Today, however, we know that none of these views alone is right *on all size scales*, and reality is very likely an unexpected mix of cyclical, one-way, *and* steady-state. On the size scales of life on Earth, the seasons are cyclical, and so are the births and deaths of generations of human beings. On the size scale of the Big Bang, the universe is changing in one direction: expanding, and we know of no good reason why it would ever contract.[35] But now we realize that if the theory of eternal inflation is right, then on that grandest of size scales the meta-universe is in a steady state: possibilities burst endlessly from every sparkpoint, yet on the whole nothing changes.

In revealing the modern universe, science has changed the nature of mystery: although our origins are still to some extent shrouded in mystery, that mystery is no longer something about which we can say nothing. Now we have theories like eternal inflation—intriguing, provocative, but untested. Eternal inflation has opened a new perspective on reality *whether or not it turns out to be true*, because if it eventually turns out to be wrong, whatever theory replaces it cannot explain less and will have to do better. In either case, a new standard has been set for creation stories. Yet science can only make this kind of progress step by step. For those who demand Ultimate Truth, there is no way to take even a single step beyond what other people have already thought. No cosmology, ancient, scientific, or otherwise, is the Ultimate Truth. What is required for a satisfying, centering cosmology is that it be big *enough* and uplifting *enough* to awaken a new level of insight, hope, and creativity.

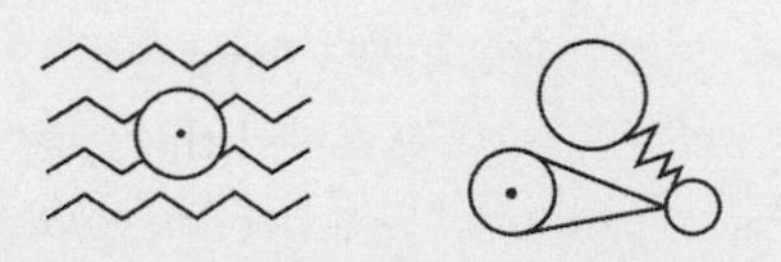

EIGHT

Are We Alone?

THE POSSIBILITY OF ALIEN WISDOM

IN OUR EXTRAORDINARY PLACE in the cosmos, "we" means different things on different size-scales. "We" of course can mean human beings, but sometimes "we" means the Milky Way, or luminous matter, or life. We're all of these and more, but perhaps the most intriguing of all our identities is "intelligent life." Thinking about our place as intelligent life in the universe immediately leads to wondering about the place of all intelligent beings. We have named this place the Sovereign Eye of the universe—but what is the Sovereign Eye? What would it mean for others to inhabit it? Do others in fact exist? Many of our identities are abstract and perhaps almost too big or strange to comprehend at first, but contact with intelligent aliens—even if you think it'll never happen or that they don't exist—is nevertheless *understandable.* Thinking about intelligent aliens helps tie us Earthlings into the universe.

Looking at ourselves in the context of all intelligent life, we begin to see our own evolution in a new light. When we ask how likely the existence of intelligent aliens is, we are also asking how likely it is that we ourselves would have evolved. How did the Sovereign Eye open on

Earth—and why Earth? By looking not only at ourselves in the context of other possible intelligent beings but at our planet in the context of other planets, we literally bring the universe down to Earth and begin to appreciate how special we and our blue planet really are. As T. S. Eliot wrote:

We shall not cease from exploration
And the end of all our exploring
Will be to arrive where we started
And know the place for the first time.[1]

This chapter thus brings us from our largest symbolic connections in eternal inflation to our much closer planetary connections, in preparation for the last chapters, which will bring us back—now in possession of cosmic perspective—to everyday life.

Many people assume that since the universe is so big, there's bound to be intelligent life out there somewhere, maybe on countless worlds. Some believe this because they are optimistic and want there to be aliens to share this beautiful universe, while others believe it because they are pessimistic and think that humanity can't possibly be anything special. The expectation of aliens has inspired not only the arts, including much science fiction, but also an occasional religion. But what is the likelihood that intelligent aliens actually exist? This is a question that scientists are now trying to answer.

There are two different scientific strategies for approaching the question of the existence of intelligent aliens. The first is to look for them directly—to *search the universe* for signs of life and intentional signals from aliens. The second is to *develop a deeper understanding* of our planet and other planets, and of how life evolved here. The second approach is less well known but has made far more progress. Following this second strategy, scientists have discovered some of the hidden requirements for intelligent life to have evolved in the one place we know it did: on Earth. And some have uncovered universal principles of living things that derive not only from earthly biology but from chemistry and physics, and therefore are likely to apply even to alien life on unknown worlds. From this kind of understanding we can start sketching the probable outlines of what it takes for a planet to produce intelligent life, and what aliens *can* be like, given what we know about how the universe works. The more we learn about how life evolved on Earth, the more we have to admit that it took a tremendous amount of both time and luck for evolution to produce our level of intelligence. This chapter will start with some of the discoveries of the second strategy and then come to the search for alien intelligence. But the likelihood of alien existence is not the only question that matters. A deeper question is: Are we alone? This is *not* a scientific question—it's a question about ourselves, and we will return to it at the end of the chapter.

Many Planets, One Earth

The first discoveries of planets outside our solar system occurred in the mid-1990s. By 2005, astronomers had already found more than 150 planets outside our solar system ("extra-solar" planets), orbiting nearby stars.[2] About 5 percent of the sunlike stars that have been studied have at least one planet, and we now know of more than ten systems with multiple planets. More extra-solar planets are being discovered every year, most less than 150 light-years from Earth. This is just a tiny part of our Milky Way galaxy—as Figure 1 shows.

Could any other planets harbor life? On all questions of life in the universe, we are forced to come back to look at Earth because we have no evidence of life arising anywhere else. Not only is Earth the only planet we know with life—all known life on Earth speaks the same genetic language. From bacteria to trees and humans, we are all descended from a single ancestor and share the same kind of genetic material: DNA. Not only do all living cells use DNA to store genetic information, they all use essentially the same code for turning this information into

Figure 1. Our Milky Way galaxy with a sphere 1,000 light-years in radius around the position of our sun. The powerful new Allen Radio Telescope now under construction will allow the SETI Institute to search for signals from intelligent aliens in this volume of space. It is a relatively small part of the Milky Way, but it contains about a thousand *times* as many stars as have been searched thus far.

proteins and living organisms. It is as if, instead of the thousands of different human languages, everything were written in the same alphabet using words with the same meaning. It's possible that completely different kinds of life arose on Earth at different times independently, but that our ancestors out-competed them and wiped them all out.[3] Alternatively, it's possible that life did not arise on Earth first, but came from Mars or even Venus. For much of the first billion years in the newly formed solar system, the planets were intensely bombarded by asteroids and meteorites from space, and pieces of each planet were knocked off and ended up on other planets. To this day there is still some active transfer of material between Mars and Earth and Venus and Earth, although far less than during the period of bombardment in the early solar system. But there is as yet no convincing evidence for life either arising multiple times on Earth or arriving from another planet.[4]

We are left with life arising exactly once, and it is never advisable to generalize from a single example and assume that because life did arise under certain conditions, it can only arise under those conditions. What we can fruitfully ask, however, is what made life possible on Earth, and what role did special conditions play? Once we determine what the factors were, we begin to get a clearer idea what to look for on alien worlds to judge the probability that they too evolved to our level.

In at least six ways, Earth is an unusually suitable planet for life.[5]

First of all, about a quarter of the extra-solar planets astronomers have found so far are "hot Jupiters," massive gasball planets like Jupiter, zipping fast in tight circular orbits around their star—closer in than little Mercury is to our sun.[6] These hot Jupiters probably formed far away and then spiraled in closer to their star in a process that would most likely have destroyed any small Earth-like planets.[7]

Second, not only has massive Jupiter not destroyed Earth by moving inward; both Jupiter's and Earth's nearly circular orbits have been favorable for the evolution of life on Earth. Ever since the end of the bombardment era in the solar system, Jupiter's gravity has helped protect Earth from being hit by comets. The importance of this protection is illustrated by the catastrophic effects of the few big impacts that have occurred anyway, the most recent of which about 65 million years ago wiped out the dinosaurs and many other species. If such events were happening much more often, life on Earth might not have had enough time between extinctions to evolve to intelligent creatures.[8]

The orbits of all the planets in our solar system except Pluto[9] are nearly circular, but circular orbits are not necessarily the norm in other planetary systems. In the extra-solar planetary systems we've found so far, planets that are as far from their star as Earth is from the sun generally follow elliptical orbits that swing them over a range of distances from their star. If Earth's orbit were highly elliptical rather than nearly circular, the seasons would be catastrophic. Earth's nearly circular orbit is at its minimum distance from the sun in January, but that is only 4 percent less than its maximum distance. We have seasons on Earth because the earth's rotation axis is tilted, and when the North Pole is tilted toward the sun it is summer in the Northern Hemisphere and winter in the Southern Hemisphere. But if Earth had a more elliptical orbit, the whole planet could suffer additional 100-degree (or more) seasonal swings in temperature. Although at least one Jupiter-like extra-solar planet has been discovered with a nearly circular orbit, the *typical* orbits of the extra-solar planets found thus far are more elliptical than even the orbit of Pluto (although this could be because planets in such orbits are easier to discover than those with large circular orbits). If Jupiter's orbit had been highly elliptical, the orbits of both Earth and Mars would most likely have been changed, and they could even have been thrown out of the solar system. Jupiter's nearly circular orbit stabilizes not only Earth's orbit but our entire solar system.

Third, Earth's distance from the sun is in what's called the "habitable zone" of the solar system—far enough from the sun for water to be liquid and not evaporate and ultimately be lost,[10] but not so far that water would be permanently frozen. Earth is the only object in our solar system that now has liquid water on its surface. What's more, the orbit of the earth has been in the habitable zone for Earth's entire lifetime. In the early solar system, when (according to the standard theory of how stars evolve) the sun was emitting about 30 percent less heat than it does now, Earth was at the outside edge of the habitable zone. As the aging sun grows hotter, the habitable zone is slowly moving outward and Earth is now nearer to the inside edge.[11] Since evolution of intelligent life here required the entire 4.5-billion-year history of Earth, a planet that was not so fortunately located with such long periods of climatic stability might not have had time. On the other hand, in any planetary system with several inner rocky planets, the spacing of such planets that is anyway required for the stability of their orbits may make it likely that at least one would be located in the continuously habitable zone.

Fourth, Earth's relatively thin crust and abundant surface water allow continued geological activity—especially plate tectonics moving the continents, forming new mountain ranges and other features, and continually recycling carbon and other elements essential for life. Neither Venus nor Mars has plate tectonics, although Venus had extensive volcanic activity as recently as 300 million years ago.[12] Data that scientists discovered after the Apollo astronauts and unmanned Russian Luna spacecraft brought back rocks from the Moon helped lead to the modern consensus that the cataclysmic crash of a proto-planet with the early earth created the Moon. This crash would have turned the earth's surface rock to liquid and destroyed any early oceans. But volcanos later spewed up water vapor, and asteroid and comet impacts brought additional water along with many of the other chemical building blocks of life.[13] The presence of all this water on Earth may be essential for plate tectonics, and it has certainly been important for the development of life, since life seems to have started under water. The timing and amount of water and other molecules delivered by impacts depends on the details of the planetary system including the orbits of the giant outer planets, so other planets in habitable zones might not have as fortunate a supply as Earth.

Fifth, the Moon, created by that chance impact, stabilizes Earth's rotation and climate. Our Moon is unusually big for a planet the size of Earth.[14] Our large Moon has kept the tilt of the earth's rotation axis essentially constant at about 23.5 degrees.[15] As we already mentioned, it is this tilt of the earth's rotation axis that causes the seasons on Earth. Thus we owe the Moon our stable seasons. Mars, with only two tiny moons that are probably captured asteroids, is susceptible to large changes in the tilt of its rotation axis, and this makes its climate much more variable than Earth's.

Our Moon also contributed to the conditions for life through the tides. Shortly after the Moon formed, it would have been much closer to the earth, producing tremendous tides. The early earth rotated much faster than it does now, but the friction of the lunar tides has gradually slowed the earth's rotation and thus also the surface winds, providing a calmer and more benign environment for life. The tides are continuing to slow the earth's rotation, while in compensation (conservation of angular momentum) the Moon is moving farther away. People find our beautiful Moon romantic but rarely appreciate that it has calmed and stabilized the earth itself, keeping our rotation on track and slowing the frantic pace of a youthful planet to a speed more conducive to the evolution of long-term complexity.

Sixth, our solar system lies in the "galactic habitable zone." Our Milky Way, like other galaxies, has a lot of stars near its center, but the stars become more spread out the farther they are from the center. As Figure 1 shows, our sun orbits in the disk of the Milky Way about halfway out from the center. Dangerous radiation is likely to have destroyed or prevented life on planets around stars that happened to lie closer to the center. Such radiation comes occasionally both from the giant black hole at the center of the Galaxy, and also from the nearby supernovas that are plentiful because of the greater density of stars closer to the galactic center. Out in the galactic suburbs where we live, the nearest supernovas were far enough away from our solar system that their radiation was weak enough for Earth's atmosphere to provide adequate protection. But if the sun had formed much farther from the center of the Milky Way where stars are even less abundant than they are nearby, there would have been too few supernovas over the history of our Galaxy to make enough stardust (heavy elements) to form rocky planets like the Earth. There may be a "ring of life" encircling the galactic center at about our distance.[16]

In all these respects, Earth has been a fortunate planet for life.

A Dance of Stability and Catastrophe Led to Life Here

How many other planets have had similar lucky conditions? No one knows how many of these unusual features are necessary for life to get started elsewhere, or for complex life to evolve. Were they even necessary on Earth? Some of the early sorts of living creatures to arise from the primordial soup are still alive and well. They're microscopic cells called prokaryotes; bacteria are familiar examples. They can live almost anywhere; they've survived every catastrophe Earth has suffered. This alone tells us that although they needed Jupiter not to destroy the Earth, and they needed liquid water, they didn't necessarily need protection from wildly varying temperatures. This kind of life has been found living in hot springs, and in rocks as deep as three kilometers below the earth's surface.

A prokaryote is one single circular strand of genetic material floating freely in a blob of protoplasm, surrounded by a cell membrane.[17] But this simplest of creatures can perform sophisticated biochemistry, including photosynthesis—it can convert sunlight efficiently into usable energy, something human chemists and physicists have not yet fully mastered. In northwestern Australia and in South Africa there are layered limestone rocks about 3.5 billion years old that were probably built by prokaryotes. Shallow seas all over the earth used to harbor vast numbers of these primitive living colonies, whose fossils are called stromatolites.[18] Prokaryotes may have been thriving and photosynthesizing as early as 300 million years after the end of the early earth's intense bombardment, and given the barren and violent state of the planet then, that is surprisingly early.[19] Since life apparently got started quickly after Earth formed and seems capable of surviving whatever goes on here, it might be that primitive life of this sort starts easily and will be found all over the universe. From a geochemical viewpoint, life is a way that a *planet* releases chemical energy; so perhaps life is almost inevitable[20]—but we're only talking about life at the level of prokaryotes or their equivalent.

It's a far cry from prokaryotes to intelligent life. For bacteria to evolve to the earliest human-like creatures took $^{999}/_{1,000}$ of the age of the earth—plus many lucky breaks. Every time we discover another such break, the likelihood of intelligent aliens decreases—but then, so does our own. Our advantages include both an extraordinary planetary habitat and a

Fruit

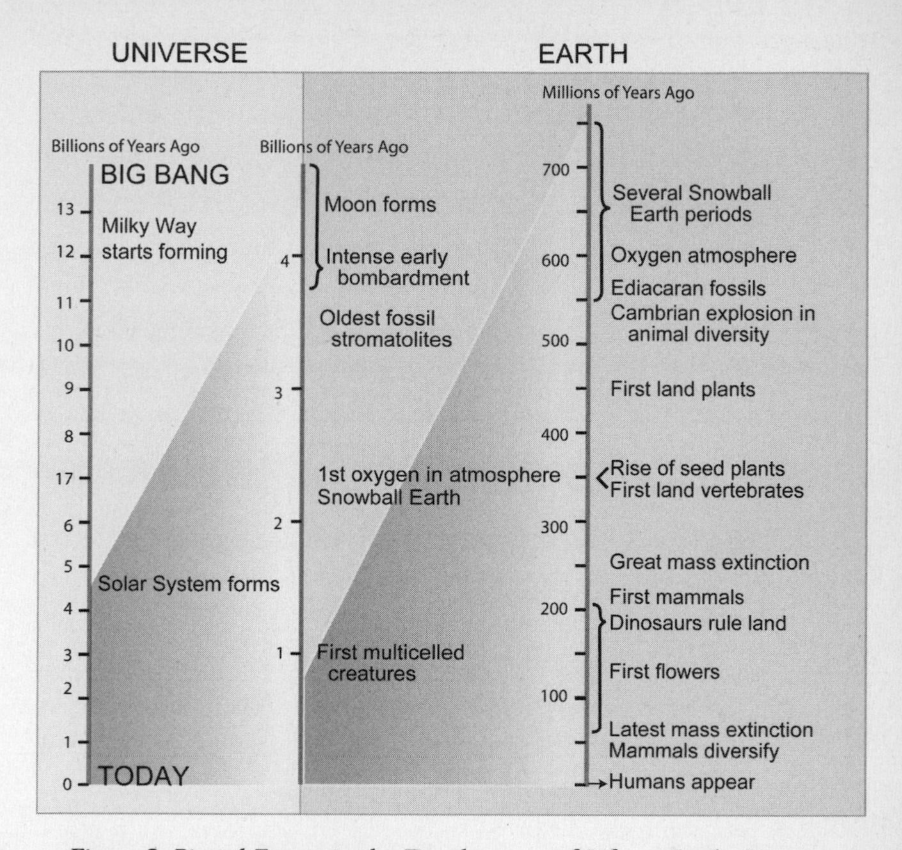

Figure 2. Pivotal Events in the Development of Life on Earth. Since so much more happened in the last 750 million years, we have expanded the timeline to show details.

series of pivotal events—shown on Figure 2—that were not always advantageous for other species.

Once the Earth-Moon system had formed, the first pivotal event in the evolution of intelligent life was the formation of the first living cell. But since no one knows if such primitive life is rare or common all over the universe, we won't even count this as a pivotal event. A billion and a half years passed before a new kind of cell, called "eukaryotes," evolved.[21] Eukaryotes have the same kind of genetic material as the prokaryotes—but about a thousand times more of it, and are thus enormously complex.[22] Their DNA isn't just floating loose but is instead wrapped around protein rods, creating chromosomes, which are contained in a special sac in the cell called the nucleus, something prokaryotes lack. In fact, eukaryotes have several kinds of little sacs, called organelles, which are miniature factories, some of which have their own separate genetic material.[23] Perhaps as a result of their special packaging of genetic material, eukaryotes invented sexual reproduction, where each parent cell contributes one of each pair of chromosomes. This was undeniably a pivotal event in evolution. Now the offspring had a much higher probability of embodying combinations of successful mutations. But evolution was still very slow because, as biologists like to say, the cardinal rule of life is "Avoid change; never evolve." If an organism is alive, it's *already* adapted, and it mainly just keeps on going and reproducing. Mutations are usually either repaired or fatal, but some not-very-deleterious ones accumulate. Evolution occurs when conditions change and some of these mutations turn out to be favored by natural selection. The evolution of eukaryotes was a tremendous step, but we're still talking about single-celled creatures. That was the extent of life on Earth for more than two billion years.

Both prokaryotes and the more complex eukaryotes could group into filaments or even colonies with more complex shapes, but each cell was still a separate organism. But then, around a billion years ago there appeared the first multicellular creatures with complicated organization—fungi, plants, and animals. The strategy of these creatures was new. Rather than trying to keep all the mechanisms of life in various sacs within the same cell, multicellular creatures divide these jobs among various specialized organs. All the cells in the creature have the same DNA, but in each organ only certain of the genetic possibilities in the shared DNA are expressed. Specialized organs not only led to far greater efficiency, they would allow entirely new possibilities—such as eyes and even brains.

The first evidence for complex life forms lies in fossils from perhaps 580 million to 600 million years ago. Although they are microscopic in size—only a few times the thickness of a human hair—these organisms were miniature versions of the more complex animals to come, with a mouth, gut, and anus.[24] Larger creatures had evolved by 570 million years ago, jellyfishlike and fanlike animals called the Ediacaran fauna. Then about 540 million to 500 million years ago a sudden riot of new forms of multicellular life appeared—the "Cambrian explosion." It is called an explosion because at this time the prototypes of the main body plans of many animals first appeared in the fossil record—including animals that developed into echinoderms (starfish, urchins, etc.), molluscs (snail, octopus), insects (six legs), arthropods (spiders, crabs, lobsters), and even vertebrates (animals with a backbone, our kind). The new creatures of the Cambrian explosion included the first ones with hard shells, which is why they are so clearly preserved in all their resplendent variety in the famous Burgess shale of Canada's Rocky Mountains, and also at Chengjiang in the Yunnan province of China.[25] Life was still underwater, but now evolution was really taking off. What could have caused this sudden increase in the complexity of living organisms after seven-eighths of the earth's history had already transpired? If it was chance mutation alone, it could happen on any alien planet where single-celled life had begun—but instead it seems to have been triggered by Earth's dance of stability and catastrophe.

Scientists recently discovered strong evidence that somewhat before the Cambrian explosion Earth was almost completely frozen several times for millions of years each time. The recent ice ages—which lasted tens of thousands of years each, during which glaciers extended only down to middle latitudes but left the tropics temperate—seem like a summer picnic in comparison. About a billion years ago the continents we know today did not exist; most land was apparently massed in the ancient supercontinent of Rodinia. But three-quarters of a billion years ago Rodinia may have begun to move, ending up near the equator. During a cooling period, much of the ocean and some land became covered with ice. The ice reflected more of the sun's light back into space, making the planet grow even colder, which made more ice form, resulting in runaway glaciation until the entire planet was frozen into what is called "Snowball Earth."[26]

With much of the ground covered by ice and life on Snowball Earth almost at a standstill, carbon dioxide began to build up in the atmosphere from volcanos. Carbon dioxide is normally removed from the atmosphere in two main ways: plants breathe it in to produce organic material (photosynthesis), and rain washes it out of the air and it combines with chemicals from rocks (weathering). Either way makes limestone (calcium carbonate) and other carbon-rich material that ends up on the bottom of the ocean. But with the planet covered in ice, neither photosynthesis nor weathering was happening. As a result, the carbon dioxide in the atmosphere reached possibly hundreds of times its present level—and this is what eventually ended these Snowball Earth periods. How? It's well known today that carbon dioxide in the atmosphere absorbs energy and heats the atmosphere; this "greenhouse effect" is the big problem caused by burning excessive fossil fuel, which releases carbon dioxide into the air. So let's imagine Snowball Earth beneath an atmosphere that was filling with carbon dioxide. The air was now well above freezing. It takes as much energy to melt ice that is 32° F (0° C) into water that is *still* 32° F as it takes to heat the water from 32° F all the way up to 176° F (80° C)! Therefore, as soon as the ice melted, the temperature on Earth would have shot up quickly to as much as 140° F (60° C) because the atmosphere was a super-greenhouse, like that of Venus. Current global warming has so far raised the earth's average temperature only about one degree F over the past twenty years, to about 59° F, but this is already causing what to us seem to be extreme swings in weather.[27]

During some of these hothouse periods there was probably a photosynthetic boom that greatly increased the amount of oxygen in the atmosphere, and this provided energy that permitted the evolution of the more complex animal types.[28] These hothouse periods would have ended when weathering of rock and photosynthesis by living organisms removed much of the carbon dioxide from the atmosphere by burying it in sedimentary layers and cooling the planet down. The effect of all this extreme climatic variability was no doubt to kill off a large fraction of the preexisting life forms, which may have prepared the way for the dramatic Cambrian explosion. This great extinction was catastrophic for earlier organisms, but it was advantageous for those of us who would succeed them.

The evolution of life cannot be explained without geology and climatology and now even astronomy.[29] For example, a crucial effect of the jump in the oxygen content of the atmosphere after each Snowball Earth period was to enhance the layer of ozone (a molecule made of three oxygen atoms) in the upper atmosphere. As the ozone layer built up, it increasingly shielded the planet from the sun's harmful ultraviolet radiation. This permitted life to thrive at the top of the ocean—and eventually to come onto land. Fungi, plants, and insects were the first groups to evolve on land; other creatures, including our ancestors the vertebrates, later came from their watery homes and adapted to living on land.

Since the Cambrian explosion, there have been at least five major extinction events, and each one was followed by major changes in the dominant sorts of living creatures.[30] The most catastrophic was about 250 million years ago, and it killed almost all life on Earth and caused the extinction of more than 50 percent of marine families of organisms, and perhaps 90 percent of all species.

This made way for both mammals and dinosaurs to evolve. The dinosaurs ruled the land for 150 million years until the extinction event about 65 million years ago wiped them out, along with a substantial number of marine families. With the dinosaurs gone, our own mammalian ancestors were able to flourish. We're not yet sure what caused the event 250 million years ago, but we do know what caused the dinosaur extinction: the collision of a large asteroid or comet with the earth, forming the enormous Chicxulub crater, which is underwater at the edge of the Yucatán peninsula of Mexico.[31] Catastrophe for them, advantage for us.

Cosmic collisions, Snowball Earths, super-greenhouse periods, asteroid impacts, extinctions, and evolutionary leaps have been pivotal events in the development of intelligent life on Earth, but through it all the original single-celled creatures that have always represented the majority of the "biomass" (the total mass of all living things) appear to have just kept going. Evolution may progress beyond primitive life only in the most unusual circumstances. It may be that the evolution of higher organisms depends on the existence of both long periods of stability and severe but not universally fatal extinction events.

Is Intelligent Life on Earth a Fluke?

We could also be reasoning completely backward: it may be that intelligence is a total fluke, and by assigning importance to the so-called pivotal events that by chance happened along the way, we may be giving them a lot more credit than they deserve. One way biologists have dealt with this possibility is to ask themselves whether, if evolution on Earth were starting all over again at the Cambrian explosion, would anything like humans ever evolve again? They are adamantly divided into two camps on this question.

The first camp argues that it's impossible. The evolutionary biologist Stephen J. Gould argued that the diversity of life was greatest at the time of the Cambrian explosion, when many body types were tried out but only a few became the basis for subsequent evolution. He said, "Replay the tape a million times . . . and I doubt that anything like *Homo sapiens* would ever evolve again."[32] Emphasizing how dependent on chance the history of complex life has been, Gould wrote, "A historical explanation does not rest on direct deductions from laws of nature, but on an unpredictable sequence of antecedent states, where any major change in any step of the sequence would have altered the final result.

This final result is therefore dependent, or contingent, upon everything that came before—the unerasable and determining signature of history." Many biologists agree with Gould. According to Loren Eiseley, "Every creature alive is the product of a unique history. The statistical probability of its precise duplication on another planet is so small as to be meaningless. . . . Nowhere in all space or on a thousand worlds will there be men to share our loneliness."[33]

Even if they didn't physically resemble human beings, what would be the chances that there would be any other intelligent creatures at all? Evolutionary biologists speak in terms of "niches" as opportunities for life. Just because a niche exists, for example the niche for intelligent creatures, does something always evolve to fill it? Jared Diamond says no. He uses the analogy of woodpeckers, a family of birds that has uniquely exploited a bountiful niche.[34] All two hundred species of woodpeckers evolved from the same ancestor, and no other birds dig holes in live trees to extract insects. According to Diamond, even on remote islands that woodpeckers never reached, such as Australia, New Guinea, and New Zealand, nothing else evolved to exploit the gourmet opportunities made available by the woodpecker lifestyle. Certain biologically useful innovations such as eyes, bioluminescence, and flight did evolve several times *independently*, but this sort of convergent evolution is not guaranteed. Diamond says, "vanishingly few animals have bothered with much of manual dexterity or intelligence. No animal has acquired remotely as much of either as have we; . . . the only other species to acquire a little of both (common and pygmy chimpanzees [bonobos]) have been rather unsuccessful." Ernst Mayr agrees. He says, "We must conclude that if high intelligence had as high a fitness value as eyes or bioluminescence, it would have emerged in numerous lineages of the animal kingdom. Actually, it happened only in a single one of the millions of lineages, the hominid line."[35] Well, that's four distinguished votes for the unlikelihood of intelligence evolving again on Earth.

But the other camp argues that it's inevitable: intelligence would evolve again. Simon Conway Morris responds to the arguments of the Impossible camp in a recent book.[36] Regarding woodpeckers, Conway Morris says that although there are no woodpeckers on the island of Madagascar, another group of birds have become true substitutes; elsewhere, three groups of mammals have converged on a woodpecker-like habit. He thus argues that other animals did evolve independently to occupy that ecological niche. But the question we're really interested in here is whether something would have evolved again to fill *our* niche—the intelligence niche. And would it have evolved with the original distinguishing characteristics of *humans*—not only intelligence but manual dexterity and toolmaking?

Conway Morris argues yes, and he gives several examples of convergent evolution to show that these basic "human" characteristics actually have already evolved independently in other animals, although not all in one. The octopus has the ability to learn and remember, as well as independently evolved camera-like eyes.[37] Birds, toothed whales (including dolphins, which until 1.5 million years ago had the biggest brains for their body size on the planet), and primates that have larger brains show more innovative behaviors. Some have large populations and range over broad areas and are thus biologically successful. In the case of the elephants, whales, dolphins, and higher primates, their dynamic social organizations—involving networks of as many as one hundred individuals in the case of dolphins—were at least partly the cause of the evolution of their excellent memories and high intelligence. Dolphins have a complex vocal communication system as well as echolocation ability, their brains are highly convoluted with the left and right sides remarkably independent (one side can sleep while the other keeps in contact with the rest of the school), and they have demonstrated the ability to learn. They can even recognize their own images in a mirror,[38] as can chimpanzees.[39] Chimpanzees and bonobos, and also some of the Central and South American monkeys such as capuchins (whose line diverged from that of the chimps and hominids about 30 million years ago) show extensive ability to make and use tools, with capuchins displaying a strong preference for right-handedness. Some crows that live in New Caledonia are known to make various standardized tool types such as hooks. Bipedality (a four-limbed animal walking on two legs) appears to have evolved not only among the hominids but also independently among the apes of the genus *Oreopithecus.* These creatures flourished about seven million years ago on islands in the Mediterranean where Tuscany is now located. They even evolved hands with a precision grip, but then they were wiped out by carnivores when their island homes were reconnected with the mainland during an ice age.[40] It may be that the role of chance, which the Impossible camp believes determines which characteristics evolve, may only determine when and where, but not whether, good survival characteristics such as intelligence and tool use will evolve.[41] Conway Morris concludes, "Rerun the tape of life as often as you like, and the end result will be much the same."

Impossible or Inevitable? We can't say yet whether human-type intelligence would evolve all over again on Earth, but a closely related question may have an answer. Once early hominids appeared, their brains—and correspondingly their intelligence—increased astonishingly rapidly in size and ability during the last two million years. This was practically overnight by Earth's standards. Why did it happen so fast? Darwin proposed the first plausible explanation for this, and it may help us see from a new angle on the alien intelligence question. Darwin explained not the total evolution of intelligence but this last enormous spurt as being a result of what he called "sexual selection."[42] The idea is that the human mind is like the peacock's tail: it evolved because it was selected in mating.[43] In other words, humans are *attracted* to intelligence when choosing mates. Contemporary evolutionary biologist Richard Dawkins agrees that sexual selection may very well explain human braininess, as well as bipedalism and hairless skin.[44] Maybe aliens need to think intelligence is sexy, too, in order to have it.

What We Already Know About Intelligent Aliens

Suppose there are indeed other worlds in our Galaxy with hospitable planetary systems and a history of fortunate events so that intelligent creatures have evolved. We already know some things about those creatures. They're not truly tiny or gigantic organisms, because, unless they're large enough to be complex, they can't be intelligent, and unless they're small enough so that the speed of internal communication is virtually instantaneous, they won't be able to think fast. That leaves a range of sizes right around our size on the Cosmic Uroboros—so intelligent aliens are likely to be about the same size as we are, although communities or integrated networks of somewhat smaller or larger creatures are also possible. Furthermore, as we explained in discussing the Cosmic Density Pyramid, stars make the same chemical elements throughout the universe. Like us, aliens are likely to be made of the most abundant reactive elements—hydrogen, oxygen, carbon, and nitrogen—which we know have a uniquely complex chemistry.

But a recent discovery has revealed an unexpected way that aliens are likely to resemble us: they are likely to have similar circulatory systems, and therefore also to have similar rates of using energy and perhaps even similar lifespans. This is more than just idle speculation. Biologists have known for many years about simple scaling relations that apply to all complex life on Earth, but they have only recently begun to understand how these regularities of life follow not from biology but from fundamental principles of physics that are likely to be applicable to life even on other worlds.

The idea of size scales is the heart of this biological discovery. Since cellular life was so rapidly successful on Earth, it may be the first step of life elsewhere, establishing the cell as the basic building block of life. Cells of *all* plants and animals on Earth are about 10^{-4} cm—regardless of the size of the plant or animal.[45] Thus large plants and animals have a lot more cells. Since every cell needs nutrients, any large organism needs a circulatory system to get the nutrients to the cells and carry away the cells' wastes. In humans the circulatory system carries oxygen to our cells and removes carbon dioxide and other wastes. The circula-

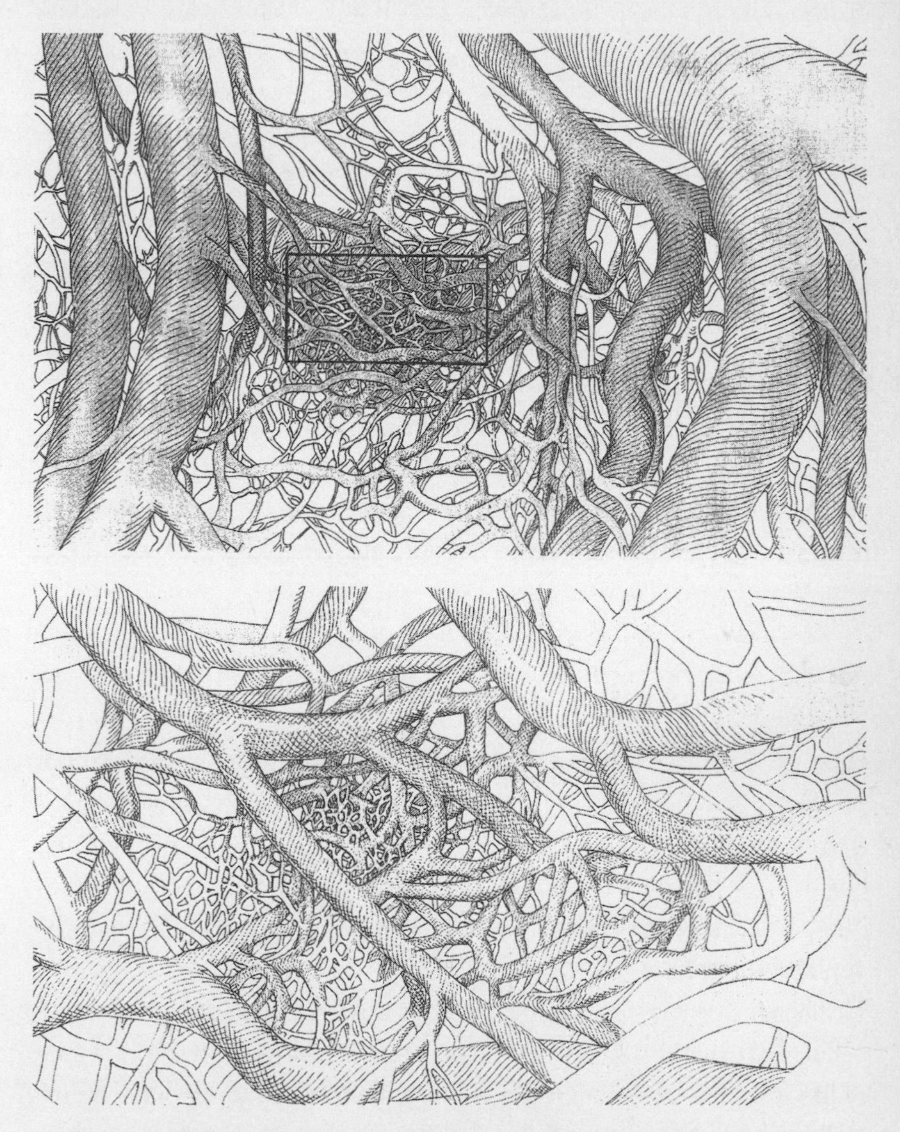

Figure 3. The fractal nature of the blood vessels of the human heart. The blowup of the central part of the top image looks like the full image.[46]

tory system is basically a network of tubes where each branch branches, and each of those branches again, down to the tiniest capillaries that supply individual cells. No matter what size scale you zoom in on, from the largest arteries to the smallest capillaries, the circulatory system looks pretty much the same. Any object that looks the same across size scales is called a "fractal."[47]

Many people have seen computer-generated fractals, which are perfect, infinite regressions. Nothing in the real world is going to be an infinite regression, since the world is made of atoms. We're not talking about these theoretical fractals but about real, biological ones, which are thus fractal across many size scales, but of course only up to a limit. In the pictures of the circulatory system above, you can't tell what scale you're looking at. The details don't look the same—it's not as though where there is a horizontal tube in one picture there will be a horizontal tube in the blowup—but the overall pattern looks the same.

All biological circulatory networks have to meet two requirements: reach every single cell (or the unreached cells will die), and do so using the minimum possible energy to circulate the blood. Why the minimum possible energy? Because organisms that use energy more efficiently will survive and reproduce more successfully, and therefore be chosen by natural selection. Physics makes a fractal the only structure that can satisfy both conditions.[48]

A consequence is that each cell in larger organisms must use energy at a slower rate than the cells of smaller organisms. Larger organisms live correspondingly longer than small ones. Mammals, for example, range from tiny tree shrews, only 5 centimeters in length and with a mass of as little as 3 grams, to blue whales as much as 30 meters long and with a mass of more than 100 tonnes.[49] Thus the length of mammals spans about three orders of magnitude, and their mass spans nearly eight. The fractal theory predicts how long each type of mammal will live on average, based only on its mass: for every four orders of magnitude that the mass increases, the heartbeat time decreases by one order of magnitude and the lifetime increases by one order of magnitude.[50] Thus the shrew, whose mass is eight orders of magnitude smaller than that of the whale, will have a heart rate about 100 times faster, and a lifetime about 1⁄100 as long. Scientists have known about such scaling relationships since the 1930s—after evolution, they are perhaps the most general facts known about complex life—but no one understood *why* mass determines lifespan until this new work revealed that it follows directly from the physics of fractals and the fractal nature of the circulatory system.[51]

That the speed of energy use in each cell depends on the mass of the whole organism continues to be true all the way down to the level of individual mitochondria *inside* cells and even down to the basic molecules in these mitochondria.[52] It even holds for human adults compared to human newborns: the cells of a fetus use energy at the slow pace of an adult while it is still effectively an organ of its mother, but after birth it becomes a separate small creature and within about thirty-six hours all its cells have sped up to about twice that rate.[53]

One stunning implication of this law is that every mammal, no matter what its size, gets roughly the same number of heartbeats in a lifetime—about 1.5 billion.[54] They're just spread out over longer lifespans (for larger mammals) or shorter ones (for smaller mammals). Other sorts of animals show similar scaling of lifespan with body mass, although birds have longer lifetimes than mammals of the same mass.[55] For example, although rats (mammals) and pigeons (birds) both have about the same mass, pigeons can live as much as ten times longer. Perhaps not coincidentally, the mitochondria (energy-processing organelles) in the cells of pigeons produce only a tenth as many "free radicals"—cancer-causing chemicals that are also present in cigarette smoke and charred meat, for example—as in rat cells. This suggests that it is such chemistry that ultimately controls lifespan, which is currently a hot research topic.

Any animal is likely to have a fractal circulatory system no matter what planet it originates on, and natural selection will probably control evolution anywhere, so the scaling laws and perhaps also lifespans among aliens may be similar to ours. However, the chemistry that all plants and animals use inside their cells to process energy may be different. Since all Earth life evolved from a single ancestor, we can't assume that alien life, which evolved from different ancestors, will process energy the same way. On the other hand, we humans are just beginning to understand and perhaps eventually will be able to control the genetic machinery of life, and it is hard to predict what we or aliens with similar abilities can do. The scientific arguments here are less certain than those we discussed earlier.

SETI—The Search for Extraterrestrial Intelligence

The other strategy for discovering how likely it is that there are intelligent aliens is to look for them. As the great physicist Richard Feynman liked to say, a problem worth working on is one where the ease of doing it multiplied by the importance of the discovery if you succeed is maximized. Since the search for extraterrestrial intelligence (SETI) is relatively easy and inexpensive compared to many other scientific pursuits, and the discovery of intelligent aliens would be among the greatest discoveries of all time, even Stephen Jay Gould, who personally assigned SETI a success probability approaching zero, agreed that it's worth doing.[56]

For us to be able to detect aliens on distant planets around other stars, the aliens need to have emitted some sort of radiation, or at least to have modified their environment in ways that we can perceive remotely. The earliest searchers began by looking at some of the nearest stars, trying to detect radio waves at a few special frequencies that seemed to be plausible choices for communication. But aliens might not choose those frequencies, so SETI scientists have invented cheap, clever ways to scan a really wide range of frequencies. They can't possibly analyze all that data for signs of intelligence—not alone; so they came up with an amazing solution. They have signed up over four million volunteers all over the world who are willing to donate their home computers' idle time to help look for signals from intelligent life. Every volunteer has a chance to be the one who finds it, and it costs the volunteer nothing.[57] But to dedicate your professional life to SETI, you have to be a dreamer.

Fortunately a substantial number of people are. Most big science is funded by governments or corporations, but SETI is mainly supported by private donations. SETI is a popular science movement, something almost unique in our time. It did get support in the beginning from NASA, but that was ended by Congress in 1992. The SETI Institute is a private organization in Mountain View, California, that describes its purpose as "to explore, understand and explain the origin, nature and prevalence of life in the universe."[58] Right now it's collaborating with scientists at the University of California to construct a new radio antenna system in a remote area of northern California. This system will extend the search from a thousand stars to as many as a million—which is still only the nearest tiny fraction of the stars in our home Galaxy. SETI keeps developing new ways to search. Since humans are now using laser light to transmit information, SETI has started to look for laser signals from space.[59]

In this mini-icon, we are at the point, which represents the region in which we are currently searching for signals from intelligent aliens. Since this mini-icon is based on the actual geography of our Galaxy, it is the only one in which our place is not central—but it is the center of the galactic habitable zone, the galactic ring of life.

SETI has not yet discovered any sign of extraterrestrial intelligence, but as its proponents like to say, absence of evidence is not evidence of absence. SETI is only fifty years old, so it's not surprising that it hasn't yet heard a signal, given the probabilities. Suppose that there are a thousand earthlike planets in our Galaxy that have intelligent, technologically savvy creatures who want to communicate with other worlds.

Suppose furthermore that we could detect signals from any of them with present technology. Given that the Galaxy is 100,000 light-years across, if they could be scattered anywhere, then the nearest of our hypothetical thousand planets would be roughly a thousand light-years away. However, these aliens are not going to be sending signals out for the entire multibillion-year lifetimes of their planets—aliens need time to evolve intelligence and communications technology, too. But let's generously suppose that the aliens on each of our thousand planets send out signals continuously for a million years—ten times longer than our own species has existed. Since they would start at different times possibly billions of years apart, most would have finished before now, or not started yet. In the immensity of cosmic time, they would have had to be sending out signals precisely as many years ago as they are light-years from us in order for us to be receiving their signals now. In that case perhaps we could detect just one signal from the entire Milky Way. Our discussion suggests that detection of signals from aliens might be highly improbable—but highly improbable events do occur, and any day the search could pay off. Maybe intelligent creatures on some planets have discovered how to live with one another and within the physical constraints of their planet(s) and have become effectively immortal—they may send out signals for many billions of years.[60] The best argument for SETI is still the one with which the original proponents concluded their first paper: "The probability of success is difficult to estimate, but if we never search, the chance of success is zero."[61]

What would happen if SETI detected a signal that appeared to be from intelligent life? The first thing they'd have to do would be to let certain other observatories around the globe in on the news so that together they could pick up the complete signal as the Earth turned. Then they'd have to figure out what it meant, if anything. It could be something very basic—just a "Hello, we're out here, are you listening?" Or it could be some kind of code packed with information that could take humans years or lifetimes to decode, if decoding of alien thought processes is even possible. Suppose under the best of circumstances the message were full of information and humans were able to decode it. We would surely want to answer them. Now another problem arises. Suppose the message was sent from a planet a thousand light-years away. That means the aliens who sent it lived a thousand years ago. It also means that it will take a thousand years for our answer to reach them at the speed of light, and another thousand years for their reply to reach us—if they're still there and interested in a thousand years, and we're still here and interested in two thousand. After all, little that was of interest scientifically to human beings a thousand years ago—for example, alchemy—is still of interest to us today. Interstellar communication requires beings who think on a long time-scale.

But now suppose that the aliens have invited us to visit them and given us directions in their very first message. How do we get there? In discussing the Cosmic Spheres of Time, we explained that in our relativistic universe, time slows down for the fast traveler compared with the speed at which time passes on Earth. "Fast" means approaching the speed of light. If we could figure out how to travel at such speeds, people on such a spaceship could quite easily travel a thousand light-years in only a few years of their own lifetime, since their internal clocks would slow down to a crawl from the point of view of us here on Earth. If the travelers got there, visited, and turned around to come home, they could possibly make it back to Earth in their lifetimes, but when they got here they would find that on Earth over two thousand years had passed since they left. This is the meaning of time travel. Time travel is indeed possible according to relativity, but only into the future, as our travelers would have done. You can't time-travel into the past. If you can get your spaceship almost to the speed of light, you can travel pretty much anywhere in the Galaxy in just a few years of your time, but you can never come home again.

Lots of people would still be willing to go. A big obstacle, however, is the fuel bill. To get even a small spaceship up to speeds close to the speed of light would take many thousands of years' worth of all the energy used by the human race at our current rate.[62] But there is another kind of obstacle that could sabotage us even if we managed against all odds to get to the alien planet. What kind of people would interact best with aliens? Scientists? Artists? Adventurers? Children? We have no idea. How would the humans on this momentous journey collaborate with each other in utterly unpredictable circumstances? What would be their shared purpose? Their highest value? How would they overcome their petty differences to rise to the tremendous challenge of the first meeting with aliens? These human problems transcend all technical solutions. They force us to confront whether we modern humans are the kind of beings capable of interstellar exploration, and if not, how we can become such beings.

Once any stable technological species—human or otherwise—embarks on space flight in a determined way, something like a billion years would be required to explore and perhaps colonize an entire galaxy.[63] A billion years is an incomprehensibly long time to us, who are awed by any civilization that lasts a few thousand years. A billion years is far longer than the lifetime of a typical *species,* which is only a few million years. Homo sapiens like us have only been around for about a hundred thousand years. The great physicist Enrico Fermi posed the question in 1950 that if alien intelligences exist, and if they have extremely long-lived civilizations and a taste for long space voyages, then "don't you ever wonder where everybody is?"[64] The answer may be that they're not here because those are very big ifs. Even if intelligence does evolve on many worlds, it may be that a long-lived civilization is rare.

The question of the chapter so far has been, "How likely is it that intelligent aliens exist?" But that may not be the most important question. The evolution of intelligence may not be the be-all and end-all but rather just one of the lucky breaks, like a large moon, on the way to what is far rarer than intelligence: the creation of a wise and long-lived civilization. When a species is as intelligent as we are—always judging from our one example, of course—its very complexity makes it volatile and unpredictable; these characteristics, however, are probably essential to the creativity and passion necessary to sustain the continued development of intelligence. Therefore, volatility and unpredictability constrained by a shared commitment to harmony with the universe is probably one key to a long-lived global civilization. Such a civilization would be the crowning achievement of evolution. This is the fire behind the Sovereign Eye of the universe roaring into life. *This is a possibility we owe to our descendants.*

Are We Alone?

Many years ago a friend of ours, who is both a Catholic priest and a philosopher of science writing books on astronomy, visited our home for several days. Every few hours he interrupted whatever he was doing and went off to read his Bible. One day Nancy asked him what he thought was real: the scientific story or the biblical stories. "The Bible is the word of God," he responded. "It's universal truth." "Then, do you think there could there be aliens on other worlds?" she asked. "Of course," he replied. "What a waste of a gorgeous universe if all those trillions of planets are uninhabited!" "If the Bible is universal truth," she puzzled, "how can it be true for aliens we know nothing about?" Our friend's reply has been an inspiration to us for years. "Universal does not mean ultimate," he said. "The Bible could have the same relationship to alien morality as Newton does to Einstein." Moralities, he was saying, could encompass one another. In the same way that Newtonian physics remains true everywhere in the universe on special size-scales, a biblical understanding of morality could in his opinion remain true under certain circumstances even if humans discover alien wisdom that is far deeper and more advanced, because alien wisdom could encompass, rather than overthrow, that understanding.[65]

Perhaps we or our descendants will someday come into contact with aliens whose long-lived civilization possesses as much wisdom as scientific knowledge, unlike our own culture where scientific knowledge seems to be far outstripping wisdom. These aliens may have nurtured themselves over millions of years without depending on material growth, instead powering their culture largely by creativity and shared commitment. They may have discovered a morality so effective that, if we knew it, it would resolve apparent inconsistencies among humans by encompassing our beliefs about morality the way Einstein's physics encompasses Newton's. But we Earthlings are unlikely to learn alien wisdom from them. We'll probably have to figure it out ourselves if we want to last long enough to encounter them. A Galactic-connected civilization must be stable and long-lived. We are not yet such a people, but we could be.

It is possible, on the other hand, that the complicated set of conditions that led to the appearance of intelligent life on Earth is so rare that we are the only intelligent creatures now alive on all the planets around the hundred billion stars in the Milky Way—perhaps even in the entire visible universe.[66] If we are, then we may be the first creatures who have begun to understand the cosmos. If so, then despite the existential feelings of lonely insignificance prevalent in our culture, we are awesomely significant.

The question "Are we alone?" is almost always asked as if it were about the existence of aliens, but it's really a question about ourselves. Are you alone when you're near an insect? Most people would say yes. How about a dolphin? Fewer people say yes. What characteristics does an alien life-form need to have before we humans will agree that by knowing it exists, we are not alone? What qualities, what compassion, what emotional potential, what ability of self-reflectiveness must they have? Any? Would we still be alone if on some alien world we discovered machines that had been created and left running by a now extinct race, but the machines were still operating and renewing themselves and superb in all kinds of artificial intelligence—would that do? This is not just science-fiction speculation, and here's why. *Whatever it is* that we require in an alien race before we'd be willing to say that the existence of such aliens has dissolved our cosmic aloneness—*that* is the essence of humanity. That is what it is in ourselves we most identify with, and value. *The qualities that we would require of such aliens are what a long-lived civilization on Earth should aim to cultivate in ourselves.* Dealing wisely with aliens or simply contacting them may be a distant goal, but understanding what it would mean can have an immediate and powerful effect. It makes clear what truly matters today: to *be* the kind of human beings we aspire to be in the long run, and to adopt this perspective *now.* The best way to get through the short run is to focus on the long run.

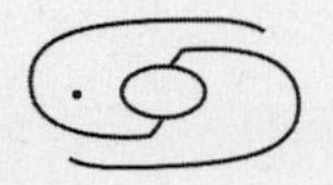

Part Three

THE MEANINGFUL UNIVERSE

NINE

Think Cosmically, Act Globally

WE HUMANS ARE at the center of a vast, cosmic adventure. Earth is four and a half billion years old, and its future stretches out for *billions* more years. Although the sun is slowly getting hotter, it will provide Earth with a perfectly livable amount of heat and light for at least several hundred million years[1]—an almost unimaginably long time. As the sun slowly heats up, our descendants will have plenty of time—millions of generations—to move to another suitable planetary system or move Earth farther from the sun.[2] Eventually, in about six billion years, the sun will evolve into a red giant star that will swallow the inner planets Mercury and Venus. Around that time, our Milky Way will be colliding with the other big galaxy in our Local Group, the great galaxy in Andromeda. But since the distance between individual stars is so great compared to their sizes, almost no stars will actually collide as these two large galaxies join over about a billion years to form a new elliptical galaxy.[3] New stars will continue to form in the united galaxy of Milky Andromeda, although more slowly. The smaller stars will continue to shine for many hundreds of billions of years, and at least some of their planets will be habitable. Thus our descendants could have many billions of years to live together—if we can just get through the next few decades without disaster.

This is the challenge to the human species today: it is as though we are on a great migration across a huge and treacherous mountain range. To get through these mountains we must gain control of human impacts on the earth and develop a sustainable relationship with our planet. The higher we are all forced to climb, the more dangerous it becomes and the more people will fall into crevasses and off cliffs or die from lack of oxygen. We need to find the lowest possible openings between the rocks. We have neither experience nor a reliable map, and we're dragging tons of baggage. We may have to jettison some of it and tie ourselves together to survive. But the goal of a sustainable global civilization is worth it.

This chapter is about re-envisioning the world through new metaphors that can help us begin to think cosmically about the future of our planet. "Thinking cosmically" doesn't require zipping around the Galaxy visiting aliens. It simply means *integrating* the new cosmic reality into our thinking whenever we try to understand what's going on in *our* world. Earth as a planet is integrated into the cosmos, but our current thinking about it is not, and therein lies the root of many problems: we are out of tune with our planet and our universe. People today are still picturing a Newtonian universe, or in some cases even a medieval earth-centered universe, while exploiting technologies based on relativity, quantum mechanics, and other new science. *The major threats to human survival today—world environmental degradation, extinction of species, climate destabilization, nuclear war, terrorists with weapons of mass destruction—result from unrestrained use of such new technologies without a cosmology that makes sense of the nature and scale of their power.*

Today many people whose destructive acts may reverberate for thousands and possibly millions of years are thinking on minuscule timescales and therefore only see short-term costs and benefits as real. One reason for this is that human beings can only perceive threats that make sense in their cosmology. For example, no one in medieval Europe recorded the sudden appearance in 1054 of a new star in the sky that burned so brightly it could be seen in daytime. In their cosmology of eternally unchanging crystal spheres such a new star was inconceivable. We know there was a supernova then because Chinese astronomers and others recorded it; they were living in a different cosmology.[4] With modern technologies, we are exercising power the long-term effects of which are as invisible to us as the 1054 supernova was to medieval Europe.

As the figure below shows, a typical person in the United States uses his or her weight in materials, fuel, and food *every day.* For every-

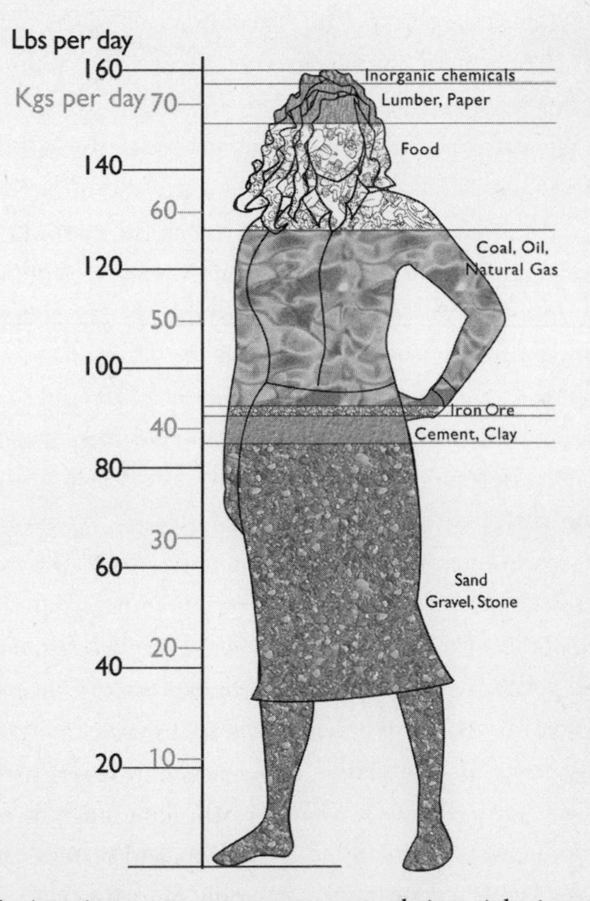

Figure 1. Americans on average consume their weight in resources every day.[5]

one in the world to reach present U.S. levels of consumption, as billions aspire to do, would with present technology require four more planet Earths.[6] Increases in greenhouse gases are now causing worldwide climate changes,[7] the effects of which we are already seeing in the form of record-breaking hurricanes, deadly heat waves, and the melting of the polar icecaps. We are running out of fresh water and topsoil worldwide.[8] We have destroyed more than half of the earth's forests and wetlands, and we are appropriating for human consumption a large and increasing fraction of the biological productivity of the entire earth.[9] Our actions are killing not just individual organisms but wiping out entire species at the greatest rate since the extinction of the dinosaurs and many other species 65 million years ago. We are only adding height to the mountains we must cross.

There is also a second reason why people today don't see long-term consequences: the still-dominant Newtonian cosmology implies that humans are of no particular significance in the universe. This reinforces our collective irresponsibility, because if we are of no significance, how much of a problem can our actions really be? Some people embrace time-myopia for religious reasons, because they think "the end" is coming soon. Some do it for financial reasons, because the quarterly bottom line is their highest standard. Some do it in pursuit of power, because nothing matters beyond the next election. But most people simply don't know yet how to think any other way.

Thus our popular cosmology cannot explain our world, and we think it doesn't matter anyway. We try as a culture to ground ourselves in each other, rather than in the earth and the universe, but we might as well try to stand still in a riptide. We humans need to ground ourselves in something real that is greater than we are. The new universe is as real as anything can be. It's happening here on Earth, as it is everywhere—that's why our technologies work! But cosmic truths will be useless if they are merely Post-it notes stuck on a Newtonian image of reality. If we want to survive and thrive, we must factor such truths, to the extent we understand them, into our policies, plans, and actions. Thinking cosmically can change our behavior globally, but to think cosmically we must begin to see through cosmic metaphors. By "cosmic metaphors" we don't mean just figures of speech but *mental reframings of reality itself.*

We Think in Metaphors

Our new cosmology is a source of new metaphors—and metaphors matter. People often assume that metaphors are merely optional figures of speech whose purpose is to enliven expression and make it more poetic and appealing. The common assumption is that we could speak literally, but it's more colloquial and comfortable to use imagery—unless we're trying to be precise, in which case metaphors muddy up the idea being expressed. But according to research in neuroscience, cognitive psychology, and linguistics, metaphors are not just words or images that help describe a concept that already exists in the mind. Instead, metaphorical connection is the way the human brain understands anything abstract. The deepest metaphors are not optional or decorative: they're a kind of sense, like seeing or hearing, and much of what we consider to be reality can be perceived and experienced only through them. We understand almost everything that is not concrete (even "concrete" is a metaphor) in terms of something else. In short, the expansiveness of our metaphors determines the expansiveness of our reality.

Those words or figures of speech that people call "metaphors" are only the last flourish of expression of an unconscious connection called a *conceptual metaphor,* which is built into our thinking. For example, when we say that an unfeeling person is ice-cold, "ice" is only the superficial metaphor; the underlying conceptual metaphor is "affection is warmth." Affection-warmth is a connection that every normal infant learns to make. While being held closely by a parent, the child's brain is activated in both the regions devoted to emotion and those devoted to temperature. As neuroscientists say, "Neurons that fire together wire together." The repeated pairing of these experiences causes the child's brain to build *physical* connections, embodied in synapses in its brain, creating the conceptual metaphor "affection is warmth."[10] When people want to describe affection, they may automatically seek a word implying warmth—love sizzles—without knowing why.

Hundreds of conceptual metaphors become hardwired during childhood as we move around in a human body on a planet with sunlight, plants, gravity, and other people. What is considered good or bad may differ between cultures, but "Up is good" and "Down is bad" are fundamental conceptual metaphors everywhere. In English this can be seen in phrases like "The economy is picking up," "She's rising in the ranks," "He really dropped the ball," and "She's feeling down." The use of conceptual metaphors is unconscious; they structure our thinking and can determine what we are able—and unable—to see. They don't act like figures of speech; they don't provide the spark, charm, or insight that makes us appreciate a genuine literary metaphor.[11] They are instead the unnoticeable medium of thought itself. The fact that many conceptual metaphors are bound into the wiring in our brains is amazing news, because to the extent that meaning is grounded in our bodies, it is as "real" as we are.

Conceptual metaphors are not all biological in origin. Cultural ones also influence our unconscious thinking. For example, "Time is money" governs how we envision and talk about time—we "spend" time or "save" time, we "invest it wisely," we "squander it," we "budget" it, we "run out" of it and "never have enough" of it.[12] Yet "Time is money" is not hardwired like "Affection is warmth." "Time is money" did not exist before the introduction of the mechanical clock in the late Middle Ages made it possible to measure hours and minutes fairly accurately. The metaphor didn't really take over our brains and our language until the Industrial Revolution reorganized all of life around timekeeping by starting to pay people not for what they produced but for their time. "Time is money" is merely a few centuries old, a heartbeat in human evolution, yet in this surprisingly brief time it has transformed the world. This illustrates the potential power of a cultural metaphor—the category into which new cosmological metaphors will also fall.

Reasoning is largely hardwired in the circuitry of our brains, which evolved from those of animals. We talk about abstractions as if they were things ("Give me liberty or give me death") because animal brains developed to seek and to navigate among solid things, and we modern humans are still using the well-tested structures of thought that animal brains developed, substituting metaphors for things.[13] If we want to expand our thinking, we have to expand the scope and richness of the metaphors we reason with. We humans have an apparently *unlimited* ability to create new combinations of imagery, and most of it is happening below awareness in what linguist George Lakoff and his coauthors have named the "cognitive unconscious."[14] The "cognitive unconscious" is thus the creator of meaning itself.

Recombining metaphors into new metaphors to build the elaborate structure we take as reality is the great unacknowledged artwork of every human brain. The natural world has always been humanity's main source of metaphor, but with science, our "natural world" includes not only each other and our local oceans, rivers, mountains, and stars, but everything between the size limits of the Planck length and the cosmic horizon. This vast conceptual universe is now available to all of us. As humanity confronts the present global challenges, we are not up against fundamental limits; it's not as though physics says we must fail. Our limits are self-imposed—ignorance, inertia, greed, fear, fanaticism, and fatalism—and thus essentially in our minds. No one has to use the symbols or names invented in this book, but everyone can start thinking more creatively from a cosmic perspective about what we humans are doing. In the next three sections we present examples of cosmic metaphors that can help us to reimagine the economic, political, and environmental challenges that we face.

Gravity and Wealth

Gravity is one example of a cosmic metaphor that suggests a new way to think about the large-scale behavior of the economy and the distribution of wealth. In astronomy, a "rich" region of the universe is defined as one that has more matter than average; a "poor" region has less matter than average. Gravity slowly magnifies subtle differences in the expanding universe.[15] Denser regions expand more slowly, and less dense regions more rapidly, than average. Gravity *always* makes the rich regions of the universe comparatively richer and the poor regions poorer, and thus gravity is the Ultimate Scrooge Principle. Eventually, as the universe expands under the force of gravity, galaxies form in denser places, and the low-density regions become cosmic voids. Wealth left to its own devices in some ways works like gravity: the rich tend to get richer and more powerful. This is the basis for our Scrooge metaphor.

The important thing about gravity for our purposes here is not simply that it concentrates matter, but that it concentrates matter *just enough* for the universe to become interesting, but not so much that everything falls into black holes. If gravity didn't concentrate matter in the rich regions, matter would have evenly thinned out as the universe expanded, and galaxies like our own Milky Way could not have formed. Similarly, if wealth were so evenly distributed that everyone in the world had the same amount, there might be little progress, since only concentrated wealth can take big risks.

In the universe, the inexorable force of gravity always gets counterbalanced—except at black holes. If we think cosmically, this suggests that the natural tendency of wealth to concentrate must also be counterbalanced *if* we want to achieve an economic stability comparable to the extremely long-term gravitational stability of our solar system and our Galaxy. What is it that counterbalances gravity? Motion. Everything is in motion. The earth and other planets are orbiting the sun, and the sun with its planetary entourage is orbiting the center of our Galaxy. *The counterbalance between gravity and motion maintains our universe.* It is worth understanding how this works, because it can perhaps tell us something important about wealth.

There is a quantity in physics that measures all such curving motion, and that quantity, called "angular momentum," is constant—or, as physicists like to say, it is *conserved.* Common sense often misleads people to assume that motion runs down, that friction eventually slows it down; but in fact angular momentum is lasting and unchangeable and thus a permanent counterbalance to gravity.[16] Conservation of angular momentum is what makes an ice skater spin faster if she pulls in her arms. Angular momentum is simply mass times radius times rotational speed. When the skater pulls in her arms, her mass stays the same but her radius decreases, so conservation of angular momentum dictates that her speed of rotation must increase. Similarly, when a galaxy forms, ordinary matter falls toward the center, and as it gets closer it rotates faster and faster about the center until its angular momentum shapes it into a disk and prevents it from falling any farther. Gravity is then counterbalanced by stable circular motion, stars like our sun form in the galactic disk, and the result is a spiral galaxy like our own Milky Way. Circular motion in the sense of angular momentum is as real and measurable as energy, and like energy it can be transferred but not lost. A balance was quickly achieved between gravity and motion in the formation of our Galaxy thanks to "violent relaxation," and the Galaxy thereafter remained stable for billions of years. It was only after that stability was achieved that generations of stars created the elements heavier than hydrogen and helium out of which planets like the earth could form, and life and intelligence could evolve.

Without something equally powerful to counterbalance the inexorable tendency of wealth to concentrate, there is a serious danger that most of the world's wealth will end up in just a few hands—which would be the economic equivalent of black holes. This is, in fact, the present trend in some countries. From the 1930s to the early 1970s, the distribution of income in the United States was stable. More recently, however, that has changed. Although the United States has the highest *average* income of any large industrial country, it also has the greatest income inequality and the largest fraction of its population in poverty. Over the past decade, most economic growth has gone to the upper 5 percent of families, and salaries and bonuses of top executives skyrocketed, but the average worker's inflation-adjusted hourly wage has not changed much in thirty years. According to the U.S. Internal Revenue Service, the wealthiest four hundred taxpayers received more than 1 percent of all income in the United States in 2000, more than double their share eight years earlier. The upward economic mobility that has long been part of the American dream has almost disappeared; very few children of the lower class now make their way even to moderate affluence.[17]

Worldwide, the inequalities are stark and increasing in much of the world. The fifth of the population living in the richest countries accounts for 86 percent of consumption, the poorest fifth a mere 1.3 percent, according to the UN Human Development Report of 1999. Although income levels are rising in some developing countries, especially in East Asia, consumption per person is lower today than it was twenty years ago in seventy countries with a total of nearly a billion people. According to the World Bank, 2.8 billion people currently live on less than $2 per day (in 1993 purchasing parity terms), compared with 2.5 billion people in 1987. Fortunately, much can be done to improve the lot of the world's poorest people, improve the education of women, and decrease the population growth rate.[18] But if current trends continue and the rich get richer while the poor get poorer, either in the United States or internationally, it may become impossible to persuade most people to cooperate in solving the problems of population, resource use, and pollution that we discuss below.

Among the rich, as among the poor, there are certainly some corrupt people and criminals, but in many cases it is no more the fault of the rich that they get richer than it is the fault of the poor that they get poorer: it's in the nature of wealth. The responsibility of a *society* with higher values than money—such as life, liberty, and the pursuit of happiness—must be to cultivate and uphold those economic forces that can counterbalance this natural tendency toward concentration of wealth and power, and to play a role analogous to motion in opposing gravity.

What could be the economic force that counterbalances the Scrooge Principle of wealth? We can perhaps get an idea if we look at motion more carefully. There are two main kinds of motion that withstand gravity in the universe: *circular* (organized, predictable) and *random* (unorganized, unpredictable on the individual level although generally predictable on large collective levels). Planets in our solar system and stars in the disk of our Galaxy are in nearly circular orbits, but in the dark matter halo of the Galaxy it is the random motion of the dark matter particles as they fly every which way that keeps the dark matter spread out. In elliptical (no-disk) galaxies and in the bulges in the centers of spiral galaxies, the motion of the densely packed stars is also random, not circular. If we zoom in to the much smaller size-scale of a single star, the matter that that star is made of stays together because of gravity, but the star doesn't collapse into a black hole except under very special circumstances, because the random thermal motions of the atoms making it up (called "pressure") prevent it from collapsing.[19] The implication for the economic analogy is that the counterbalancing force needs to be a combination of random and organized motions on various size-scales.

What might this mean in practice? Circular-motion redistribution would be predictable, and everyone in a certain category could count on it—for example, free public education for all children, police and fire protection, and public health and health care in almost all industrialized countries. In contrast, random-motion redistribution would be less uniform.[20] Human abilities are distributed somewhat randomly, so scholarships based on merit and grants for the arts or scientific research could be one way to keep money in useful non-uniform motion. Admittedly, it is difficult to see how these different kinds of motion can further illuminate the prevention of overconcentration of wealth, especially on the international level, where there is no government and few shared ideals. But if the cosmic analogy is apt, then these different kinds of motion may indeed be relevant, and if issues concerning the concentration of wealth are almost as inevitable as gravity, then the analogy can encourage us to confront them pragmatically rather than get bogged down in questions of morality and blame. The fact that the cosmos effectively counterbalances gravity, the Ultimate Scrooge Principle, can be an inspiration for us all.

The special challenge and opportunity of thinking cosmically is to take advantage of cosmic ideas like gravity, or scale thinking, or cosmic inflation, and use them as new lenses through which to view our world. We're rusty as a culture—we haven't seen reality through a new set of cosmic metaphors for centuries. But the time has come to try.

Scale and Politics

The size of each thing in the universe determines how it works. A horse can only be about the size of a horse; there can't be a horse the size of a mouse. A shining star can't be small like a planet or huge like a galaxy; an atom can't be even the slightest bit larger or smaller than it is. The symbol of the Cosmic Uroboros stands for this defining role of size in the universe. The cosmic serpent swallowing its tail reminds us that each size scale is in seamless continuity with all others, and yet every few powers of ten, wham, there's a qualitative change, possibly even bringing into play hitherto irrelevant laws of physics or emergent phenomena like temperature or even consciousness. This choppiness in the way complexity grows is characteristic of the entire Cosmic Uroboros, and thus qualitative change every few powers of ten in size is a *universal* pattern. This also applies to increasing numbers of people.

Groups of people in fact do not behave like individuals, and larger groups like countries behave differently than do smaller groups like clubs or communities. It has long been noticed that there are special group sizes—for example, ten to fifteen is a sympathy group, a jury, a sports team, the apostles, a government cabinet. Recent research has demonstrated that the limiting size of a group whose members are *personally* connected, not only in humans but many animals, is determined by the size of the neocortex of the brain compared to the whole brain.[21] Humans have the largest neocortex compared with brain size, and the corresponding number of individuals we can personally have relationships with is about 150. The earliest farm villages, clans among Australian Aborigines, businesses that can be organized informally, the company size in armies since the seventeenth century, the number of people in four generations of an extended family—all of these groups are about 150. But the decisions some people are making today affect many orders of magnitude more people than they can personally care about or even conceptualize. Rather than developing adequate new ways of understanding such new (from an evolutionary stance) size groups, many people are still thinking about global politics or economics with rules, understandings, and moral judgments that are appropriate to smaller size-scales such as a family or tribe. They use metaphors that reinforce this blindness, such as a "family" of companies or the nation as a "family." Nations and companies don't work anything like families. You might as well say a "nation of horses." No matter how many horses you put together, you will never get a nation.

Some optimists think corporations need to become stewards of the land, and that the problem is that "we haven't yet learned how to stay human when assembled in masses."[22] But a group cannot *in principle* act like a human being because it's not human, even though it's made of humans. Each person is made of elementary particles, but we don't act like elementary particles. People collectively, as committees, cities, or countries, are almost never kind, generous, or wise. Families, guilds, crowds, mobs, corporations, governmental entities—none have the same needs, purpose, or psychology as their individual human members. They may have a national constitution or a corporate mission statement, but they don't feel human values like honor, honesty, or compassion. Their indefinite life spans and ability to be in many places at the same time mean that they don't even follow all the laws of biology and physics to which we human beings are subject.

Facing this reality, we must find metaphors that *accurately* frame the behavior of the specific size group that concerns us. Some argue today, for example, that the world is broken up into "civilizations," and that what seem like smaller-scale conflicts or terrorist acts are actually part of a very large "clash of civilizations."[23] If "civilizations" is the right metaphor for understanding world politics, then it greatly simplifies the problems of predicting the interactions of some two hundred countries by reducing these to seven or eight chief actors, and there are probably cases where this is useful. But since civilizations cannot behave like individuals and vice versa, to describe individual acts as civilizational may be a kind of Scale Chauvinism (the logical fallacy in which a favorite size-scale is considered more fundamental than the others). Furthermore, there are groups trying to use this metaphor to instigate a clash of civilizations, and the danger exists that the metaphor, right or wrong, could itself become a rallying cry and a terrible, self-fulfilling prophecy. The main strategy of such instigators seems to be to encourage people to violently impose their local rivalries and narrow religious notions on the entire world—in short, to think locally, but act globally. This is backward: one's thinking should always be on a *larger* scale than one's actions if those actions are to be meaningful. To act wisely globally, we must think cosmically.[24] Furthermore, the events of our time that will have the longest-term impacts on the world are probably the destabilization of climate by global warming and the extinction of many of the world's species because of destruction of natural habitats. But these are hardly relevant to the concept of civilizations.

Putting too much weight on the "civilizations" metaphor tends to obscure the fact that in many ways the young everywhere have more in common with one another than with the old who live in their country

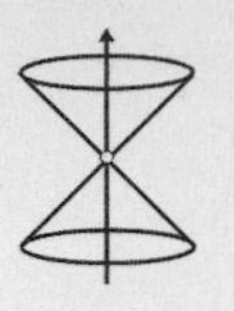

or even in their own house. Young people do not share a single religion or language, but they share an epoch in cosmic and terrestrial history that their elders will never know. "Generations" are at least as fundamental a category as civilizations into which to divide the world's people. Given the looming abrupt end of inflationary growth on the earth, it is the young everywhere who have the most at stake and who in their own interest should transcend any civilizational divisions of their forebears. All young people today are moving *as a group* up the worldline of Earth's lightcone. They are bound together by time to travel through Earth's future the same way a galaxy is bound together by gravity to travel through the expanding universe. The cohort of twenty-year-olds in the world today will never have new members, and they can only lose one another. How tragic that they often kill one another at the behest of older and less innocent cohorts. If twenty-year-olds looked at global politics through the "generations" metaphor, they might get a very different perspective.

Clearly we need many metaphors—but in an overarching context of meaning that makes sense of the variety of metaphors necessary for the many sizes of human groups. Such a context would let us consider which are appropriate, which mislead, and which ones crush our imagination or shunt it down narrow or unnecessarily dangerous corridors. Metaphors are powerful and can be perilous, but the danger can't be avoided by locking them in a drawer. Our best defense against their possible misuse is to encompass them in a higher understanding. Scale thinking gives us a superstructure in which we can evaluate and use metaphors like "civilizations" to the extent that they are accurate.

Cosmic Inflation and the Environment

Like gravity and scale thinking, cosmic inflation as a metaphor can help us understand our own world—specifically, that the dizzying rate of growth of the past century or so must end soon. Below is a graph of the

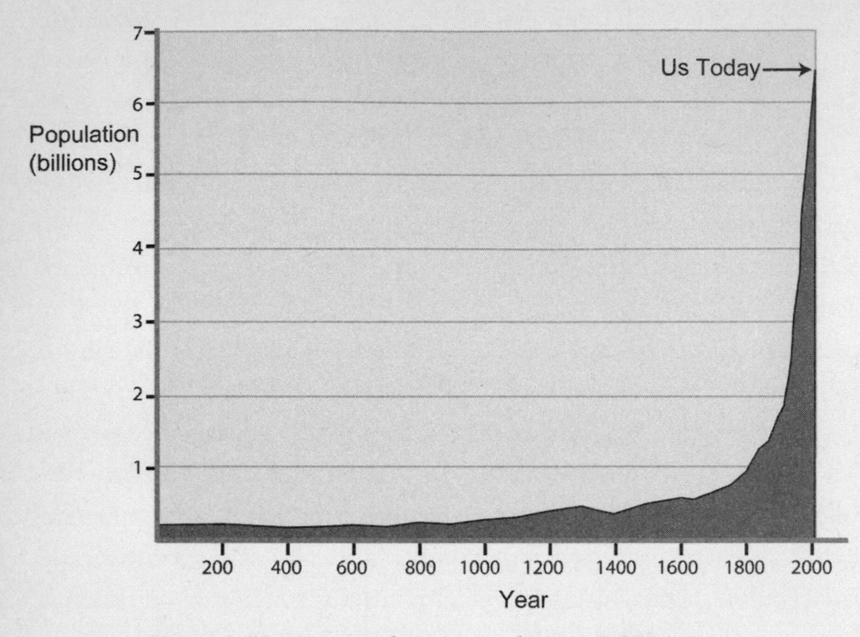

Figure 2. Human population over the past 2,000 years.

human population, which grew slowly for almost 2,000 years and then in the past century exploded. Before the twentieth century, no one ever lived through a doubling of the entire human population. But during the twentieth century the human population *quadrupled*. Experts dispute how many people the earth can support, but no one seriously proposes that the earth can sustain another population doubling.[25]

When a number doubles over a certain period of time and doubles again in the next equal period, this is exponential growth, and it is the way our universe began. Cosmic inflation was the instant lasting perhaps 10^{-32} seconds just before the Big Bang, when the size of the universe exploded exponentially, so wildly fast that the part now inside our cosmic horizon grew by as large a factor in that tiny fraction of a second as it has in the 14 billion years since. The way the universe looked on the quantum level at that instant was frozen into the blueprint for the future universe. The Big Bang was the transformation from wildly accelerating expansion to the slow and steady expansion that followed for billions of years. The time between doublings was incomparably shorter for the inflating universe than for our inflating human population, it's true, but the curve on a graph looks pretty much the same for both; only the time intervals along the bottom are different.

All exponential growth in a finite environment follows similar patterns. Exponential growth reaches a physical limit and ends—perhaps smoothly, perhaps abruptly. When growth overshoots a limiting resource, the population collapses.[26] The exponential growth in human population and resource use must end one way or another in the current generation. That makes our present era unique in human history.

Almost all the population growth is now in poorer countries, but human aspirations are similar around the world. If everyone wanted only what their parents had, resource use would be inflating at the same rate as the population. But resource use is inflating much faster than population because people's *expectations* soar as TV sets find their way into every human habitation with electricity, and people see glamorized versions of what others possess in countries like the United States that produce entertainment for export. The average technological power of each individual and the rate of pollution production are also inflating. Our finite Earth cannot fulfill the resource-profligate dreams of so many people.

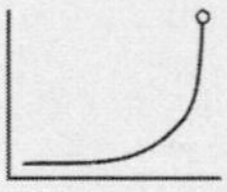

In this mini-icon of inflationary growth on Earth, we are the point. The question is, Where will the curve take us from here? We are at the current peak, but its future remains to be drawn.

The human species is reenacting the story of "The Sorcerer's Apprentice," a poem by Goethe (later animated in the Disney movie *Fantasia*). The apprentice was supposed to haul water to the house while the master was gone, but by using one of his master's incantations the apprentice cast a spell over a broomstick to haul the water for him. The apprentice was quite proud of himself, until water was flooding the house and he realized he had not learned how to break the spell. Desperately, he chopped the broomstick in two, but now each half kept bringing water. Each chopping doubled the problem. He had unleashed an inflationary situation without understanding how to control the magic, and that is precisely where we are today.

Collapse has never happened globally in human history, although it's happened locally in the collapses of empires and island cultures.[27] *If we wake up to the cosmic uniqueness of our time,* we may find the motivation to direct our efforts toward changing the things we can right now. We can start altering the trends now. But if we just continue with business as usual, Faust will reach the end of his free ride and the devil will show up in lots of ways. No one needs cosmology to diagnose this state of affairs—millions of people already sense it. Many turn to drugs, junk food, distracting entertainment, and shopping to quell the wordless fear that accumulation of material things is not a meaningful purpose and that it can't be sustained by degrading the climate, water, and land we depend on. If you are one of these people, you have sensed the enormity of our predicament, although you may not yet have found the opportunity to remedy it. Cosmology is not necessary to the diagnosis, but cosmology—specifically, cosmic inflation—helps reveal the opportunity.

Many people genuinely trying to save the earth argue that the only solution is for growth of population and of resource use to stop, but the universe in fact did not stop dead at the end of cosmic inflation like a truck hitting a brick wall. The inflationary *rate* of growth stopped, but not growth itself. The universe slammed on the brakes, slowed to a crawl, and kept going for the next 14 billion years with no wall in sight. Inflation that is transformed to slow and steady expansion can go on for billions of years. The end of inflation is like the end of adolescence; it's a coming of age, and from then on growth must be not physical but intellectual, emotional, and spiritual.

Would creeping along be the end of human progress? In the current economic paradigm, yes—stability of this kind would be considered disastrous. But the present inflationary economic paradigm is itself the long-term disaster. At the end of cosmic inflation, the universe settled into a slow and steady expansion. *Only then did it enter its most creative and long-lived phase during which it produced galaxies, stars, planets, and life. The fundamental character of the universe has been to grow in complexity.* In human affairs, very slow growth in materials use would not prevent fast growth in complexity of ideas and cultural interactions, and if those were valued, then we would have a new paradigm for human progress. On a practical level, meaningful jobs would be created. Not all human activities have to be environmentally costly. Processing information, for example, is not, and it already occupies more and more of the world's population. Creative content—books, software, recordings and movies—are already one of the biggest U.S. export sectors.[28] Smart use of resources can keep us within sustainable limits, and competition to be smart in ways that serve us all is always exciting and morally fulfilling.

We have to stop fearing the coming changes as merely material sacrifices and start seeing them as cosmic opportunities not to *acquire* more but to *become* much more. As a Zen saying goes, "Enough is a feast." Human life can continue to be enhanced indefinitely after this current period of inflationary growth if we can find a rate of growth—let's call it "sustainable prosperity"—just slow enough so that our creativity in restoring the earth stays ahead of the effects of our resource use. Short-term crises will inevitably arise, but the key to sustainable prosperity is to refuse to sacrifice the long-term goal for a short-term benefit. One might say that faith in the future means refusing to sell our cosmic birthright for a mess of pottage.[29]

In "The Sorcerer's Apprentice," the sorcerer returns in the nick of time, but in our world, it's up to us. We humans are both the sorcerer and the apprentice. We are fighting against ourselves over the fate of the earth; our cosmic self is that of the sorcerer, and our ignorant, impulsive self is the naïve young apprentice looking for a quick fix. In the end the apprentice, shaken and humbled by his close call, defers to the master, and we must do the same—our myopic, compulsively consuming small selves must defer to our cosmic selves. We can do this. Sorcerers and apprentices differ only in knowledge, maturity, and commitment. Every sorcerer began as an apprentice.

The metaphors we share, the images that collectively shape our thoughts, will make all the difference in how we humans proceed through this very dangerous pass in the mountains. We may only get one chance, and there is no heroism in forging ahead bravely in ignorance. We need to broaden our perspective and think cosmically. We are part of something bigger than most of us realize. And yet we can easily become overwhelmed or discouraged by this enormity. How can we maintain our faith in the future?

Faith in the Future

In the year 2000, the two of us participated in a conference in Iceland called "Faith in the Future." The name had a double meaning, since the conference was in honor of the thousandth anniversary of Christianity in Iceland and was partly about the future of the Christian faith, but it also focused on the role of science in the modern world and asked whether there are grounds for optimism about the human future. In one of the workshops, many experts spoke about disastrous trends—burgeoning impoverished populations, global warming, destruction of the environment, and so on. All these speakers were personally and professionally committed to countering these dangerous trends, and many of them had devised wonderfully creative and energetic means to do so. At the end of the workshop the consensus seemed to be that the future was bleak; a palpable sense of depression filled the room. There is something wrong here, we both thought, but we could not put our finger on it. As we wrote this book, we realized that, like so much of the confusion of modern life, it was connected to a misunderstanding of size scales.

"Scale Confusion" is the error of applying a way of thinking appropriate to one size scale to a different size-scale where it is inapplicable and misleading. Applying Newtonian physics to the entire universe, for example, produced for centuries a deceptive picture of reality on both large and small size-scales. Our Iceland experience showed that Scale Confusion happens not only intellectually but emotionally: feelings of devastation that would be entirely appropriate for disasters among people we know are scale-confused when aroused by abstract understandings of statistical trends describing events on a mass scale. This can be especially dangerous when such feelings demoralize creative people into paralysis.

"Emotional Scale Confusion" seems to be the human default setting. We evolved in small tribes. Our distant ancestors felt emotions about people they knew, but today we are in a different situation, inundated every day with news of everything wrong in the world. Just as we are learning to think about more than one size scale, we also need to learn to feel—consciously—on more than one size scale. Freedom from Emotional Scale Confusion means living with love and serenity in our personal lives while feeling commitment and urgency at the collective level. Nothing could be more destructive than believing that our feelings have to be consistent on all size scales. The explorers of the universe are our intellects and imaginations, not our emotions, which can expand only much more slowly. Emotional Scale Confusion is what happened in the Iceland workshop, where participants despaired for the future even though everyone around them was a jewel. It was as though they experienced not the people around them but only the looming of distant abstractions. On the personal, emotional level, instead of being sunk in depression, we should all have been grateful for having learned more about the world and been celebrating the discovery of so many inspiring people dedicated to such high ideals.

Personal demoralization because of the world situation is not only counterproductive—it may be wrong on the facts. It assumes that we really understand the present and can accurately foresee the future.

This is often not the case. In simple situations, small changes cause proportionally small consequences. But in more complex systems, small changes can have huge effects. For example, a single vote on the United States Supreme Court can affect the lives of billions of people all over the world for decades. Sometimes the effects of minuscule causes can be amplified *exponentially* by unforeseeable conjunctions of events—commonly known as the "butterfly effect."[30] Furthermore, statistics don't tell simple truths; they always require selection and interpretation, further hampering accurate prediction. Ironically, the future on the size scale of the entire universe may be predictable because it depends on so few factors (the most uncertain of which is the nature of dark energy), and some aspects of planetary behavior too, huge though they may be, are fairly simple and predictable. But human phenomena are very complex, and they are often impossible to predict because small actions can sometimes have enormous consequences.

A slogan of the women's movement for decades has been "the personal is political." But the personal and the political are on very different size scales and often require different approaches, although we must function on both of them at once. Navigating between these worlds of different scales is a constant negotiation for any human who is trying to live consciously at the center of a meaningful universe. The seriousness of the overall world situation does not at all require endless seriousness on the personal level. Our job is to live with joy while doing everything we possibly can to improve the odds for our planet. Joy can help by increasing motivation. The Talmud intriguingly says that we will be held to account for all permitted pleasures we did not enjoy.[31] Our species' best hope is energized, not demoralized, individuals vigorously and creatively attacking large-scale problems, but guiltlessly living fun, loving lives on the small scale whether or not we succeed on the large. This does not mean that our personal choices should not reflect our larger principles, but that what we *feel* about large-scale trends is about the large-scale trends—not about who we are and the possibilities open in our personal lives.

This Worldwide Turning Point

The world is at a turning point. Not the turning point of this election cycle, not even the turning point of a lifetime, but a turning point that can only happen once in the evolution of our planet. Some may dismiss this as a ridiculous exaggeration, since it is so unlikely that such a momentous turning point would occur in *our* short lifetimes. Unlikely or not, it's here. If we take our cosmic role seriously and let our largest selves find the sanest way across the mountains, we can come down the other side having created a stable and wise long-term civilization that will allow our descendants to benefit from the amazingly benign conditions of our beautiful planet. If we don't, they may curse us forever.

Modern people are horrified at the practice in some parts of the ancient world of punishing children for their fathers' crimes. "What kind of people would do that?" they wonder. The answer is: We would, and we're doing it on a much bigger scale. We're like a bomber who calmly programs a computer to drop explosives on a city below but who would recoil in horror if he had to kill every child in the city personally with a knife or his bare hands. He shields his better emotions behind the screen of distance; we hide ours behind the screen of time.

Mention a date thirty or forty years in the future, and most people will throw their hands up and say, "That's so far off we can't possibly predict the situation." But in many cases we can. Although we may not be able to predict what any individual or group of individuals will be doing in forty years (people are the most complex of all), we can predict with increasing accuracy the consequences of physical alterations to the planet, such as ozone holes and climate changes. However, many of the people with power over those consequences are either ignorant of or in denial about them. Those people's power exceeds their comprehension, and they know not what they do. Other people are courageously trying to avert catastrophes, but although their work is essential, they cannot succeed alone. It is equally essential that many people join in a positive way to discuss where "we," in the largest possible sense, want to be in a hundred years and a thousand years. Some civilizations have lasted longer than that. Our civilization may or may not, but its destructive effects certainly will.

Right now almost the only policy-makers who talk seriously about planning for the long-term—even a few thousand years from now—are people discussing how to dispose of nuclear waste properly. These people think about ethics across millennia and what we might owe our distant descendants. They wonder, for example, how a nuclear waste dump, which will be dangerously radioactive for perhaps 100,000 years, should be identified with signs to keep people from unknowingly digging in it when no one may read our language anymore or understand the concept of radioactive waste. It's a profound question how we could warn people of the distant future not to dig up hazards we have buried. One thoughtful person proposed that instead of relying on the standard radiation symbol, which may be absurdly meaningless then, we should carve into stone at the sites of nuclear waste dumps copies of Edvard Munch's terrifying painting *The Scream*.[32] Clarity about the danger is in some sense our only possible atonement for leaving that poison behind in their world.

I have set before you life and death, blessings and curses. Now choose life, so that you and your descendants may live . . .[33] These profound words from the Bible are reminders to all of us about what we owe our descendants. Our distant descendants—science helps us understand—may live thousands and millions of years in the future. How do we choose life for them? How do we choose life for our planet? These are central questions of our time and cannot be answered by simplistic notions of "life." Food is nourishing and essential; therefore, one might conclude, more food is more nourishing and more essential. But acting on this logic we would gorge ourselves on everything in sight and die early of heart attacks. This is the logic of defending every human embryo without balancing that goal with an even greater concern for the destructive long-term implications for life itself of overpopulation. "Life" is the entire four-billion-year process of evolution on Earth, which has only in the last cosmic instant reached the Sovereign Eye. To "choose life" is to nurture and protect this great cosmic process.

Many people today simply feel nothing about those future people, no connection, no concern that consequences to them will be real. But

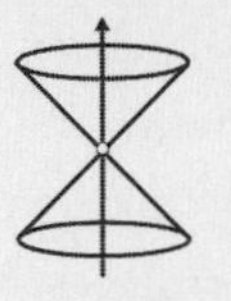

what is inevitable is real. We stand at the center of our planet's lightcone of past and future. In one direction lies everything that has ever existed or happened in the universe that can have affected us, and in the opposite direction lies everything that we can possibly affect into the most distant future. The consequences of our actions will reverberate throughout our forward lightcone, and thus the future is as real as the past—it's just not knowable yet, since we still have the opportunity to change it. From a cosmic perspective, our larger identities are bound equally into the past and future of our species and our planet. To discount the future, as though consequences that will only hit a later generation are insignificant for present calculations, is a crime against ourselves, not just against our descendants, because it distorts and truncates our concept of what *we* are.

No one needs to worry about the multibillion-year future. But right now many people are nowhere near their limit of concern—they're stuck inside a one- or maybe two-generation "responsibility-horizon," feeling they owe little to anyone beyond their own children. But it's an evasion to say, "My children or infinity, and infinity is impossible to care about." There is plenty in between, and a cosmic perspective may help to conceptualize what lies between and how our current choices are likely to affect them. If our children and their descendants keep doing always a little better than the generation before them in this one thing alone—expanding the human responsibility-horizon—the world will be fine. But only if we start now at a high enough level. We're the generation that needs to make the big jump.

Our challenge today is not only to survive the next few decades or so but, with faith in the future, to minimize our negative impacts and maximize the positive as far into the future as we can. This is not to suggest that people today try to plan the world of the future in any detail. But we should think with a long-term perspective about what *we* have to do *now*—first, to keep our own actions from leading to foreseeable disaster for our descendants; and second, to make the investments today from which the future is likely to reap the greatest benefit. We don't have to change the world by next week; all we have to do is make relatively small changes to *alter the trends*, and if people persevere in those small changes, then after a few decades the results will be huge.

Relatively modest improvements can have big payoffs in the long run. Simply setting efficiency standards for refrigerators back in the 1970s has over the three decades since then saved the United States from the need to build the equivalent of forty large nuclear power plants.[34] As a result of continuing improvements in the efficiency of electrical appliances and air-conditioning systems, which happened partly because of higher energy efficiency standards promulgated by the California Energy Commission, Californians still use about the same amount of electricity per capita as they did in 1976, while per capita electricity use in the United States as a whole has climbed by 50 percent since then.[35]

Modest changes in consumption trends could lessen the likelihood of catastrophic climate change. If we actually started now to implement improvements that have *already* been shown to work at industrial scale—such as using energy-saving hybrid cars, building more efficient buildings, restoring forests, managing soil better, and injecting waste carbon dioxide underground (rather than releasing it to the atmosphere to cause global warming)—current technology could completely stop the growth of carbon dioxide emissions for the next fifty years.[36] Big results can come from small changes. But to make the right changes, we have to plan now, since small changes in the wrong direction can have results just as huge, but devastating.

Many people comfort themselves with a pendulum metaphor of history—a very Newtonian metaphor. Today no technology dependent on accurate timekeeping would rely on pendulum clocks, and neither should any serious analysis of history. The pendulum metaphor falsely reassures us that things may swing from one extreme to the other but that there's no need to worry because they always return to the middle if we just wait long enough. From a cosmic perspective, the metaphor of history as an endlessly swinging pendulum is completely inappropriate and dangerously hypnotic. On larger scales time has an arrow and the past never returns. The current period of exponential growth in our impact on the earth is a singular point in human evolution: we who are alive today just happened to be born around the time when the curve of human population was rising most steeply.

The long-lived universe was shaped by an instant of cosmic inflation. We are living in that same kind of dangerous but fertile instant in the life of our species right now. We hold a special power that no one may ever wield again: the power to shape the very long-term future not only of our species but many others. When inflation ends, this power may be lost, and the resources and organization to carry it out irretrievable. Now is the season for planting the seeds of a long-lived civilization. If we let this season pass and do nothing, there will be no harvest, only a long, hungry winter, which many may not survive. A successful harvest is produced by human beings working in harmony with nature *on the scale at which they hope to see results.* That scale now is global. Harmony with nature is now harmony with the universe. Let's use our power not to gorge on everything in sight, but to launch the best possible future, and in doing so enhance the lives of those who will follow us.

Sustainable prosperity will not be a set of instructions for the world. This planet is so diverse that the way to deal with global problems is not to impose global solutions but to cultivate the *common ground of a large-scale goal* and encourage small-scale, decentralized solutions, appropriate to different situations, created by different kinds of people inspired by that goal. René Dubos pointed out that as more public activities and experiences become globalized, the counter-trend of identifying with a chosen neighborhood or community—what he called "local patriotism"[37]—will increase. We can in this way live on different size-scales at once without falling into Scale Confusion. A common universe can provide common ground.

The morality of sustainable prosperity, like everything else, cannot be absolute across all size-scales or time-scales. When the issues we're dealing with are on far larger scales than humans are accustomed to, both in power and in long-lasting impact, we are obliged to seek an encompassing moral understanding. What is right or wrong for the emerging global culture is going to be different in some ways from what it is inside any subgroup or any religion, but the purpose of knowing right and wrong is still the same today as it was in the days of hunter-gatherer societies: what is "right" must promote the welfare and indeed the survival of the group; what is "wrong" are self-serving behaviors at the expense of the group or of other members.[38] Today, self-serving behaviors at the expense of the group—above all, those behaviors that threaten extinction of our entire species, or of much of life—must be wrong, even if there are no laws yet against them. In the emerging global culture, our group is now the human species for better or worse, and maybe also other species.[39] Intuition and tradition can no more tell us what is right or wrong on this scale than they can tell us what is scientifically correct. To act morally today therefore requires knowledge as to what the needs of the time are, and much of that will be scientific knowledge. A moral concern in many public-policy decisions from now on would not even have made sense at earlier times: "What if this single decision actually turns out, by the falling of the blade of chance that ends inflation, to determine the pattern of the world for a thousand years? *Given this risk, is it moral?"*

Seeking sustainable prosperity through an appropriate cosmic morality need not be a distant goal. We can start today—and the mere attempt would immediately raise our probability of success. Is the very idea of a global search for sustainable prosperity a pie-in-the-sky absurdity? Is it, as our culture trains us to assume, beyond human abilities? (After all, we're so weak and insignificant . . .) *What is possible versus impossible depends entirely on what universe you're living in. Until you understand the universe you're living in, you cannot know what is possible.* Not only is it possible that we will find a sustainable rate of growth: it is almost inevitable that we will do so—sooner or later. We humans have survived because we always adapt. But right now, sustainable *prosperity* is still among the sustainable choices. Like the tremendously hopeful celebrations that took place literally around the globe on the eve of the year 2000, the global search for sustainable prosperity could unite anyone from a teenager in California to the King of Tonga.

Many of us may not yet feel that our material-growth-addicted society could drastically slow down. How could we? We don't know anything else. But even if we don't yet feel that it could stop, we can *act* as though it can, and will. Actions change feelings. Think cosmically, act globally. We must live as though we are setting the pattern for the future. At any moment, we may be. How the present period of human inflation ends will determine whether the stable period that is coming will be dark and repressive or will nurture the human spirit. It may sound terribly overwhelming and even unfair that so much responsibility for the future rides on every decision we make. But no—this way we live LARGE. *This is what it means to matter to the universe.* To save the potential of humanity on Earth may save the only intelligent life in the universe. If we destroy this great experiment on Earth, it could have the cosmic consequence, as we have warned, of extinguishing the Sovereign Eye of the universe. Like the ancients who felt there was a bridge between their acts and the invisible beyond, our generation's choices will have power over times and size-scales that we can hardly imagine. If we take on the cosmic responsibility, we get the cosmic opportunity—that rarest of opportunities for the kind of transcendent cultural leap possible only at the dawn of a new picture of the universe.

As we discover the path of sustainable prosperity, we have to follow it with faith. Not faith that some magical force will save us despite the evidence but faith *in ourselves* that we humans can understand and use the evidence science is presenting; that we are capable of growing to accommodate its implications, however strange or disconcerting; and that we can find the creativity and energy to turn these implications to the advantage of our planet as a whole. We are at the *center* of a vast, cosmic adventure—not outside it and not at its end.

TEN

Taking Our Extraordinary Place in the Cosmos

COSMIC PERSPECTIVE is the greatest gift that modern cosmology gives us. It reveals that the Big Bang powers us all, galaxies and humans alike, in different ways on our respective size-scales. Every one of us is entitled to say, "*I* am what the expanding universe is doing *here and now*." Yet this gift of perspective is not so easy to integrate into daily life. Most people's cosmic imagery is left over from earlier notions of the universe—the flat earth of the Bible, the heavenly spheres of medieval Europe, or the endless emptiness of Newton's meaningless universe. We don't live in those universes. There is real dissonance between the colorful, volatile, science-expanded world we actually inhabit and the monotonously recycled language that religions use to describe "ultimate reality." Anything described in tired metaphors from an admittedly unreal world must inevitably be accompanied by doubts and eventually boredom and indifference. The lack of a meaningful universe is a modern mental handicap. The ancient Egyptians understood that the earth reflects the cosmos, and that gave their universe meaning. *We* need to see this. Earth does reflect the cosmos. The Egyptians had a cosmic sense of time. *We,* who have the scientific ability to see so much farther, do not feel the enormity.

But if we intend to navigate Earth's coming transition from inflation to stability successfully, with sanity and justice, we will need to inspire huge creativity, intense commitment, and immense stores of enthusiasm and raw hope. To guide and use this human energy will require every powerful tool that people have ever come up with, including religious truths, ancient mythology, science, and art, plus a detailed picture of reality. Motivated by big and inspiring ideas, people without power tools built the pyramids. To perform what look like miracles, humans need big and inspiring ideas.

We are central to the universe. This belief has been the foundation of all centering cosmologies of the past, but today it is no longer merely an assumption. Now we have evidence. During the centuries between Newton and the current cosmological revolution, however, people could find no such evidence and abandoned centrality as wishful thinking. Instead, they embraced the notion that humans are insignificant, isolated beings in a vast, mostly empty space, and made the best of it by finding a kind of nobility in self-deprecation. This has led to the cultural result that the phrase "I'm human" now means basically "I make mistakes," "I have my limits," "Don't expect too much of me." Admitting our own imperfections and apologizing for our mistakes is a worthy purpose for invoking this phrase, but thinking of being human *essentially* as a limitation is a self-fulfilling prophecy and denies us our cosmic potential. In the expanding universe, human beings are not only significant—we are central, not in a simple geographic sense or through convenient choice of units, but in at least seven different ways all of which follow directly from astronomy and physics. In each of the mini-icons here and on the following pages, all of which represent the universe, *we are the point.*

1. We are made of the rarest material in the universe: stardust.

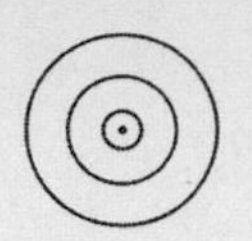

2. We live at the center of our Cosmic Spheres of Time, because every place is the center of its own cosmic spheres of time. The finite speed of light makes this inevitable in a uniformly expanding universe.

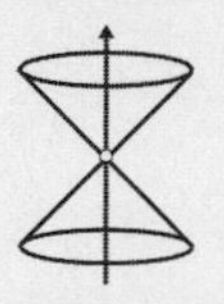

3. We live at the midpoint of time, which is also the peak period in the entire evolution of the universe for astronomical observation. Most nearby galaxies are middle-aged, past their violent youths but not yet senescent and finished with star formation. The most distant galaxies—which we have *just* acquired the technological ability to see—are beginning to disappear over the cosmic horizon now that the once-slowing expansion of the universe has begun instead to accelerate. The universe *as we are observing it today* will truly become mythic, since it will become the lost Golden Age—a fabulously rich sky that, our distant descendants will know, actually existed but will never be seen again.

4. We live at the middle of all possible sizes—in Midgard, where the possibility of tremendous variety and complexity coming in small packages keeps life interesting. Life of our complexity could bloom on no other size-scales of the Cosmic Uroboros.

5. We live in a universe that may be a rare bubble of spacetime in the infinite, seething cauldron of the eternal meta-universe. Outside our absolutely unique and isolated bubble, which we call the Big Bang, there is neither space nor time as we know it. But here inside, there is time for evolution and history, and there is space across which connections can form and structures can develop. We are not geographically central in eternal inflation, but we are very special.

6. We live at more or less the midpoint in the life of our planet. It formed, along with the sun and other planets,

about 4.5 billion years ago. It has about six billion more to go before it is roasted when our sun swells into a red giant star. We also live in the middle of the billion-year period during which Earth is most hospitable to complex life. From the point of view of our species, whose recorded history is a mere 5,000 years, today is late enough to have evolved to our present abilities while early enough still to have a potential future so vast it beggars the imagination.

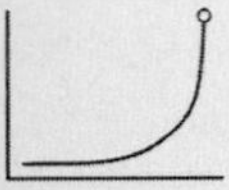

7. We live at a turning point for our species. From the point of view of the generations alive at this moment, it is late enough that we are sobering up to the scale of our problems, but not so late that we have lost all chance to solve them. This is a very special time that will never come again.

Successful cosmologies have been centering for a reason. Humans experience our own consciousness as the center of our reality—we always look *from here,* from some point of view that is characteristically us. This is why all descriptions of reality, whether based on science, logic, philosophy, or authority, that contradict this hardwired internal sense will feel unsatisfying. They may be believable and yet not believed. The only place beings with a consciousness like ours can ever feel ourselves *belonging* to the universe is at the center. But the longing to be central is not what makes us central: the structure of the universe makes us central. Cosmologists did not intend to find this and cannot tell us what it might mean. People who know some science but have not read this book may automatically assume that since there is no geographic center to the expanding universe, our claim that humans are central means nothing more than "we are of central importance to ourselves." No, it means much more. We are at the center of the *principles* that uphold the universe, and our generation is the first to know it.

The Existential Alternative

Not everyone agrees. Some excellent scientists and other scientifically literate people today look at recent discoveries and reach quite a different conclusion—that in the expanding universe "human" is a small, even pathetic identity. We will refer to the attitude they collectively express as the "existential" view of scientific cosmology, although they cover a range. "Existential" is a somewhat ambiguous word, as we explained in the context of the Copernican/Newtonian revolution, but it seems to be the best fit overall. An eloquent statement of this view was by philosopher and mathematician Bertrand Russell:

> *That Man is the product of causes which had no prevision of the end they were achieving; that his origin, his growth, his hopes and fears, his loves and his beliefs, are but the outcome of accidental collocations of atoms; that no fire, no heroism, no intensity of thought and feeling, can preserve an individual life beyond the grave; that all the labours of the ages, all the devotion, all the inspiration, all the noonday brightness of human genius, are destined to extinction in the vast death of the solar system, and that the whole temple of Man's achievement must inevitably be buried beneath the debris of a universe in ruins—all these things, if not quite beyond dispute, are yet so nearly certain, that no philosophy which rejects them can hope to stand. Only within the scaffolding of these truths, only on the firm foundation of unyielding despair, can the soul's habitation henceforth be safely built.*[1]

Humans are merely "a fortuitious cosmic afterthought," according to Stephen Jay Gould.[2] Even the mythologist Joseph Campbell, who looked expectantly to science to provide a foundation for a future centering cosmology, wrote that humanity is a "scurf on the epidermis" of a small planet of an average star.[3] The brilliant astronomer and science popularizer Carl Sagan was perhaps the best known and most tireless debunker of our cosmic centrality:[4]

> *We live on a hunk of rock and metal that orbits a humdrum star in the obscure outskirts of an ordinary galaxy comprised of 400 billion stars in a universe of some hundred billion galaxies, which may be one of a very large number, perhaps an infinite number of separate, closed-off universes. Many, perhaps most, of those stars probably have planets. In this perspective, how can anyone seriously believe that we are central—physically, much less to the purpose of the universe?*

Sagan had a purpose in insisting on the cosmic demotion of humanity from its traditional place at the center of the universe. He believed that if people want to feel important to the universe, we have to *do* something important—like end poverty, bring about equality, explore other worlds—because there is no justification for feeling important on the simple basis of our position in the universe. "We have not been given the lead in the cosmic drama," he wrote.[5] Sagan's goal was admirable, and he deserves to be honored for his immense contribution in bringing the highest values of science before the public eye. But the existential view is missing something crucial.

The difference between the existential view and the meaningful view is not just "The glass is half empty" versus "The glass is half full," because everyone knows what a glass looks like and what the alternatives mean. But if we resign ourselves to being some minor trash in the universe, we will *never see* what the universe looks like, because that can only be seen with the mind's eye, and the mind's eye works from metaphors that are inaccessible to people who hold the humans-as-trash assumption. There is nothing in modern cosmology that requires the existential view, nor anything that requires the meaningful view. The bottom line of both views is scientific accuracy: both hold that interpretations of reality where science is compromised for ideological purity should be rejected. But given this bottom line, an *attitude* toward the discoveries of modern cosmology is every person's choice. This choice is too easily made, however, based solely on what feels intuitively right or seems smart in our shortsighted, ironic culture. The existential view automatically feels more familiar and natural because the West has cultivated it for generations, and much that is beloved in art and culture reflects it. But where the existential view veers off into emotions like despair or resignation or a *feeling* of insignificance or even of dark satisfaction, those emotions are arbitrary and unnecessary. The meaningful universe encompasses the existential, in the sense that the meaningful can understand the existential, but the existential cannot see the meaningful.

The choice of attitude is not a casual one. It's all too easy to see scientific cosmology as an intellectual challenge, entertainment, cocktail banter, or even, as some cosmologists treat it, a professionally played sport, because these are the normal attitudes as long as we don't participate in what a new universe *means*. But cosmology is not a game; it has the power to overturn the fundamental institutions of society. Only by approaching these questions of the nature and meaning of the universe with an appreciation of the stakes involved are we likely to emerge unscathed. There are countless cautionary tales of hubris among the ancient Greeks, the myth of Faust, the legend of the rabbis in Pardes, and other fables of well-meaning people who dive headlong into mysteries for which they are not prepared and are in some way destroyed. All echo the same message: humility is essential in the face of higher powers.

Claiming our centrality does not imply that the universe was created for our eventual arrival or evolved with us in mind. It did not. Nevertheless, the entire Cosmic Density Pyramid *supports* us and any others who sit with us in the Sovereign Eye. It doesn't do so willingly, but simply in fact. Dark matter did not intentionally gather its forces over billions of years, pulling cosmically wandering atoms together into spectacularly crashing "violent relaxations" for our benefit. Dark matter didn't herd a dispersed, fertile mix of hydrogen and helium into a small region at the center of our Galaxy so that those primal atoms could easily interact and evolve into stars and worlds in preparation for us. Dark matter didn't commit itself unconditionally for billions of years, with never an instant off, to hold the atoms in its charge safely in a stable home Galaxy for us. Dark matter doesn't cradle the entire Milky Way—and all galaxies—in delicate, invisible hands, protecting it from the cosmic hurricane of dark energy tearing space apart outside because it cares about us. It does all these things because it has no choice. Its behavior is built into the order of the universe. Nevertheless, we benefit.

This kind of integration of science and meaning is considered by many scientists to be a danger to science, but a science that doesn't consider its own meaning can be a danger to everyone else. Interpreting modern cosmology is—if anything is—a sacred responsibility. As we move into this final chapter, we offer interpretations which are more subjective than any that have come before. This may raise in some readers' minds the question of what we ourselves actually believe. Specifically, do we believe in God? We believe in God as we understand God. Some words commonly used to describe God are "without end, eternally creative, unknowable, and the source of everything we know." But this could just as well be a description of the state of eternal inflation that may surround the Big Bang, so those can't be defining characteristics of God. The Cosmic Uroboros tells us that the universe encompasses all size scales, so any serious concept of God must *at least* do as much. "God" must therefore mean something different on different size-scales yet encompass all of them. "All-loving," "all-knowing," "all-everything-else-we-humans-do-only-partially-well" may suggest God-possibilities on the human size-scale, but what about all the other scales? What might God mean on the galactic scale, or the atomic? A God disconnected from this amazing universe that science is revealing would be a God entirely of the imagination—in fact, well worked over by many imaginations. But a God that arises from our scientific understanding is not entirely created by us. Such a God runs deeper than humankind's imagination and is speaking in some way for the universe itself.

The new scientific picture of the universe establishes a lower limit for God. For us, God can't be less or simpler, but could be more. God represents a maximum that is ever-expanding, and we are on the *inside*.

God represents the *directions* of our wonder—not the destination. The more that people discover about the universe, the faster God keeps expanding, always ahead, pulling yet teasing scientists. As God expands, God also deepens at all levels, just as our understanding of gravity deepens on the level of ordinary affairs because of Einstein's discoveries, even though we don't have to *use* relativity to calculate on that level. In this way scientific discoveries endlessly enrich the possibilities of God.

This is the universe to which we want to feel connected, and to which many more people will want to feel connected as they come to believe in its likely existence. We two believe in God as nothing less than the process of opening our personal lines of contact with the unknown potential of the universe. This process is an experience, and finding the words to describe it has been part of what we've been doing all through this book. We have a deep faith that if humans could come into harmony with the real universe, our troubled species would have its best chance to enjoy this jewel of a planet, unique in all the cosmos.

We both started in the existential camp ourselves. In Joel's widely read 1984 lectures about the Cold Dark Matter theory, he said that if the bulk of the matter in the universe is not made of atoms, "that is yet another blow to anthropocentricity: not only is man not the center of the universe physically (as Copernicus showed) or biologically (as Darwin showed), it now appears that we and all that we see are not even made of the predominant variety of matter in the universe!"[6] Humans are indeed not. But it was pure interpretation to conclude that not being made of the predominant variety of matter is somehow a blow and rules out human centrality. The very word "anthropocentricity" not only communicates "human centrality": it judges it and finds it unacceptable.

When we started teaching the Cosmic Uroboros, we realized that it might matter that humans are at the center of all possible sizes. We realized further that stardust can be seen as central, not peripheral. In fact, the entire existential façade of despair and stoicism flips inside out if we simply view the universe *from the inside,* where we indisputably are. Once we made this mental shift and opened our eyes to the view from the center of the universe, we not only kept discovering more ways that we are central: we found that doing so evoked the opposite emotions from the existential stance—not despair but hope, not resignation but excitement. These may be equally arbitrary emotions, but they lead to nonarbitrary actions.

We've actually moved into these ideas with all our furniture, and we love our new home. New Year's Day is the day we celebrate the universe and recharge ourselves by resetting our mental focus into harmony with it. Most holidays have special foods associated with them, and New Universe Day is no exception. Here is our recipe for the Cosmic Dessert, the very construction of which re-creates the universe itself.

• THE COSMIC DESSERT •

Layer ingredients in order so that when it is finished the creation looks like the Cosmic Density Pyramid, albeit a bit messier.

Dark Energy: chocolate cake (70 percent)

Dark Matter: chocolate ice cream (25 percent)

Invisible mixed atoms: chopped nuts dusted with ground coffee—crunchy but dark (4 percent)

Visible hydrogen and helium, the lightest elements: whipped cream (0.5 percent. Since this is very little whipped cream, seize the opportunity to arrange more around the base of the cake to represent the ribbon at the bottom of the Cosmic Density Pyramid.)

Stardust: the tiniest pinch of cinnamon you can see on top of the whipped cream (0.01 percent)

Intelligent Life: a cherry on top (not to scale, like the Sovereign Eye). Or substitute your favorite fruit—this is your opportunity to choose your representative.

The Price of Centrality

There's a joke among cosmologists that romantics are made of stardust, but cynics are made of the nuclear waste of worn-out stars. Sure enough, the complex atoms coming out of supernovas can be seen either way, but these atoms introduce into matter the possibility of complexity, and complexity allows the possibility of life and intelligence. To call them nuclear waste is like calling consumer goods the waste products of factories. A cosmology can be a source of tremendous inspirational and even healing power, or it can transform a people into slaves or automatons and squash their universe into obsession with the next meal or with trivial entertainment.[7] The choice of what attitude the twenty-first century will adopt toward the new universe may be the greatest opportunity of our time. The choice between existential and meaningful is still open.

The scientific discovery of our centrality tantalizes us with the prospect of finding meaning and purpose deep enough to inspire and transform our culture, and this transformation could happen. The meaningful universe offers us a home in the great scheme of things, and the feeling of confidence and comfort that come from knowing that we are central. But these benefits are not free. The price is that we have to live according to the cosmological principles that *make* us central, understand them, and continually enrich our mental vocabulary with imagery worthy of them.

A pyramid may seem to have nothing to do with a serpent, which has nothing to do with a slot machine and a blizzard, but these metaphors build upon the same physical worldview, and as ancient Egypt taught us, such metaphors can appear disconnected and even inconsistent, and yet not be mutually exclusive. Their very proliferation inspires humility and helps us approach reality. The universe is a pyramid of matter and energy. The universe is concentric spheres of time. The universe is a serpent swallowing its tail. The universe is a cosmic Las Vegas. And the universe will no doubt be much more as people free their creativity from the confines of old ways of thinking.

We have to learn how to juggle multiple universe images. *Seeing reality takes a lot of imagination*—but it takes *disciplined* imagination, which is sensitive to scientific knowledge, humble before it, and committed to consistency with it. The undisciplined imagination cannot learn because it refuses all constraints; it claims "Anything is possible!" when in fact, much of what it imagines is not possible, while most of what is possible, it will never imagine. Scientific knowledge does limit the imagination, but only in the same healthy way that sanity limits what we take as real.

The metaphors that make us cosmically central are powerful tools only if we use them. When we take the actions that make our centrality real and vital—when we cultivate and teach this new way of thinking, when we seek and create new metaphors that expand our consciousness large enough to grasp this new universe accurately, when we make an ongoing effort to apply cosmic perspective, especially about time and truth, to the big social and political decisions of our day—then we become the kind of people who are far more likely to care about and achieve Carl Sagan's high goals for humanity.

Human Is a Big Identity

When we come to understand that we are central to the universe, "we" can mean humanity, all intelligent life, the Galaxy, even luminous matter. "I'm human" can mean "I stand here on the Cosmic Uroboros, midway between the largest and the smallest things in the universe. I can trace my lineage back fourteen billion years through generations of stars. My atoms were created in stars, blown out in stellar winds or massive explosions, and soared for millions of years through space to become part of a newly forming solar system—my solar system. And back before those creator stars, there was a time when the particles that *at this very moment* make up my body and brain were mixing in an amorphous cloud of dark matter and quarks. Intimately woven into me are billions of bits of information that had to be encoded and tested and preserved to create me. Billions of years of cosmic evolution have produced *me*."

Every one of us is connected mathematically, physically, cosmically, and now also consciously with all the size scales around the Cosmic Uroboros. How exactly are we connected? It's not visually obvious. We look down at our bodies and see a finite package of living organs, sealed neatly in skin, and because we look that way, we assume we objectively are that way. The skin is the sacred organ of our culture because we allow it to define each of us cosmically. We assume that it is a wall dividing us from the rest of the universe, and that it is a cosmic absolute—"as constant as the Northern Star," as Shakespeare's Julius Caesar says. But the North Star is not constant. Earth's axis[8] will no longer point toward Polaris in a few thousand years—and similarly there should be a time limit on this old idea of ourselves.

Einstein and many other scientists have shown us that things are not the way they seem. In what other area do we take our gross perceptions as the final word anymore? After all, we enhance our senses with eyeglasses, telescopes, night-vision goggles, microphones, infrared detectors, telephones, radios, TVs, and many other technological devices to see, talk, feel, and perceive beyond what our unaided senses can detect. Yet when we look at our own bodies, we fall back into an intuitive sense of what is real. Far from being a package cosmically sealed off by skin, each of us is the tip of a great iceberg of cultural and genetic history moving among and through each other. We sit, with the other intelligent beings of this universe, in the Sovereign Eye of the Cosmic Density Pyramid—a position whose power and uniqueness cannot be threatened even by the rising tide of dark energy. As time goes on, each human becomes more precious, because of all the evolution and luck it took to get to that person. But we lose our preciousness and ignore it in each other when we lose our conscious connection with the enormity of time *embodied* in each of us. For practical purposes of spatial orientation and movement we need the boundary of skin, but we don't have to assume that therefore our earthbound, working sense of limitation and location reflects some "ultimate reality." We can enhance the sense of ourselves, as we've been successfully doing with our other senses, by means of a scientific but nevertheless metaphorical telescope—a new cosmological lens through which we can see how the expanding universe really works and how astoundingly special our place is in it.

Scientific cosmology is offering us a new picture of an awesome universe, but science provides no way of personally connecting to it. New scientific ideas as intellectual entertainment are not going to change our point of view. The scientific picture of the universe and the actual *experience* of it as reality—like mind and heart—each come fully alive in connection with the other. Cosmic ideas need to be integrated harmoniously into all we know, and that can open us up to the universe. But how in practice is that done? How can we bring our stunted consciousness into harmony with scientific reality? This is a great challenge of our time, but similar challenges have been met successfully before.

This book's history of cosmologies began with the flat earth. The ancient Egyptians and Babylonians could probably imagine distances of a thousand (10^3) miles horizontally, but only a few miles vertically. (Some Babylonians thought heaven was above the clouds, but not so far that they couldn't build a tower to it.) Then the flat earth was replaced with the sense that the earth is actually spherical but so huge that from any point on its surface it looks flat. This tremendous shift eventually became intuitive and enlarged the size scale that people could imagine to

10^8 miles (based on the popular medieval notion that it would take at least 8,000 years at forty miles per day by mule to reach the sphere of stars). The ratio of 10^3 to 10^8 expresses how much the size of the universe increased in people's consciousness from the flat earth to the heavenly spheres, and it's a factor of 10^5, or 100,000, times! But the change to the possibly infinite Newtonian universe was far larger. And the change from the Newtonian universe to a possibly eternal meta-universe spawning infinite bubble universes of which ours is one is even greater. Every cosmology since the flat earth has been counterintuitive at first, but cultures succeeded in re-envisioning reality every time on the basis of less evidence than we have today.

Midgard, the Golden City of the Expanding Universe

This book has proposed a name for the center of the size scales of the universe: Midgard. In the Old Norse mythological cosmos, Midgard was the human world; it was an island representing stability and civilized society in the middle of the world-sea, the Norse universe. The world-sea was large, and there was room not only for Midgard but for the land of the giants and the land of the gods. This is an excellent description—metaphorically, of course—of Midgard as the center of the expanding universe. Our Midgard is the island of size scales that are familiar and comprehensible to human beings. But beyond the shores of Midgard in one direction—outward—into the expanding world-sea is the land of incomprehensibly giant beings, like black holes a million times the mass of the sun and galaxies made of hundreds of billions of stars.[9] In the other direction from Midgard—inward, toward the small—lies a living cellular world, and beyond that the quantum world, and these microlands are the evolutionary and physical sources of everything we are. That may not make them gods, but compared to us they are more prolific, more ancient, universal, and omnipresent. Like the Norse Midgard, our Midgard is not isolated from these other lands in the world-sea. It is connected by the Cosmic Uroboros.

People disagree on just about everything that has to do with spirituality, but the one thing they do tend to agree on is that whatever the spiritual may be, it's not physical. However, the concept of Midgard helps us understand why this physical/spiritual dichotomy is illusory. Backing up a bit, in medieval cosmology heaven was understood to be physically enveloping the sphere of the fixed stars at a finite distance

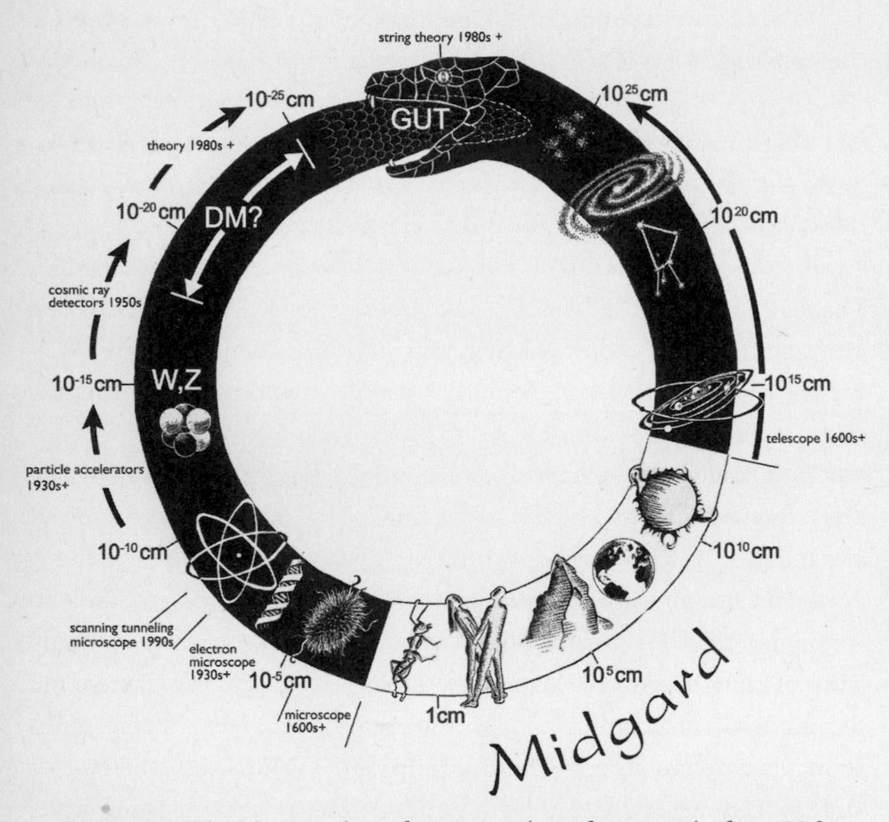

Figure 1. Midgard spans about fourteen orders of magnitude, from 10^{-2} cm to 10^{12} cm, holding everything for which people have intuition. The figure also shows the approximate decade and technology by which scientists discovered the rest of the Cosmic Uroboros.[10]

away from the earth (so close that in Dante's *Paradiso* it was possible from the height of Paradise to see the shoreline from Asia to Cadiz). But after medieval cosmology was replaced by the Newtonian picture, space was understood to go on forever, leaving no location for heaven. God was said to be "outside the universe" or "in the heart." Today most people have the idea that the spiritual, if it exists at all, is mysteriously other than the physical or material world and "transcends" the physical universe. The concept of Midgard erases this not by telling us what the spiritual is but by clarifying what the physical is.

As we discussed earlier, the largest structures inside the cosmic horizon, superclusters of galaxies, are expanding apart, and in billions of years they will disperse; they're "bound together" not by gravity but by our dot-connecting minds. In the opposite direction from Midgard toward the very small, there are elementary particles that are not really "physical" particles but rather quantum-mechanical ones that are routinely in two or more places at once. The strange truth is that what we usually think of as "physical" is a property of Midgard, perhaps the defining property, and thus Midgard is what people generally think of as the "physical" universe. Beyond Midgard, however, lies most of the Cosmic Uroboros.

The popular idea that the spiritual is a realm outside the universe that "transcends" the universe is a holdover from an earlier picture of the universe. But the concept of transcendence is not meaningless—it is merely misunderstood. Transcendence is not an imaginary jump to some place "outside" the universe. Transcendence is what happens many times within this universe, every few powers of ten. For example, on the atomic and subatomic scales, "human" means nothing. There is no humanness to our atoms. Whether atoms are inside us, inside a rock, or drifting through space, is all the same to them. On the atomic scale, therefore, even inside our own bodies we do not exist. "We" are something that transcends atoms. In the same way the universe as a whole transcends familiar Midgard. It is possible that from the perspective of some size-scale substantially larger than Midgard, we humans are playing a role as far beyond our imagination as that which atoms and cells are playing for us. The experience of transcendence, understood this way, might be described as the feeling of exponentially increasing complexity suddenly making sense to the imagination.

Let us dig a little deeper. By the "spiritual" we mean *the relationship between a conscious mind and the cosmos*. It's not the study of the cosmos—that's science. It's the way we relate to it. Given this definition of spiritual, *the Cosmic Uroboros is a bridge between the spiritual and the physical*. The Cosmic Uroboros gives not only meaning but a context to those exotic size-scales of the universe that no one ever had a connection with before. Amazingly, in this interpretation the difference between spiritual and physical becomes—in an approximate way—quantifiable with powers of ten. Things larger than about 10^{12} cm, or smaller than about 10^{-2} cm, can only be known through science and only experienced spiritually. This includes most of the universe.

Understanding Midgard changes many things. The great novelist D. H. Lawrence, writing in 1931, felt himself attracted to the metaphors of the Biblical Apocalypse because they presented the only escape he could imagine from the cold Newtonian universe:

> *Some of the great images of the Apocalypse move us to strange depths, and to a strange wild fluttering of freedom: of true freedom, really, an escape to* somewhere, *not an escape to nowhere. An escape from the tight little cage of our universe; tight, in spite of all the astronomist's vast and unthinkable stretches of space; tight, because it is only a continuous extension, a dreary on and on, without any meaning . . .*[11]

Midgard reveals the profound fallacies in Lawrence's thinking: the universe is not a dreary on and on without any meaning—and Apocalypse is not an escape to somewhere. This poetic language, so hauntingly beautiful, is in the service of a murderous idea based on ignorance of the universe. This is a very real danger today.

Midgard also challenges an aspect of what people think is human nature. Many people insist that human beings need to organize around the principle of "us versus them" in order to bond with other people and get motivated to act. This, the idea goes, is why wars and prejudice are inevitable. Even if this is true, it doesn't follow that us-versus-them has to be a division among human beings. This, in the long term, is the most counterproductive possible choice. Nor does "them" have to be an enemy—it just has to be different enough to put us on our guard in dealing with it and to help define who we are. What other kind of division is there? "Us" might at first glance include whatever supports our lives—not only other people and animals but all aspects of the planet, its atmosphere, and the web of life. But there's a problem with this definition of "us" and "them"—it requires endless judgments about who and what is on our side: does that feature of Earth support "us" or doesn't it? How important does its support have to be, before we include it as "us?" And every judgment is another opportunity for conflict. There's a better, completely impartial way to define "us," which enlarges not only who's with us but what we are ourselves. "Us" is Midgard. "Them" is the other size-scales—both the giants and the wee gods. Midgard is familiar throughout the universe. You can travel a billion light-years to a distant galaxy, but wherever you find seas, there will be sandy beaches.[12] Earthbound intuition would be helpful on such alien worlds, although fallible. But "them" is absolutely Other. Mentally step outside Midgard and you don't have to travel across the universe—what is going on *right here and now* is utterly outside our intuition. That's the real "them." When it comes to the future of our species, we humans are scattered around the globe, but we're all on the same side of the negotiating table. What's on the other side is the laws of physics.

The fact that Midgard *today* lies in the middle of the Cosmic Uroboros is another way that we are central to the universe. Like everything else, Midgard is evolving. In the beginning there was no Midgard. At the Big Bang, as mentioned earlier, the Cosmic Uroboros consisted of only the tip of the serpent's tail in the mouth. There was no body in between because the smallest scale was also the largest. "Head swallowing tail" was all there was at first: the serpent acquired a body as expansion created larger size-scales. As time passed and the universe kept expanding, larger size-scales continued to come into existence at the head. But if the current acceleration of expansion continues, the universe on the scales larger than our Local Group of galaxies will empty out as more distant galaxies race away from us. Over billions of years, the universe will become very different from what it is now. The universe is always becoming. In ancient Egypt it was written of Atum, the Creative Principle, "You came into being in this your name of 'Becoming One.' "[13] In the Bible when Moses asked God His name, God said, "I am Becoming What I am Becoming."[14] In the timeless, bottomless cauldron of eternal inflation, our universe is an evolving bubble of spacetime, becoming what it is becoming.

Integrating Cosmic Ideas into Our Lives

A great obstacle to experiencing ourselves in cosmic space and time is the habit of considering the universe to be what's out there, with ourselves as somehow objective observers. We don't normally think of reality as funneling from great galaxy clusters into us and spreading cell to cell, then soaring inward to the molecular level, the atomic, the quantum levels—and our humanness the fulcrum at the center of the entire process. But we need to. We need to experience the universe from the *inside*. We have to imagine ourselves in our proper place, *inside* the symbols, *part* of the symbols, the *point* of the symbols. Until we find our symbolic place in the universe, we will always misinterpret ourselves, feeling as though we are outside, sensing the familiar existentialist isolation, and looking at a universe in which we play no part.

For example, the Cosmic Spheres of Time are shown in Chapter 5 as if they are there and you, the reader, are outside them, holding this book in your hands and looking at them on a page. Jump in! In your imagination *take* your place there in the center of the symbol, at "Today," and then close the spheres around yourself. You are immersed in the history of the universe. You're at the center of your past. The past is not "over"—it's racing away from you at the speed of light like ripples from a pebble thrown into a pond but in spheres, not circles. The *era* when our sun was forming is still out there, four and a half billion light-years away, spherically enfolding our solar system, our Galaxy, and our Local Supercluster. Far, far beyond, the era of the earliest galaxies engulfs us, and beyond that is a deep sphere of utter blackness that theory tells us is the Dark Ages of the universe before the first galaxy had formed. Earlier still lies the sphere of the cosmic background radiation and at last the cosmic horizon. The Cosmic Spheres of Time are real because the past is real—evidence of it in the form of light and other radiation is perpetually arriving.

Since we're all immersed in the universe and much of it is immersed in us, there is indisputably a relationship between the universe and us. A real relationship is not something you view from outside but something you participate in. But how do we relate to the universe? All our relating tools are small. A million years of evolution honed them for connecting to other people and special animals and maybe plants. But for thousands of years people exploited these same relating tools to take their place in a powerful, larger, invisible world by talking about elements of that world *as if* they were like people or special animals; this helped them know how to relate to the invisible world. People have always "personified" gods with creative imagery because that's how human relating tools work; we too can do that as long as we don't fall into the trap of taking metaphors as real and assuming that in some independent spiritual realm gods are actually persons and have human characteristics. There is no *independent* spiritual realm. The *uni*verse is One. But if we modern humans want to experience our relationship with this newly discovered cosmos—which is the only way we will be able to discover our own fully expanded selves—we have to take those same tools evolution gave us and get creative.

Our constraint, however, as well as our inspiration, must always be scientific knowledge. That's what guarantees that if we do experience a connection, it's to the real universe and not to some disembodied fantasy rattling around in our minds. There is nothing wrong with "personifying" nature in an attempt to get access to our tool kit, as long as we do so knowing that nature is not a lot of persons or gods resembling persons, but that *we* are persons and that is the way *we* can exploit the power of language and metaphor to connect to the cosmos. Therefore, to the extent we use language that personifies aspects of the universe, we are doing so metaphorically, but for our kind of intelligence metaphors are not optional, and ones that use personification are among the most effective anyone has discovered. This is why many ancient spiritual ideas still resonate—they were not arbitrary inventions but reflect discoveries about how humans think.

We need somehow to recognize our ancient ancestry *living in ourselves* through imagery that speaks to our time and our universe. After all, our earliest ancestors, the elementary hydrogen nuclei that came out of the Big Bang, are 10 percent of our body today, by weight. We are made of history as surely as we are made of matter. We have to find imagery that takes us there and then dare to jump in and claim our place in the universe and experience its counterintuitive nature in our lives right here.

Taking Our Extraordinary Place in Time

The Cosmic Spheres of Time can be thought of as the homes of our ancestors. This no doubt sounds strange today, when people generally think of ancestors as human relatives older than grandparents, or perhaps an ethnic group or country, but never as objects or phenomena. Yet the Huichol Indians of Mexico, discussed at the beginning of this book, think of their ancestors as including the forces of nature responsible for the very existence of the world. This may be a kind of language we need.

Thinking of nonhuman things as "ancestors" bothers many Westerners. We have been taught to be wary of superstition or idolatry. But consider the assumption implicit in our outlook: that the term "ancestors" must be cut off at some arbitrary point between grandparents and the primordial soup. We didn't start at some arbitrary point. It is not idolatry but science to say that we are the result of an unbroken chain going back to the Big Bang. Those ancestors that Huichols recognize and honor are actually our ancestors, too.

A few years ago we read *The Seven Daughters of Eve*[15] and discovered that almost all women of European stock are descended from one of only seven women who lived a few tens of thousands of years ago. Nancy became curious from which daughter she was descended and sent in a DNA sample. Much to her surprise she found out that she is descended from none of them but instead shares the mitochondrial DNA of most Native Americans. She is descended from the founder of one of the four major clans that colonized both North and South America by crossing the land bridge from Northeastern Asia to Alaska about 12,000 years ago. Suddenly Nancy felt part of that tremendous history—but that is only a drop in the bucket on the time-scale we're talking about here.

From a cosmic viewpoint, it doesn't matter who our recent ancestors were. Our *distant* ancestors are what everyone on Earth shares. To connect to our distant descendants, who will take over the world in the future, to care what happens to them, to take responsibility for the actions of our own time that will have enormous impact on them, we can start by connecting to our past, shifting to a cosmological sense of time. Can we begin to appreciate from the Huichols and others like them that cultivating a feeling of honor and respect for the past, not just knowledge about it, makes the possibility of personal connection more attractive and meaningful? Can we borrow any of their techniques to help us connect to our universe?

Suppose as a thought experiment, we try out the Huichol myth that "Grandfather Fire," was the original light, the original wisdom, and the universe's own memory. Scientifically this triple association is actually accurate. Fire produces light, and light carries cosmic memory. All of astronomy is based on this fact. Everything we see in the universe, we see only because something burning produced light, which sometimes traveled through space for billions of years before reaching us, carrying the memory of its source at the time it was emitted. As we have explained, the first light—the cosmic background radiation, the ubiquitous heat radiation from the Big Bang—is inundating spacetime with baby pictures of the universe, and those historical photos can be decoded by the "eyes" and instruments of any intelligent being anywhere in the universe who is able to understand them.

Grandfather Fire in the modern universe would mean all the forms of fire we know of today—not just combustion but electricity, fission, fusion, and the Big Bang itself. Thus Grandfather Fire would include all the generations of stars that have ever burned, including all those that exploded in supernovas and other star deaths, seeding the universe with the stardust we and our planet are made of. How could anyone feel closer to the stars that created us and the star, our sun, that daily sustains us, than to think of them as Grandfather?

What more appreciative way to think of the evolution of life than to honor it, as the Huichols do, as Grandmother Growth? Ever since early life on our planet, billions upon billions of creatures have struggled so that their children would survive, and those children have led to us. There may be no better way to compress the untold ages of time between ourselves and the beginning—emphasizing the enduring, personal relationship *across* that immense gap—than to think of those original creative forces as "grandparents." Scientists perform a somewhat similar compression of incomprehensible amounts of time by using exponents rather than a linear scale to make vast expanses of space and time manageable, but for them the impact is mathematical rather than personal.

Cosmologists have discovered the age of the universe, but to many people numbers in themselves mean nothing, and a million might as well be a trillion. We can learn from the way the Huichols emotionally connect to the fundamentals of their universe through the simple concept of Grandparent that *from the point of view of experience* the numbers are in fact not important. What matters is to develop imagery that will bridge the incomprehensible gap between us and our 14-billion-year-old source. Cosmology thus depends on artists—those who create such imagery not to manipulate but to inspire.

The terms "Grandfather" and "Grandmother" are perhaps unlikely to catch on in the West as names for forces of nature, but the impulse toward connection that is reflected in this naming must somehow find a place in our own language. Otherwise, we sophisticated people may continue to have little connection with anyone except those we have personally known, or with any time frame longer than a human lifetime or two. We will continue to inhabit a cramped but culturally correct bell jar, directing the immense mental energy our human ancestors expended in cosmic awareness into trying to make middling-scale pursuits move faster or middling-scale possessions accumulate faster. By making every effort to connect, we will do even more than expand our sense of cosmic identity: we will actually make our scientific understanding of the universe more accurate by integrating ourselves into the big picture—and we are certainly here.

What we are, however, is not so certain; our personal identity runs only as deep as our awareness of where we come from.[16] This in turn depends on our level of scientific understanding and our willingness to experience that understanding metaphorically. Our own thinking thus sets the size-scale for our possible connection to the universe.

Somewhere in the Deep[17] of our brains, in the metaphor-ocean of

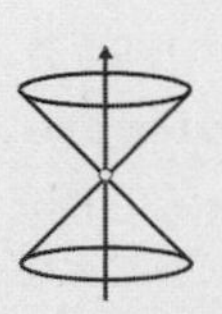

the cognitive unconscious, there may be images that can help us open a mental window to our earliest ancestors—those phenomena that acted out the drama of creation as we now understand it. Our consciousness can then open a mental window on the other side of the room to an equally vast future, and transform even our distant descendants into our "grandchildren." Seeing and living on all levels at once is what cosmic connection is all about. It is not mystical; it is as practical—and essential—as the visualizations that athletes do before a competition, or concert pianists before they go out on stage. It situates us in reality at our best.

If we can learn from indigenous peoples to honor among our ancestors the forces of nature in the far deeper way that we now scientifically understand nature, it may at last dawn on us that we, too, are ancestors, and our actions today will have consequences possibly as broad and far-reaching as the sum total of all the influences that have led to us.

From this perspective it becomes clear that no human goal today is more pressing than the search for sustainable prosperity. It encompasses all other purposes, including peace, justice, health, science, economic development, and spiritual pursuits, and without it these others will prove transient and illusory. Saint Augustine enunciated the Christian doctrine: "The deliberate sin of the first man is the cause of *original sin*." Whether you believe that or not, failing to protect our species and destroying the promise of the only intelligent life that may exist would surely be a *final sin*. It could extinguish the Sovereign Eye of the universe.

We can't create sustainable prosperity by an act of will. It must be founded on harmony with the universe. We've got to discover it—we've got to hear it. Scientific cosmology is making possible the most careful kind of listening, and through it the basic rhythm of sustainability seems faintly audible. We can all drown it out yelling about our old differences, trying to control matters by fiat, or we can close our mouths and listen and try to follow. The sheer importance of what is happening at the end of this human inflationary period on Earth, whether the results are brilliant or terrible, is likely to make our age a mythic one to our distant descendants. Only realize it! We matter beyond our wildest imagination.

This Sacred Opportunity

The ancient Egyptians saw themselves upholding the cosmos itself by upholding Order, Harmony, and Truth through their rituals; the Hebrews saw themselves upholding the moral order of the entire universe, at God's behest. These people had no doubt that they mattered to the universe. Modern people instead scoff, "It's a nice illusion the ancients had—that they were upholding the universe—but there is no way human beings 'uphold' the expanding universe." But we actually do. We matter to the universe as we know it, because the universe *as we know it* dies without us.

We uphold the new universe—but only if we too, like the ancients, consciously do uphold it in our thoughts and actions. The way to uphold the universe is to embrace scientific reality to the extent the evidence supports it, and commit ourselves as a culture to develop its meaning collectively. Scientific cosmology is not the last word: it's the first word. Scientific accuracy has to be our minimum standard. What is the extent of the evidence of the new cosmology? Here is a comparison of the relative scientific certainty of the theories we've discussed in this book:

Very Well Verified Scientific Theories
- Physics: Thermodynamics, Quantum Mechanics, Relativity
- Biology: Evolution
- Geology: Plate Tectonics

Strong Consistent Evidence / Multiple Independent Tests
- Standard Model of Particle Physics
- Standard Model of Cosmology (Double Dark Theory)

Good Evidence
- Cosmic Inflation
- Snowball Earth
- Fractal Theory of Biological Scaling

Very Modest Evidence For / No Evidence Against

Supersymmetry

No Tests, but Many Experts Find Them Attractive

Eternal Inflation

String Theory

No earlier cosmology could frankly assess how likely its many notions are to be true. Belief that rests on authority and the feeling of certainty is, as we have seen, earthbound. We need so much more today, but it is becoming available. Scientific cosmology is emerging at this moment in history because of the felicitous combination of thousands of brilliant, curious human beings, collaborating from the widest range of cultural backgrounds in history, exploiting unprecedented sums that governments have been willing to spend, amazingly, on superb new scientific instruments with no practical use whatsoever except to understand the universe. International scientific cooperation is part of modern ritual. It represents standards that serve us well. But we still need more: *we need, collectively, to become the kind of people capable of using science to uphold a globally inclusive, long-lived civilization*. How do we get there? By facing up to the imperative that we have to face up to. As H. G. Wells put it, "Human history becomes more and more a race between education and catastrophe."[18]

A new universe picture alone is not destiny. Egyptians, Hebrews, and Babylonians shared the flat-earth picture, but they developed radically different attitudes toward life: inheritors of the gods, teachers of the world, or in the case of Babylon, serfs and vassals of the gods' feudal domains.[19] This is why today it matters so much—and is entirely legitimate—to choose to interpret the new universe in a positive way. We who are alive today are the first to get the chance to shape the coming civilization's attitude, foreseeing that some new attitude toward life must accompany so enormous a shift in our picture of reality as the one now occurring.

We are at a turning point in history as the rising curve of human inflationary growth on Earth approaches its maximum. What direction it will take from here we do not know. The next decades—the active lifetimes of the people reading this book and our children—will create the blueprint of the future Earth. We represent an age on Earth comparable to the age of cosmic inflation at the beginning of the universe: brief, but about to set the pattern for the long-term future. We have a chance to be heroes if we have the courage to do what the knowledge implies, or to be reviled as ignorant, selfish, and hugely destructive if we do not. This is a once-in-all-of-life's-time opportunity, but we can only take advantage of it if we understand what a unique moment this is and how crucial it is to get things right, and to do it now. For our entire species, today *matters*. Our society's current choices—and interpretations—might be the ones that reverberate for millennia, out of all proportion to the thought that went into them. We already have knowledge we need to use, and we have to find more, and the right kind, fast.

The end of inflationary growth on Earth may cause global disruptions—economies collapsing, resource wars, plagues, climate destabilization, mass migrations. But we have an extraordinary opportunity that has arisen only twice before in the history of Western civilization—*the opportunity to see everything afresh through a new cosmological lens.* We, in our generation, are witnessing a full-blown scientific revolution in cosmology—a field where the unrestricted, dataless fantasizing of theorists has been succeeded by reliable theory that has been tested against the entire visible universe. We are the first humans privileged to see a face of the universe no earlier culture ever imagined. Perhaps not the last face, whatever that may mean, but an exciting and counterintuitive new one. We dare not undervalue this immense privilege, even though it is happening at the same time as some of the most barbaric and self-defeating behavior our species has ever exhibited. This is all the more reason we need it. We must absorb the immense realization that there is an overarching truth encompassing all our religions: we are at the center of a new universe.

Whatever scientific truth may be, there is no defensible choice but to accept it. The greatest scholars and teachers, the most creative artists, and the noblest intentions in the world, all working together, can never make sense of our lives in the old metaphors of the Newtonian or earlier universes. But if all people build on the groundwork of recent discoveries, they will be able to do what scientists alone can never do—create a visual and poetic language through which the universe as we scientifically understand it can speak accurately and meaningfully to a global, science-based culture. Artists, take heed: realistic art is considered old-fashioned, but art that portrays the new universe realistically could be the most subversive art in history.

These are the simple elements that can promote the human species' success:

1. Accept the new universe.
2. Commit to a meaningful, not an existential, view of the universe.
3. Open your mind and heart to a long time-horizon both behind us and before us.
4. Make choices that support the long-term future now.

Humanity is still capable of having a long-lived civilization, but in order to become the kind of people who can sustain such a civilization, we need to change. Ancient Egypt had a goddess, Ma'at, who represented a profound fusion of qualities: Order, Harmony, and Truth. Ma'at could be an inspiration for us, because there is in fact an organic fusion of these three in the meaningful universe. Cosmic order is discovered through cosmology; harmony is how we hope to relate personally to the cosmic order and appreciate what is at stake; and truth is the standard by which we can most safely judge our success—truth, that is, as the whole process of approaching the unknown with humility, skepticism, scientific method, curiosity, and equality of all such willing people. Our own "Ma'at" would encourage us to seek an integrated understanding of the scientific principles of cosmology with a commitment to the still-to-be-discovered principles of right behavior on large scales. But even intellectually discovering this new Order/Harmony/Truth will not be enough unless it changes us.

People alive today have the choice of giving up and going the existential route, mocking each step of the demise of our planet, resigning ourselves to it, despairing about it, or even, as in some dangerously twisted ideologies, welcoming it. Or we can grasp this cosmological revolution as a miraculous opportunity *we will not get again*. We need to feel like the stardust beings we are. We need to take our place in the glowing company of all intelligent life in the Sovereign Eye of the Cosmic Density Pyramid. We need to cherish our lush homeland in Midgard, the golden city at the center of the expanding universe. We need to overflow with gratitude that our universe—of the infinite bubble universes lost to each other in eternal inflation—is filled with light and possibilities and billions of years still to develop them.

We will change when we *accept* our cosmological truths, when we identify with and honor the forces of the early universe as our oldest ancestors. We will change when we see *ourselves* moving outward into the Cosmic Spheres of Time to become ancestors worthy of honor a thousand or a million years from now. When with all our minds and hearts we grasp that we are central to the expanding universe, we will have connected. Then we too, like our ancient ancestors the world over, can say once again with confidence and commitment that we uphold the universe.

Acknowledgments

In looking back, we are overwhelmed that all these wonderful people helped us in so many ways:

First we thank our brilliant daughter, Samara, who edited every chapter several times with the eye of a dramaturg and the sparkle of a muse, and Rena Ellenbogen Abrams and Larry Abrams, who have been our most loving and steadfast supporters. Immeasurable thanks to Doug Abrams, our agent, "idea architect," and dear friend, who has been like a third parent to this book from conception to delivery, and whose editing and structural suggestions have been indispensable. For helping keep us alive and sane throughout this process, we owe love and gratitude to Dave Dorfan, Brant and Barbara Secunda, Cindee Jeski, Larry Blumberg, and Robin Lynn, whose lifelong friendship and support we cherish. We thank in particular Nicolle Rager Fuller, our talented young illustrator, who brought many of our ideas to life and endured our requests for endless revisions with astonishing patience.

For carefully reading earlier versions of various chapters and giving us invaluable comments, corrections, and/or essential new information, we are indebted to Bob Balopole, John Faulkner, Mary-Kay Gamel, Ray Gibbs, Gildas Hamel, Tsafrir Kolatt, Leonard Lesko, Daniel C. Matt, Michael Nauenberg, James Redfield, Robert Ritner, Virginia Trimble, and especially Owen Gingerich (on historical and cultural material), and Jim Gunn, Greg Laughlin, Andrei Linde, and Kevin Zahnle (on scientific issues). We are tremendously grateful to Anthony Aguirre, David Babcock, Sandra Faber, Tara Firenzi, Ricardo and Consuelo Flores, Hal Hansen, Mel Honowitz, Joe Kochan, and Deva O'Neill for detailed comments that helped us improve the text. For research assistance we thank Amanda Mehl.

For their encouragement when this book was just a dream, we thank again many of those just mentioned, and also John Barrow, Audrey Chapman, Philip Clayton, Hossain Danesh, Carl Djerassi, Alan Dressler, John Harte, Roald Hoffmann, Rocky Kolb, Michael Lerner, Vilhjalmur Ludviksson, Jim Miller, Martin Rees, Mark Richardson, Robert J. Russell, Zalman M. Schachter-Shalomi, Stephen Schneider, Joe Silk, Al Teich, Frank von Hippel, Steven Weinberg, Jozef Zycinski, and several close friends who prefer to remain anonymous but who know who they are. For excellent advice and for introducing us to people who turned out to be essential, we thank Marc Chamlin and Seth Gellblum. For helping make possible the interdisciplinary course from which this book arose, we thank the University of California, Santa Cruz; the American Council of Learned Societies' Center for Contemplative Mind; the Templeton Foundation; and especially Triloki Nath Pandey, who cotaught the course the first year.

Finally, we thank Riverhead/Penguin and Wendy Carlton for their enthusiasm for this book, and particularly our editor, Jake Morrissey, whose careful reading and quiet but perceptive suggestions have greatly helped clarify and shape it.

Notes

INTRODUCTION

1. Blaise Pascal, *Pensées* (posthumously published 1670), III, 206. That this feeling was never expressed in medieval literature is discussed in C. S. Lewis, *The Discarded Image: An Introduction to Medieval and Renaissance Literature* (Cambridge University Press, 1964), pp. 98–100.
2. Carl Jung, *The Spirit in Man, Art, and Literature* (Bollingen Series XX; Princeton University Press, 1966), p. 77.

CHAPTER I

1. A book that has inspired many in this kind of endeavor, although not at the cosmological level, is Thomas Berry, *The Dream of the Earth* (Sierra Club, 1988). Berry, a visionary and "geologian," followed this with a book coauthored with Brian Swimme, *The Universe Story: From the Primordial Flaring Forth to the Ecozoic Era—A Celebration of the Unfolding of the Cosmos* (HarperSanFrancisco, 1992), which is perhaps the first attempt to mythologize the evolution of the modern universe.
2. There are perhaps three thousand scientists around the world who either observe cosmological phenomena or are theorists working with cosmological

data. Young scientists are increasingly doing both. The vast majority of cosmologists are in the United States, Europe, and Japan, but China, India, and other countries are also becoming centers of cosmological research. Several thousand other scientists who consider themselves mainly physicists also do some research on cosmological questions.

3. The shortest wavelengths are called gamma rays, and X-rays and ultraviolet light also have shorter wavelengths than visible light. Infrared light and radio waves have longer wavelengths than visible light. All of this radiation is carrying different yet mutually consistent information that needs to be decoded. (If the information isn't mutually consistent, either the supposed data are inaccurate or the theory used to interpret them is wrong.)
4. Steven Weinberg, *The First Three Minutes: A Modern View of the Origin of the Universe* (Basic Books, 1977), pp. 131–132.
5. Edward R. Harrison, *Masks of the Universe* (Macmillan, 1985), pp. 1–2.
6. Clifford M. Will, *Was Einstein Right?*, 2nd ed. (Basic Books, 1993). For more recent details, see Clifford M. Will, "The Confrontation Between General Relativity and Experiment," *Living Reviews in Relativity* (2001), www.living reviews.org.
7. Karl R. Popper, *The Logic of Scientific Discovery* (Basic Books, 1959). Robert Nozick, *Invariances* (Harvard University Press, 2001), especially pp. 102ff, summarizes and criticizes Popper's views (what Nozick calls "the standard model of science").
8. Thomas S. Kuhn, *The Structure of Scientific Revolutions* (University of Chicago Press, 1962; 2nd ed. 1970; 3rd ed. 1996). See also Paul Hoyningen-Huene, *Reconstructing Scientific Revolutions* (University of Chicago Press, 1993).
9. Thomas S. Kuhn, *The Copernican Revolution: Planetary Astronomy in the Development of Western Thought* (Harvard University Press, 1957).
10. A major postmodern journal published a special "science wars" issue including responses to the book *Higher Superstition: The Academic Left and Its Quarrels with Science* by biologist Paul R. Gross and mathematician Norman Levitt, which had criticized postmodern attacks on science, and an essay by physicist Alan Sokal, "Transgressing the Boundaries: Toward a Transformative Hermeneutics of Quantum Gravity," *Social Text* (Spring/Summer 1996), pp. 217–252. Sokal's article appeared to support the postmodern view, quoting reputable writers about science (including the present authors), but it was a hoax "liberally salted with nonsense," as Sokal himself admitted in *Lingua Franca* (May/June 1996), pp. 62–64. See also Steven Weinberg, "Sokal's

Hoax," in *Facing Up: Science and Its Cultural Adversaries* (Harvard University Press, 2001).

11. According to Steven Weinberg, "The birth of Newtonian physics was a mega–paradigm shift, but nothing that has happened in our understanding of motion since then—not the transition from Newtonian to Einsteinian mechanics, or from classical to quantum physics—fits Kuhn's description of a paradigm shift." "The Non-Revolution of Thomas Kuhn," in *Facing Up*, p. 205. Niels Bohr said in 1931, "The language of Newton and Maxwell will remain the language of physicists for all time." Quoted in Abraham Pais, *Niels Bohr's Time: In Physics, Philosophy, and Polity* (Oxford University Press, 1991), p. 426. There have been several revolutions in biology since Darwin, from Mendelian genetics in the nineteenth century to the discovery of the molecular basis for genetics and the recent mapping of the human genome. But John Maynard Smith, a leading theorist of evolution, says, "The major scientific revolution of my working life has been the rise of molecular biology, which has all the characteristics of a new 'disciplinary matrix' in Kuhn's sense—new scientists, new problems, new experimental methods, new journals, new textbooks, and new culture heroes. But where was the incommensurability? . . . there were few conceptual difficulties and no paradigm debate." "Science and Myth," in Niles Eldredge, ed., *The Natural History Reader* (Columbia University Press, 1987), pp. 222–229. Ernst Mayr, perhaps the leading neo-Darwinist, analyzed many revolutions in biology, and concluded that none of them was of the type Kuhn described, not even the Darwinian revolution itself; see his "The Advance of Science and Scientific Revolutions" (1994), reprinted as "Do Thomas Kuhn's Scientific Revolutions Take Place?" in Ernst Mayr, *What Makes Biology Unique* (Cambridge University Press, 2004).
12. The encompassing theory "reduces" to the encompassed theory in the sense that the two theories make the same predictions (to within some specified accuracy) under certain specific conditions. Such theories are sometimes said to obey a "correspondence principle." Note that it is the predictions, and not the words that scientists use to describe the theories, that correspond. The words are often completely different. But physicists all learn that such apparent differences in description are not necessarily significant. Newtonian mechanics can be expressed as equations of motion or by saying that particles move so that a certain quantity (called the "action") is as small or as large as possible. Despite the very different sound of these two descriptions, they are mathematically equivalent and make identical predictions. Maxwell's me-

chanical analogies and his luminiferous ether have been discarded, but his equations of electromagnetism express fundamental truths about nature that have been tested countless times.

Kuhn was a physicist, and he was of course aware that relativity and Newtonian mechanics make essentially identical predictions under many circumstances, and that most physicists do not agree with him that Newtonian and Einsteinian dynamics "are fundamentally incompatible in the sense illustrated by the relation of Copernican to Ptolemaic astronomy" (*The Structure of Scientific Revolutions*, 2nd ed., p. 98). He argued however that the meanings of the quantities in the equations, such as mass, are fundamentally different: "Newtonian mass is conserved; Einsteinian is convertible with energy. Only at low relative velocities may the two be measured the same way, and even then they must not be conceived to be the same" (*ibid.*, p. 102). This focus on words rather than predictions is what we just argued against.

13. Charles W. Misner, "Cosmology and Theology," in W. Yourgrau and A. D. Breck, eds., *Cosmology, History, and Theology* (Plenum Press, 1977), pp. 75–100. Misner is coauthor of one of the leading textbooks on general relativity.
14. See Nancy L. Etcoff, *Survival of the Prettiest: The Science of Beauty* (Doubleday, 1999), and Geoffrey Miller, *The Mating Mind: How Sexual Choice Shaped the Evolution of Human Nature* (Anchor, 2000).
15. The great theoretical physicist Paul Dirac (1902–1984) said that "it is more important to have beauty in one's equations than to have them fit experiment." P. A. M. Dirac, "The evolution of the Physicist's Picture of Nature," *Scientific American* **208** (5), 45 (1963). Dirac was discussing Erwin Schrödinger's first attempt to derive the famous equation of quantum mechanics now named after him. Dirac was guided by these ideas in discovering the equation that describes electrons and their antiparticles (positrons), and he received the 1933 Nobel Prize after his prediction of positrons was experimentally confirmed. However, scientists have also been misled by their quest for beauty. For example, for essentially aesthetic reasons both Claudius Ptolemy and Nicholas Copernicus expected the planets to move along circles, but it was not until Johannes Kepler abandoned circular for elliptical orbits that he was able to fit Tycho Brahe's observations. Newton's mechanics subsequently showed that elliptical orbits are the rule, and circles an exceptional special case.
16. The terms "cold dark matter" and "hot dark matter" were first used in a talk Joel gave in March 1983 and in J. Richard Bond's talk at the same conference.

The first detailed description of the CDM theory and comparison of its predictions with observations was in George R. Blumenthal, S. M. Faber, Joel R. Primack, and Martin J. Rees, "Formation of Galaxies and Large-Scale Structure with Cold Dark Matter," *Nature* **311**, 517–525 (1984). The website TheViewfromtheCenteroftheUniverse.com contains additional relevant references and links.

17. Joseph Campbell, *The Inner Reaches of Outer Space: Metaphor as Myth and as Religion* (HarperCollins, 1988).
18. We are grateful to Brant Secunda, Huichol shaman and founder of the Dance of the Deer Foundation for Shamanic Studies, Soquel, California (www.shamanism.com), for our understanding of and experience with Huichol shamanism.
19. People in many traditional earth-centered cosmologies trace their genealogy back to the beginning through their real and imagined ancestors. Gregory Schrempp, *Magical Arrows: The Maori, the Greeks, and the Folklore of the Universe* (University of Wisconsin Press, 1992), explains how the Maori (native people of New Zealand) do this. That book and Robin Horton's *Patterns of Thought in Africa and the West: Essays on Magic, Religion and Science* (Cambridge University Press, 1993) discuss and critique anthropological and philosophical views of cosmology.
20. Jean Clottes and David Lewis-Williams, *The Shamans of Prehistory: Trance and Magic in the Painted Caves* (Abrams, 1998); Randall White, *Prehistoric Art: The Symbolic Journey of Humankind* (Abrams, 2003).
21. Lynn Gamwell, *Exploring the Invisible: Art, Science, and the Spiritual* (Princeton University Press, 2002).
22. Stephen Hawking, *A Brief History of Time: From the Big Bang to Black Holes* (Bantam, 1988), p. 174. See also Kitty Ferguson, *The Fire in the Equations: Science, Religion, and the Search for God* (Eerdsmans, 1994).

CHAPTER 2

1. This selection of examples does not imply that any new cosmology is going to be "Western," since science is international, and nature is the same the world over. Other examples of historical cosmologies could have been chosen and would have been fascinating, but given the need to limit the number discussed, we chose to explore those that led directly to our current modern worldview.

2. Your authors are not historians or professional experts in the individual cultures we describe here, and for more complete views of those cultures we will refer you to authors who are. However, in the subject of this chapter—how human pictures of the universe developed from ancient times to the present, and how the various versions affected people's lives and cultures—there are no professional experts because it is vast and inherently multidisciplinary. We have been synthesizing this cultural history for the past decade in our course Cosmology and Culture at the University of California, Santa Cruz, asking the questions that this and the following chapter explore.
3. The goddess Nut (pronounced "Noot") was often painted on the inside of coffin lids. There is a particularly beautiful example in the British Museum, pictured in Geraldine Harris and Delia Pemberton, *Illustrated Encyclopedia of Ancient Egypt* (British Museum, 1999), p. 110.
4. This figure is redrawn and simplified from a scene showing the creation of the world according to a Heliopolitan myth, from the *Book of the Dead* of the priestess Nesitanebtashru, daughter of High Priest Pinudjem I, in the British Museum (the Greenfield papyrus), reproduced on p. 122 of David P. Silverman, ed., *Ancient Egypt* (Oxford University Press, 1997). See also the British Museum website, http://www.thebritishmuseum.ac.uk/compass/.
5. John A. Wilson, "The Nature of the Universe," in Henri Frankfort et al., *The Intellectual Adventure of Ancient Man: An Essay on Speculative Thought in the Ancient Near East* (University of Chicago Press, 1946, repr. 1977), pp. 34, 43.
6. The following diagram shows the relations among the Egyptian gods we discuss:

The scarab is a beetle that rolls dung into large balls from which life appeared to emerge spontaneously. (In fact, scarabs lay their eggs in dung, which they have formed into balls to feed the larvae.) Atum was portrayed as a scarab pushing the black sun out of the primal darkness of Nun (pronounced "noon") into the sky. Thus Atum was later identified with the sun during its dark phase in its nightly journey through the innerworld (the Duat) as it fought the forces of chaos in preparation for its morning rebirth as Re, the shining sun.

7. Some polarities that had to be kept in harmony were intellectually much more demanding than these. For example, "existence" included "that which is" and "that which is not."
8. All of Shu (space) was Egypt. The domain of the Pharaoh was said "to extend beneath the whole reach of the sky, to the limits of the eternal darkness." R. T. Rundle-Clark, *Myth and Symbol in Ancient Egypt* (Thames & Hudson, 1959), p. 35.
9. From the New Kingdom on (i.e., after about 1570 B.C.E.), outsiders began to be considered people in formal Egyptian religious texts.
10. "O Atum, when you came into being you rose up as a High Hill." Utterance 600, the Pyramid Texts. These are prayers, praises, stories, and incantations inscribed on the inside walls of Fifth and Sixth Dynasty (2494–2181 B.C.E.) pyramids of the Old Kingdom, although it is probably safe to assume that most of the religious ideas expressed in them were even older. The Pyramid Texts are the largest single collection of religious documents from the Old Kingdom. Scholars usually divide the period of ancient Egypt into the Old, Middle, and New Kingdoms. The Great Pyramids of Giza date from Fourth Dynasty (2613–2494 B.C.E.) of the Old Kingdom. Rundle-Clark, *Myth and Symbol in Ancient Egypt*, p. 37.
11. "Leonard H. Lesko, "Ancient Egyptian Cosmogonies and Cosmology," in Byron E. Shafer, ed., *Religion in Ancient Egypt: Gods, Myths, and Personal Practice* (Cornell University Press, 1991).
12. Wilson, "The Nature of the Universe," p. 33.
13. In the year 1357 B.C.E., the new pharaoh, Amenhotep IV ("the one who satisfies the god Amen"), age sixteen, had a revelation that in lieu of the many Egyptian gods there should be only one god, "Aten," and that Aten required a new city, Akhetaten (whose ruins are called Tel al Amarna), be built far from all other cities, dedicated only to him. Amenhotep changed his name to Akhenaten, which was a hugely significant act, since to the ancient Egyptians,

your name was who you were. Pharaoh Akhenaten diverted all resources to the construction of his new city, and forced all his retainers and craftspeople to move with him. The name "Amen" was erased from the old monuments. Hundreds of illiterate people were given chisels and copies of the god's name, and they removed all words with those characters, including the name of Akhenaten's own father, Amenhotep, which included the god's name. Since the name was the reality, the god Amen disappeared. The great god now was Aten, usually identified with the energy and light of the sun, represented in drawings as a circle with rays coming out and a hand at the end of each ray. This picture might have been a kind of explanation of how the god could affect human life, since his hands were everywhere. But there were no myths about the creation of Aten. He had no consort, and there was little attempt to humanize him for the sake of the people. Only Pharaoh Akhenaten could communicate with this god, and people had to worship through him. Since Aten had no divine family tree, he was presented as part of a trinity, with Akhenaten and his queen, the famous Nefertiti, raised to almost his equal. Festivals and feasts to other gods were reduced or suppressed, replaced with events glorifying the royal couple. The old idea of piety was replaced by loyalty to the king. After Akhenaten died, the young Pharaoh Tutankhaten took the throne and changed his name to Tutankhamen (better known today as King Tut), returning to the name of the old god Amen. Egypt returned to polytheism, the order of the world was reestablished, the temples reopened, and the capital moved back to Thebes (near modern Luxor), and in a few years the new city of Akhetaten was largely demolished. See Donald B. Redford, *Akhenaten: The Heretic King* (Princeton University Press, 1984).

14. "Do not be evil, for kindliness is good. More acceptable is the character of a man of just heart than the ox [i.e., offering] of the evildoer," reads a text from the period. Wilson, "The Nature of the Universe," p. 106.
15. This phrase is used in a different context by Daniel Matt, *Essential Kabbalah* (HarperSanFrancisco, 1995), p. 10, and Gershom Scholem, *Major Trends in Jewish Mysticism*, 3rd revised ed. (Schocken, 1961), p. 35.
16. Zahi Hawass, the archaeologist in charge of the monuments at Giza, believes that the Pharaoh Khufu (called Cheops in Greek) contrived the Great Pyramid as a monument intended to consolidate and preserve royal power by linking it to the primary myth of Egyptian kingship—the command of celestial forces and cosmic order. See E. C. Krupp, *Skywatchers, Shamans & Kings: Astronomy and the Archaeology of Power* (Wiley, 1997), p. 289.

17. This discovery was made by Egyptologist Alexander Badaway in collaboration with Virginia Trimble (then an undergraduate, now a well-known astrophysicist). The Great Pyramid was built about 2500 B.C.E. Earth's axis pointed then toward a fairly dim star named Thuban (Alpha Draconis). The axis of the spinning earth precesses like the axis of a spinning top that wobbles, completing one complete precession (wobble) every 26,000 years—see the figure:

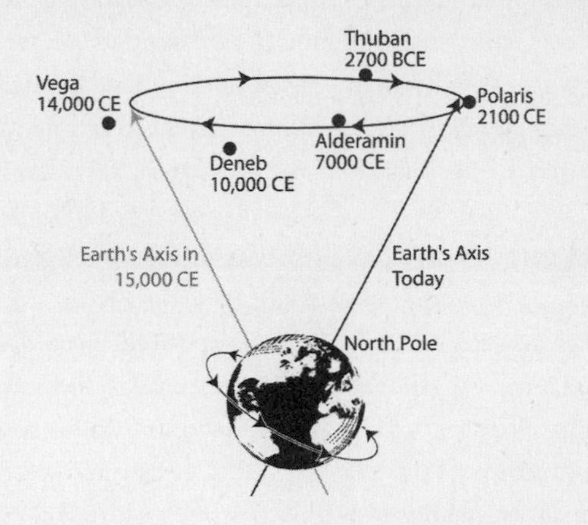

Polaris, which we call the North Star today (and which will be even more accurately over the pole in the year 2100), was not in line with the earth's north-pointing axis 4,500 years ago. Stellar north could also be found by observing the points on the northern horizon where a more southern star rises and sets and finding the exact midpoint between these points; for alternative methods, see K. Spence, "Astronomical Orientation of the Pyramids," *Nature* **412**, 699 (2001). Any buildings oriented then toward stellar north would now point toward Polaris.

18. Krupp, *Skywatchers, Shamans & Kings*, pp. 287–288. A popular book, R. Bauval and A. Gilbert, *The Orion Mystery: Unlocking the Secrets of the Pyramids* (Crown, 1994), claims that the Great Pyramid and the two smaller pyramids of Giza were intended to reflect the belt stars of Orion on earth—but then why would the Egyptians, who were so committed to orienting their spiritual buildings to the four directions, have reversed north and south in such a reflection? See E. C. Krupp, "Pyramid Marketing Schemes," *Sky & Telescope*, **93** (2) 64 (February 1997); also Krupp, *Skywatchers, Shamans & Kings*, pp. 288–290.

19. Psalms 104:2–3. Other examples: Genesis 11:2–9; Deuteronomy 5:8, 13:7, 28:64; I Samuel 2:10; Proverbs 30:4; Psalms 48:10, 61:2, 65:5, 98:3, 104:6, 135:6–7; Proverbs 30:4; Job 22:14, 37:3; Amos 9:6; Ezekiel 7:2.
20. *Genesis*, introduction, translation, and notes by E. A. Speiser (The Anchor Bible, Doubleday, 1964), pp. 9–10. The Babylonian myth is called *Enuma Elish*, after its opening words. In it the Primeval Waters were the great goddess Tiamat, who was vanquished in a tremendous, mythic battle by the high god of Babylon, Marduk. After killing her he divided her waters in two, constructed something to hold the upper waters up, and then created the world and humans. *Myths from Mesopotamia: Creation, The Flood, Gilgamesh, and Others*, trans. and ed. Stephanie Dalley (Oxford University Press, 1989), pp. 233ff. This story has survived on largely intact clay tablets from about 1800 B.C.E. Babylon, but aspects of the story seem to come from Sumer (a Mesopotamian civilization that preceded the Babylonians).
21. Robert Graves and Raphael Patai, *Hebrew Myths* (Anchor Books, 1989), pp. 21–25.
22. Richard Elliott Friedman, *Who Wrote the Bible?* (Prentice Hall, 1987), pp. 150–160.
23. Psalms 74:13–14. A few more examples: "Thou dost rule the raging of the sea; when its waves rise, thou stillest them. Thou didst crush Rahab [a sea demon] like a carcass, thou didst scatter thine enemies with thy mighty arm" (Psalms 89:9–10). "He gathered the waters of the sea as in a bottle; He put the deep in storehouses" (Psalm 33:7). "By his power he stilled the sea; by his understanding he smote Rahab. By his wind the heavens were made fair; his hand pierced the fleeing serpent" (Job 26:10–13). These quotations are from *The Oxford Annotated Bible* (1965).
24. For example, Claudius Ptolemy states that the Babylonians compiled practically complete lists of eclipses from the reign of Nabonassar (747 B.C.E.), although their information on the planetary motions was incomplete. See Otto Neugebauer, *The Exact Sciences in Antiquity* (Princeton University Press, 1952; repr. Dover, 1969), p. 98.
25. The psychologist Julian Jaynes, in an intriguing but controversial theory, dates what he considers a profound mental shift from god-driven consciousness to modern, questioning self-consciousness to late-second-millennium Mesopotamia, when rulers, becoming unsure of the gods, turned to omens and divination. Jaynes, *The Origin of Consciousness in the Breakdown of the Bicameral Mind* (Houghton Mifflin, 1976).

26. The attacks in the Bible against those Jews who worshipped the sun, moon, or planets (e.g., 2 Kings 23:5, Jeremiah 8:1–2, 44:17–19) show that the practice was nevertheless widespread.
27. Daniel 2.
28. Neugebauer, *Exact Sciences in Antiquity*, p. 36.
29. As we have explained, the Genesis story attempts to describe the origin of the flat earth. No matter how great or inspired such a story was, its authors did not dream of explaining the universe we now know about. Several modern books—for example, Gerald Schroeder, *Genesis and the Big Bang: the Discovery of Harmony between Modern Science and the Bible* (Bantam, 1992)—have attempted to argue that the six days of creation in Genesis are a prophetic description of the Big Bang. This argument can be made if an author stretches words enough, but the effort is pointless. Not only is it scientifically absurd, it does no justice to the Bible, according to Richard Elliott Friedmann, *The Disappearance of God* (Little, Brown, 1995), p. 231.
30. See Michael David Coogan, *Stories from Ancient Canaan* (Westminster, 1978), pp.15–23.
31. Robert Merton, *On the Shoulders of Giants* (Harcourt, 1965) traces the origins of this phrase, used by Newton, to Bernard of Chartres in the twelfth century.
32. This is according to both Herodotus (fifth century B.C.E.) and the much later neo-Platonist mathematician Proclus (411–485). The Egyptians had developed geometry in order to resurvey the land after the annual flood wiped out boundaries. Without geometry they could never have built the pyramids, but their geometry was highly practical; they didn't prove general theorems. Marshall Clagett, *Greek Science in Antiquity* (Collier, 1955), pp. 36–37.
33. G. E. R. Lloyd, *Early Greek Science: Thales to Aristotle* (Norton, 1970), pp. 7–15.
34. Jacques Brunschwig and Geoffrey E. R. Lloyd, eds., *Greek Thought: A Guide to Classical Knowledge* (Harvard University Press, 2000), pp. 434–435.
35. Francis M. Cornford, *From Religion to Philosophy: A Study in the Origins of Western Speculation* (1912; repr. Dover, 2004).
36. Benjamin Farrington, *Greek Science* (Penguin, 1953), p. 50.
37. People today who mix up the words "cosmology" and "cosmetology" have a good reason—there was an original connection. Ancient Greece had many alternative stories about the origin of the earliest gods, the Titans; one of them was that Form (male) married Matter (female), and Cosmos was the robe that

he gave her as a wedding gift. Cosmos—that is, the material world of heaven and earth—was thus the ornamentation of the primeval bride. James M. Redfield, *The Locrian Maidens: Love and Death in Greek Italy* (Princeton University Press, 2003), p. 406. This story of the cosmic marriage was told by Pherecydes of Syros, who was the teacher of Pythagoras, and who could supposedly predict earthquakes from changes in well water (p. 394).

38. Gregory Vlastos, *Plato's Universe* (University of Washington Press, 1975), p. 3.
39. Heraclitus reasoned like this: almost everything burns. Thus fire must be latent inside all things and only needs to be kindled. Fire is even latent in water, since oil is a liquid (a "water" in his scheme), yet it can burn. Water can put out fire, fire can be smothered by earth or else consume it, and water can wear down earth. He concluded that fire is the universal substance, and that it is constantly turning into earth and water, and they into each other. The ever-changing world could be explained as an eternal process of transformation and remixing of the elements earth, water, and fire. Vlastos, *Plato's Universe*, p. 6.
40. Plato is one source of the assumption concerning science that we warned against earlier: that truth must be by definition unchanging and eternal.
41. Lloyd, *Early Greek Science*, p. 84.
42. Each sphere in Eudoxus's system rotated about an axis through the center of the Earth. The rotation of this axis was determined by points fixed on another rotating sphere. Each planet required four spheres, the sun and moon three each. The stars were on the outermost sphere, and the whole system rotated about the earth each day. By this complicated scheme, Eudoxus was able to account fairly well for the observed motions of the planets and stars, including their occasional retrograde motions. The Eudoxian, Ptolemaic, and Copernican models are animated at http://faculty.fullerton.edu/cmcconnell/Planets.html#3.
43. An ellipse is the oval-shaped closed plane curve consisting of points the sum of whose distance from two fixed points (the "foci") is constant. Kepler found that the sun lies at one focus of any planet's orbit. A circle is the special case where the two foci coincide.
44. Plato described the five possible solid objects all of whose sides are identical polygons ("Platonic solids"). These solids are the tetrahedron (four identical equalateral triangles), cube (six identical squares), octahedron (eight identical equilateral triangles: two pyramids glued together), icosahedron (twenty faces which are identical equilateral triangles), and dodecahedron (twelve equilateral pentagons). He associated four of them with Empedocles's four

"elements" earth, water, air, and fire. (In *Timaeus*, 56B, for example, Plato identifies the tetrahedron, the one with the fewest faces and sharpest edges, with fire.) The fifth (subsequently sometimes called "quintessence" or "ether") he associated with the heavens. See Francis M. Cornford, *Plato's Cosmology: The Timaeus* (Routledge, 1935), p. 222. Plato's theory is in some ways similar to the atomic theory of his philosophical rival Democritus, whose atoms came in many different sizes and shapes. Vlastos, *Plato's Universe*, pp. 93–94, asks: "If we were satisfied that the choice between the unordered polymorphic infinity of Democritean atoms and the elegantly patterned order of Plato's polyhedra was incapable of empirical adjudication and could only be settled by asking how a divine, geometrically minded artificer would have made the choice, would we have hesitated about the answer?"

45. Protons and neutrons are made out of more elementary particles, quarks and gluons. Both the characteristics of such elementary particles and the nature of the forces between them are apparently completely described by a branch of mathematics called *group theory*. The electromagnetic, weak, and strong interactions are governed by the three simplest groups of a certain basic type.
46. Eratosthenes learned that at noon in midsummer the sun was visible in a deep well at Syene (near modern Aswan), and therefore directly overhead. The only additional information that he needed was (1) the angle that the sun made at the same time in Alexandria, which could be determined accurately simply by comparing the length of the shadow of a vertical stick with its height, and (2) the distance that Alexandria is north of Syene, which Eratosthenes arranged to have accurately measured. Dividing the distance by the angle, he thus determined the distance corresponding to a degree of latitude. Multiplying by 360, the number of degrees in a circle, gave the circumference of the earth.
47. This Greek translation is known as the Septuagint, after the Latin word for seventy. The first five books, the Torah or Pentateuch, were supposedly translated by a team of seventy (or seventy-two) Jewish scholars sent by the High Priest in Jerusalem at the request of Philadelphus II, the ruler of Egypt, in 282 B.C.E.. The remaining books were translated later.
48. Brunschwig and Lloyd, *Greek Thought*, pp. 877–878.
49. In Ptolemy's system, described in his book *Mathematical Treatise* (or *Syntaxis* in Greek; the Arabic translation was known as *Almagest*), each planet moves on a small circle (epicycle) riding on a large circle (deferent). His system is contrasted with Copernicus's in Figure 1 of Chapter 3 of our book. Ptolemy

developed it further in a later book, *Planetary Hypotheses*. See G. J. Toomer, "Ptolemy and his Greek Predecessors," in Christopher Walker, ed., *Astronomy Before the Telescope* (British Museum, 1996), p. 90; and Liba Taub, *Ptolemy's Universe* (Open Court, 1993).

50. The last of the Egyptian rulers, Cleopatra, was defeated by Rome in 30 B.C.E., and Egypt became a Roman province.

51. "The effort to harmonize Athens and Jerusalem [on cosmological issues] had not originated with Christianity but with Judaism, more particularly with Philo of Alexandria," according to Jaroslav Pelikan, "Athens and/or Jerusalem: Cosmology and/or Creation," in James B. Miller, ed., *Cosmic Questions, Annals of the New York Academy of Sciences* **950**, 17 (2001). See also Jaroslav Pelikan, *What Has Athens to Do with Jerusalem? Timaeus and Genesis in Counterpoint* (University of Michigan Press, 1997), and David T. Runia, *Philo of Alexandria and the Timaeus of Plato* (E. J. Brill, 1986). Philo (c. 20 B.C.E.–c. 50 C.E.) was a Hellenized Jewish philosopher. He may have influenced the author of the Gospel of John, according to C. H. Dodd, *The Interpretation of the Fourth Gospel* (Cambridge University Press, 1963). Saint Jerome (345–420), who translated the Bible into Latin, even lists Philo as a Church Father. Echoing Plato's *Timaeus*, Philo argued in *On the Creation* (26) that time did not exist before the world was created by God. This was also the view subsequently adopted by Saint Augustine (see, for example, his *Confessions*, XI, 15). Augustine (354–430) also argued that different interpretations of scripture are possible "in matters that are obscure and far beyond our vision. . . . In such a case, we should not rush in headlong and so firmly take our stand on one side that, if further progress in the search for truth undermines this position, we too fall with it." Augustine, *The Literal Meaning of Genesis*, trans. John H. Taylor, S.J. (Newman, 1982), p. 41. Augustine warned Christians not to make fools of themselves by taking the Bible literally on such questions as the shape of the earth. Rather, they should take references to the sky as a tent, for example, as metaphors. "Usually, even a non-Christian knows something about the earth, the heavens, and the other elements of the world. . . . Now, it is a disgraceful and dangerous thing for an infidel to hear a Christian presumably giving the meaning of Holy Scripture, talking nonsense on these topics" (p. 43). Subjects like the form and shape of heaven "are of no profit for those who seek beatitude. . . . What concern is it of mine whether heaven is like a sphere and the earth is enclosed by it and suspended in the

middle of the universe, or whether heaven is like a disk above the earth and covers it over on one side?" (p. 59).

CHAPTER 3

1. Lynn White, Jr., *Medieval Technology and Social Change* (Oxford University Press, 1962).
2. C. S. Lewis, *The Discarded Image: An Introduction to Medieval and Renaissance Literature* (Cambridge University Press, 1964). Much of the following discussion is based on this brilliant little book. An eminent medieval scholar, Lewis also wrote the *Narnia* stories for children and other books, including *The Screwtape Letters.*
3. Arthur O. Lovejoy, *The Great Chain of Being* (Harvard University Press, 1936).
4. Johan Huizinga, *The Waning of the Middle Ages* (Doubleday, 1956).
5. For example, there is a prediction in the New Testament that God will eventually destroy "the heavens and earth that now exist" with fire (2 Peter 3:5–7), and medieval Christian theologians were faced with explaining this prediction despite their equally certain belief that the heavenly spheres were eternal and perfect and therefore could not be destroyed. The reconciliation they reached was that the Biblical prediction must have been referring only to the part of the heavens beneath the sphere of the moon, because God would never destroy the spheres themselves. Medieval people had little historical understanding of how drastically cosmology had changed since Biblical times and that this prediction was probably written by people who believed in a flat earth and who had never considered the distant future problem of whether the prediction should cover the spheres or not. Lewis, *The Discarded Image,* pp. 120–121.
6. Chaucer, *Hous of Fame,* II, 730ff, quoted ibid., p. 92.
7. Muslims had their own explanations of the spheres. Some relevant Koranic quotations: "And He it is Who created the night and the day and the sun and the moon; all travel along swiftly in their celestial spheres" (Koran 21:33). "Neither is it allowable to the sun that it should overtake the moon, nor can the night outstrip the day; and all float on in a sphere" (36:40). "Do you not see how God has created the seven heavens one above the other, and made the moon a light in their midst, and made the sun as a lamp?" (71:15–16). The seven heavens (or spheres) corresponded to the sun and moon and the

five naked-eye planets (Mercury, Venus, Mars, Jupiter, and Saturn); the sphere of Saturn was thus the "seventh heaven."

8. Dante, *Paradiso*, XXX, 38.
9. Christian theologians in general accepted the idea of planetary influences. The Church only objected when these ideas led to actual worship of the planets, to denial of free will, or to the lucrative practice of prediction. Lewis, *The Discarded Image*, pp. 103–4; cf. E. M. W. Tillyard, *The Elizabethan World Picture* (Vintage, 1959), pp. 52–60.
10. "Kabbalah" means tradition, and its origins are uncertain. Some adherents claimed it began in the Garden of Eden; others held that it came from secret teachings that God gave Moses at the same time as the Ten Commandments, and that from that time onward it was passed down orally through the most religious Jews. Most scholars trace Kabbalah to Jewish texts from eleventh- and twelfth-century Spain. A good introduction to Kabbalah is Daniel C. Matt, *The Essential Kabbalah: The Heart of Jewish Mysticism* (HarperSanFrancisco, 1996).
11. Joseph Campbell, *The Inner Reaches of Outer Space: Metaphor as Myth and as Religion* (Harper & Row, 1988), p. 20.
12. Building on this foundation, Muslim scientists made significant advances in fields such as astronomy and mathematics—still commemorated in the common Arabic names of most of the bright stars (Aldebaran, Betelgeuse . . .) and in mathematics (algebra, algorithm . . .).
13. In fact, the causes were complex both for the decapitations of the two monarchs, and for the subsequent restorations of the monarchy in England and France. On the complexity of historical reasoning and the necessity but also limitations of historical analogies, see, for example, Bertrand de Juvenal, *The Art of Conjecture* (Basic Books, 1967), pp. 64–65.
14. Nicolaus Copernicus, *De revolutionibus orbium coelestium* (*On the Revolutions of the Heavenly Spheres*), Book I, chapter 10. For the story behind this historic book, see Owen Gingerich, *The Book Nobody Read* (Walker & Company, 2004).
15. Actually, in ancient Greece not long after Aristotle's death, Aristarchus of Samos (310–230 B.C.E.) proposed that earth is not the center of the universe, as Aristotle had claimed, but instead revolves around the sun. (We know of this from *The Sand Reckoner* by Archimedes, and from the Greek biographer Plutarch.) Aristarchus beat Copernicus to this realization by 1,800 years, but Aristarchus's heliocentrism was so far ahead of his time that even the most

forward-thinking philosophers in Greece could not absorb this idea. This is a perfect illustration of the Primack Principle, which holds that a discovery is named after the last person who discovers it. The astrophysicist who credited Joel with inventing this principle was Virginia Trimble, who discovered (see note 17 of Chapter 2) that the shafts starting in the King's Chamber inside the Great Pyramid of Giza pointed toward Orion and Thuban. (She cautioned, however, that the name of this principle could change if someone else discovers it.)

16. Copernicus's system was actually no more accurate than Ptolemy's—it was essentially a remapping of the Ptolemaic scheme to make the center of the system the sun rather than the earth. Copernicus also needed epicycles in his heliocentric system, in order to represent the planetary motion using only circles. It is commonly believed today that observations of improved accuracy required additional epicycles in the Ptolemaic system, and that Copernicus needed fewer epicycles than Ptolemy, but both of these are wrong. (See Gingerich, *The Book Nobody Read*, pp. 55–60.) There were no significant improvements in European observations of the stars and planets until Tycho Brahe.
17. The fact that the preface had been written by Andreas Osiander rather than Copernicus was not generally known until Kepler published it in 1609.
18. Bruno's views on the infinity of the universe echoed those of the English astronomer Thomas Digges (1546–1595). But according to Arthur O. Lovejoy, *The Great Chain of Being* (Harvard University Press, 1936), p. 119, and Alexandre Koyre, *From the Closed World to the Infinite Universe* (Johns Hopkins Press, 1957), pp. 39ff, although Digges put his stars into a theological heaven (also occupied by God and angels), it was Bruno who first championed the idea of an infinite and infinitely populous universe.
19. Cardinal Nicholas of Cusa (1401–1464) wrote on mysticism, infinity, and cosmology, and taught some surprisingly modern ideas including that the center of the universe is everywhere and its circumference nowhere. *Nicholas of Cusa: Selected Spiritual Writings*, trans. H. Lawrence Bond (Paulist Press, 1977), p. 161; see also Koyre, *From the Closed World to the Infinite Universe*, chapter 1.
20. Both these drawings are simplified. For example, in the Ptolemaic system the planets each move at constant speed with respect to a point, called the equant, which is not at the center of the earth.
21. Hold a finger vertically in front of your nose: parallax causes your finger to

appear to move with respect to distant objects when you compare the views with your right or left eye closed.

22. Quoted in Thomas S. Kuhn, *The Copernican Revolution: Planetary Astronomy in the Development of Western Thought* (Harvard University Press, 1957), p. 190.
23. The figure below shows Venus as seen from the earth in the Copernican scheme. When Venus is near the earth, it appears to be a crescent. When it is nearly on the opposite side of the sun it appears almost circular, but smaller. In the Ptolemaic scheme, however, Venus is always between the Earth and the sun; it can never be on the opposite side of the sun, as seen from the earth. Therefore, it should only go through phases from a crescent on one side to a crescent on the other side. Since Galileo saw Venus through his telescope go from circular to crescent-shaped as it grew larger in his telescope, that disproved Ptolemy's scheme, since the fact that the whole face of Venus is seen to be lit shows that its orbit takes it to the far side of the sun. Galileo did not announce his discovery of the phases of Venus in his book *Sidereus nuncius* (1610), since he had not yet seen its full cycle of phases. However, by early 1611 he had already seen Venus change from a small circle to a hemisphere and then become sickle-shaped, with the horns pointing away from the sun, as expected in the Copernican heliocentric system. See Owen Gingerich, *The Great Copernicus Chase and Other Adventures in Astronomical History* (Cambridge University Press, 1992), pp. 98–104, 110.

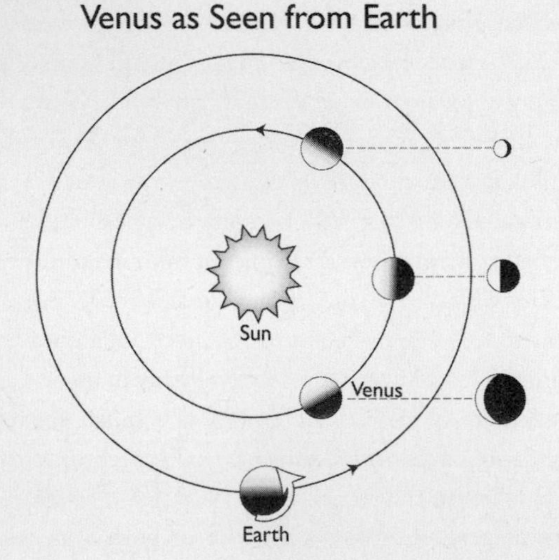

24. Galileo had not, however, proved that Copernicus was right, because it was still possible that, as Tycho Brahe had proposed, all the planets orbited the sun, while the sun revolves around the earth. This was essentially the same as the Copernican picture except for the point of view, and equally inconsistent with the medieval crystalline spheres, but it avoided the problems of the earth moving.
25. John Donne, *Anatomy of the World, The First Anniversary* (first published 1611), lines 205ff. Marjorie Nicolson, "The 'New Astronomy' and English Literary Imagination," *Studies in Philology* **32**, 428–462 (1935), repr. in Marjorie Nicolson, *Science and Imagination* (Cornell University Press, 1956), pp. 30–57, shows that Donne was following the latest developments in astronomy, and that these lines in his poem were responding to Galileo's *Sidereus nuncius*.
26. Meaning that the medieval sphere of fire below the moon doesn't really exist.
27. The Catholic Church in 1992 partially repudiated its condemnation of Galileo. Pope John Paul II said that "It is a duty for theologians to keep themselves regularly informed of scientific advances in order to examine if such be necessary, whether or not there are reasons for taking them into account in their reflection or for introducing changes in their teaching. . . . The error of the theologians of the time, when they maintained the centrality of the earth, was to think that our understanding of the physical world's structure was in some way imposed by the literal sense of Sacred Scriptures." Quoted from the English translation of the Pope's address to the Pontifical Academy of Sciences, which appeared in *L'osservatore romano*, November 4, 1992.
28. When Descartes learned of Galileo's arrest, he was finishing a book on cosmology. He wrote, "I was so astonished that I have resolved to burn all my papers, or at least not show them to anyone." The version of this material published in his *Principles of Philosophy* ten years later was carefully rewritten to avoid provoking the Church. See Martin Gorst, *Measuring Eternity* (Broadway, 2001), chapter 4; and Stephen Toulmin and June Goodfield, *The Discovery of Time* (University of Chicago Press, 1965), pp. 82–83.
29. This common statement is not entirely correct. Galileo died on January 8, 1642, according to the Gregorian calendar already in use in Italy. Newton was born on December 25, 1642, according to the Julian calendar then still in use in England—but this was January 4, 1643, according to the Gregorian calendar.
30. For the scientific revolution to occur required both a succession of brilliant

scientists (including Copernicus, Brahe, Kepler, Galileo, and Newton) and these social institutions within which they could flourish. The miracle is that it happened at all, not that it happened in Europe. See H. Floris Cohen, *The Scientific Revolution: A Historiographical Inquiry* (University of Chicago Press, 1990), pp. 506–525. In addition to official scientific organizations like the Royal Society of London, there were many less formal groups—see, for example, Jenny Unglow, *The Lunar Men: Five Friends Whose Curiosity Changed the World* (Farrar, Straus & Giroux, 2003).

31. As Newton described this incident in 1726 to his friend William Stukeley, he realized on seeing the apple fall straight downward that "there must be a drawing power in matter: and the sum of the drawing power in the matter of the earth must be in the earths center. . . . If matter thus draws matter, it must be in proportion to its quantity. Therefore the apple draws the earth as well as the earth draws the apple. That there is a power, like that we here call gravity, which extends its self thro' the universe." This quotation, from William Stukeley, M.D., F.R.S., *Memoirs of Sir Isaac Newton's Life,* ed. A. Hastings White (Taylor and Francis, London, 1936), pp. 19–20, encapsulates key ideas of Newtonian mechanics. The "quantity" of matter is its mass, which is a property of any object. Newton's mechanics can be summarized in his three laws of motion and his law of gravity:

    1. Inertia—things keep moving in a straight line at the same speed unless acted on by a force;
    2. Acceleration (rate of change of speed and direction) is proportional to force divided by mass; and
    3. The force of A on B is equal and opposite to the force of B on A.

    Newton's law of gravity says that the gravitational force of A on B is attractive (directed toward A) and proportional to the product of their masses divided by the square of the distance between them. However, Stukeley's report of Newton's claim that he understood all this in 1666 has been shown to be false; Newton only understood it as he was writing his great *Principia* twenty years later—see I. B. Cohen, "Newton's Discovery of Gravity," *Scientific American* **244** (3), 166–179 (March 1981).

32. Newton would have been surprised by some aspects of this "Newtonian" picture. Newton himself believed that God had an essential role in the universe, for example setting the planets on their paths and correcting their orbits every now and then. See, for example, Stephen Toulmin and June Goodfield,

The Fabric of the Heavens (Harper & Row, 1961), pp. 252–253. Newton's belief in absolute space and time is based on God's omnipresence and eternity, according to E. J. Dijksterhuis, *The Mechanization of the World Picture* (Oxford University Press, 1961), p. 487.

33. See Charles Coulston Gillespie, *The Edge of Objectivity* (Princeton University Press, 1960), p. 159. Voltaire's companion Gabrielle Émilie de Breteuil, the Marquise du Châtelet, translated Newton's *Principia*; and Voltaire's own *Elements of the Philosophy of Newton* appeared in 1738.
34. Newton, "Letter to Richard Bentley," in Milton Munitz, ed., *Theories of the Universe* (Free Press, 1957), pp. 211–212.
35. This is the basis of "Olbers's paradox," which was known to Kepler and first solved by Edgar Allan Poe. See Edward Harrison, *Darkness at Night: A Riddle of the Universe* (Harvard University Press, 1987).
36. The German philosopher Immanuel Kant (1724–1804) had suggested this "nebular hypothesis" in his first book, which had the ambitious title *Universal Natural History and Theory of the Heavens: An Essay on the Constitution and Mechanical Origin of the Whole Universe According to Newton's Principles* (1755). But the original publisher went bankrupt before the book could be distributed. Although Kant published a lengthy excerpt from it in 1791, after he became a famous philosopher, it was not widely known until the mid-1800s. Laplace described the nebular hypothesis in his popular book *Exposition du système du monde* (1796), and worked it out in mathematical detail in his *Traité de mécanique céleste* (1799). The nebular hypothesis was the first successful naturalistic picture of the origin of the solar system, and along with Darwin's theory of evolution it helped to spread scientific views of origins among the general population in nineteenth-century Europe and America, according to Ronald Numbers, *Creation by Natural Law: Laplace's Nebular Hypothesis in American Thought* (University of Washington Press, 1977). The modern scientific pictures of the origin both of the solar system and of spiral galaxies are generalizations of the nebular hypothesis.
37. Caveat: The "Existentialists" mostly disavowed that term and disagreed with each other on lots of things. But we're going to generalize about them anyway, because they read and responded to each other, and their influence has spilled over into literature and art of all media; indeed, some of them were just as much artists as philosophers. See also Chapter 10.
38. Karl Jaspers, quoted in Walter Kaufmann, *Existentialism from Dostoevsky to Sartre* (Meridian, 1956), pp. 33–34.

39. See, for example, Barbara Sproul, *Primal Myths: Creation Myths Around the World* (Harper & Row, 1979; HarperCollins, 1991); David Leeming, *Dictionary of Creation Myths* (Oxford University Press, 1994).

CHAPTER 4

1. About 61 percent of your weight is oxygen (mostly in water molecules), 23 percent carbon, 10 percent hydrogen, 2.6 percent nitrogen, 1.4 percent calcium, and 1.1 percent phosphorus, according to John Emsley, *Nature's Building Blocks* (Oxford University Press, 2001), p. 6. The last percent or so is other kinds of atoms. The top five elements in the sun are hydrogen, helium, oxygen, carbon, and nitrogen. The proteins out of which our bodies are made are composed almost entirely of carbon, hydrogen, oxygen, and nitrogen (CHON for short)—the same elements that are most abundant in the sun except for helium. (Helium plays no role in our bodies since it is chemically inactive.) Proteins are made of twenty amino acids, and only one of these is made of any other element besides CHON (cystine has one atom of sulfur). The nucleic acids that store our genetic information are based on CHON, plus phosphorus. For more on stardust, see John R. Gribbin, *Stardust: Supernovae and Life: The Cosmic Connection* (Yale University Press, 2000), and Hannah Holmes, *The Secret Life of Dust: From the Cosmos to the Kitchen Counter, the Big Consequences of Little Things* (Wiley, 2001).
2. Most stars in the night sky are relatively close by. For example, Sirius, the brightest star, is about eight light-years away. Deneb, the brightest star of the constellation Cygnus the Swan, is by far the most distant of the twenty brightest stars in the sky. It is about 1,400 light-years away, but that is only a twentieth of the distance to the center of the Milky Way. We cannot see very far in the disk of the Milky Way using ordinary light because our view is obscured by galactic dust. But we can get much clearer views using infrared radiation and radio waves.
3. Many such photos with brief explanations can be found in Timothy Ferris, *Galaxies* (Sierra Club, 1980) and other popular astronomy books, and on the Web—Google "Astronomy Picture of the Day" or "Galaxy Catalog" or "HubbleSite Galaxies."
4. Alan Dressler, *Voyage to the Great Attractor: Exploring Intergalactic Space* (Knopf, 1994).
5. The terms "particle" and especially "elementary particle" do not have fixed

meanings. The term "particle" is usually used in physics to mean something that comes in discrete units with fixed properties (such as mass), in contrast to "waves." But in quantum theory, objects sometimes behave like particles and sometimes like waves. The Greek philosophers who introduced the idea of atoms thought that they were indestructible particles, but a century ago the constituent particles of atoms were discovered—the electron in 1897 by J. J. Thomson, the proton in 1918 by Ernest Rutherford, and the neutron in 1932 by James Chadwick (working with Rutherford). For a while it was thought that these were elementary particles, but then in the 1970s it became clear that the neutron and proton are made of particles called "quarks" and "gluons." It is possible that the electrons, quarks, and gluons are elementary particles, along with the photon (the particle of light) and other particles that are not made of any particles more elementary than they are. But it is popular among theoretical physicists to speculate that all these particles are in turn made of microscopic pieces of "string" that can oscillate in unseen dimensions (see Chapter 6).

6. Radioactive iron-60 from a core-collapse supernova that exploded tens of light-years away about 3 million years ago has been found on Earth in ice cores and ocean sediments.
7. A rare, heavier sort of hydrogen, called deuterium, has a neutron as well as a proton in its nucleus. A small amount of deuterium was produced in the Big Bang. Atomic nuclei with the same numbers of protons but different numbers of neutrons are called *isotopes*; thus deuterium (pn) is a heavy stable isotope of hydrogen. Tritium is an even heavier isotope of hydrogen, with two neutrons and one proton in its nucleus (pnn). Tritium is radioactive, with a half-life of 12.3 years, and it decays to helium-3 (ppn), which is lighter than the common helium isotope helium-4 (ppnn).
8. For example, consider electrons that can travel to a screen through two slits. It is found that the number of electrons arriving at some points on the screen is actually *decreased* when both slits are open compared to what happens when one slit is closed. This is exactly as predicted by quantum theory, in which the electrons (behaving more as waves than as particles) can effectively go through both slits simultaneously. See R. P. Feynman, *The Character of Physical Law* (MIT Press, 1965), or chapter 37 of *The Feynman Lectures on Physics* (Addison-Wesley, 1963), reprinted as chapter 6 of *Six Easy Pieces* (Addison-Wesley, 1995).
9. When electrons are changing energy levels they oscillate between the levels,

radiating light, until the transition is complete. The oscillation frequency (which is faster the farther apart the levels are in energy) corresponds to the frequency of the light radiated.

10. The electron has a property called "spin" that can have two possible orientations, and if both electrons in a helium atom are in the lowest energy state (like the leftmost illustration in Figure 2) then the exclusion principle requires that their spins be oppositely oriented. Lithium has three electrons, but just as with helium only two oppositely oriented electrons can be in the lowest energy state. The third electron must be in the next-higher energy state, which means that lithium has quite different chemical properties from helium.
11. A neutron star has a radius of about 10 kilometers, the size of a city, but it has a mass comparable to that of our sun. A teaspoon of neutron star would have a mass of nearly a billion tons. A black hole has no meaningful size at all, but it is surrounded by a sphere of invisibility called the event horizon. According to Einstein's relativity, no material or information of any kind can get out of this sphere, which is about 3 kilometers (1.9 miles) in radius for a black hole with the mass of our sun, and ten times larger for a black hole ten times more massive.
12. A supernova was observed in 1987 in the Large Magellanic Cloud, the Milky Way's nearby satellite galaxy, which is visible from the Southern Hemisphere. Two separate neutrino detectors on Earth each registered the expected neutrino burst—which gives us confidence in the theory of how the core collapses in a supernova. Most likely Kepler's supernova of 1604, the last one definitely seen inside the Milky Way, was also this core-collapse type, called by astronomers Type II supernovas. The first Astronomer Royal of England, John Flamsteed, may have observed the Cassiopeia A supernova in 1680, which in that case was the most recent supernova observed in the Milky Way. It is the brightest astronomical radio source in the sky, and it was probably also of Type II.
13. These heavy elements are made by bombarding lighter nuclei with neutrons rapidly (the "r-process"), most likely in the outgoing debris from the supernova of a massive star. For popular accounts of the origin of the elements, see Timothy Ferris, *Coming of Age in the Milky Way* (Anchor, 1988); Ken Crosswell, *The Alchemy of the Heavens: Searching for Meaning in the Milky Way* (Anchor, 1995); or Virginia Trimble, "The Origin and Evolution of the Chemical

Elements," in B. Zuckerman and M. A. Malkan, eds., *The Origin and Evolution of the Universe* (Jones and Bartlett, 1996).

14. Planetary nebulas are visible for only about 100,000 years, after which they disperse. There are beautiful pictures of them in S. Kwok, *Cosmic Butterflies* (Cambridge University Press, 2001), and on the Web.
15. Subrahmanyan Chandrasekhar showed this in 1931 while he was still a graduate student at Cambridge University. He received the Nobel Prize for this work in 1983.
16. White dwarfs explode with an energy of about 2.5×10^{30} megatons, or about 10^{53} ergs. The material ejected includes up to about half the mass of the sun of radioactive nickel, which decays to radioactive cobalt and then to stable iron, and such supernovas are the source of most of the iron in our blood. The companion star to the white dwarf that exploded as Tycho's supernova was recently discovered—see Dennis Overbye, "How Supernovas Happen," *New York Times*, November 9, 2004. White dwarf supernovas are known as Type Ia supernovas.
17. There is astronomical evidence that the heavy elements out of which very long-lived stars formed indeed came from earlier core-collapse supernovas, and that the fraction of iron in the universe grew later as more white dwarfs had time to form and later explode.
18. "What's out there" is measured in terms of density. The total average density of the universe has been measured several ways, most accurately using the heat radiation of the Big Bang. To within a measurement uncertainty that is now only about 1 percent, the result is that the universe has "critical density." According to general relativity, critical density means that, on large scales, the universe is Euclidean—i.e., parallel lines neither meet nor diverge. It is numerically equal to about 9.5×10^{-30} grams per cubic centimeter. The amount of various types of matter and energy is often expressed by giving it as a fraction of critical density, which is represented by the Greek capital letter omega (Ω). For example, visible matter makes up about half a percent of all that's there, so $\Omega_{visible} \approx 0.005$.
19. One method is based on the amount of deuterium produced in the first few minutes after the Big Bang. (The less deuterium remains, the larger the total amount of atomic matter. This is because the more atoms there were at this early time, the more deuterium nuclei would have found each other and been fused to make helium.) It has been possible to determine the amount of pri-

mordial deuterium by studying the absorption of the light from distant quasars by nearer gas clouds in which very few stars had yet formed. Such absorption also gives an independent estimate of the total quantity of atomic matter. A third method of measuring the amount of ordinary matter is based on the sizes of the hot and cold spots in the heat radiation from the Big Bang. A fourth method is based on clusters of galaxies, whose tremendous gravity should retain most of their initial gas. Because it is very hot, the cluster gas emits X-rays, and this has allowed astronomers to measure how much of it there is. It is impressive that these quite different physical phenomena all give the same result for the amount of atomic matter: 4 to 5 percent of the cosmic density.

20. Protons and neutrons, the particles that make up atomic nuclei, are called "baryons," from the Greek word for heavy. The electron and its sister particles (muon, tau, and the corresponding neutrinos) are called "leptons," from a Greek word for small, since the electron has only about 1/2000 the mass of the proton or neutron. Because almost all the mass of atoms is in the tiny nuclei at their centers, ordinary or atomic matter is often referred to as baryonic matter. Thus non-atomic dark matter is also called "non-baryonic."
21. The term "density" in this context means the total amount of matter and energy in a representative volume of space. Because matter and energy can be converted into each other, as Einstein first showed in 1905, it has become the standard practice in astronomy to add them together, with mass multiplied by the quantity c^2, where c = 300,000 km/s is the velocity of light. This of course means that a small quantity of matter is equivalent to an enormous amount of energy—for example, one gram of matter can be converted into 25 million kilowatt-hours of energy. This happens not just in nuclear reactions but also in combustion, since a carbon and two oxygen atoms have less mass than the carbon dioxide molecule produced by their combustion. Although matter and dark energy both contribute to density, they have quite different effects on the expansion of the universe.
22. Zwicky had observed that galaxies in a cluster were moving at such high speeds that the gravity of the cluster could not confine them, if all the matter was just the visible stars in these galaxies. He therefore hypothesized that most of the mass was in the form of *"dunkle Materie"*—dark matter.
23. Newtonian gravity should apply because Newtonian and Einsteinian gravity agree at the relatively small scales and slow speeds of objects in galaxies (a few hundred kilometers per second, only about a thousandth of the speed of

light $c = 300{,}000$ km/s). The most serious effort to develop an alternative in order to avoid the need for dark matter is called Modified Newtonian Dynamics, or MOND. Although MOND can account for the motions of stars in many galaxies, it fails completely when it is applied to larger things like galaxy clusters. It also fails to account for the evidence for dark matter from gravitational lensing, discussed in note 27 below, and from observations of galaxy and cluster collisions.

24. Today we also have independent evidence of the existence of dark matter gathered by X-ray telescopes in space. Giant clusters of galaxies are always observed to contain high-temperature gas, which emits X-rays. The gas temperature is a measure of the high-speed random motions of the gas particles, and gives an independent check that there must be much more gravity holding the cluster together than could be produced by the visible stars and gas.
25. If the distant galaxy is directly behind the center of the cluster as seen from Earth, then on the sky around the cluster center there is an almost perfect circle of light, called an "Einstein ring." But if the distant galaxy is off-center, which is more common, we see fragmented arcs. Measurements of the light from the arcs always show that the light comes from much farther away than the foreground cluster. The redshift (explained in Chapter 5) of the light in the arcs is always higher, often much higher, than that of the cluster that is acting as the gravitational lens. (If even a single counterexample were discovered, it would undermine our entire picture of the expanding universe!)
26. Einstein first became a great celebrity when his prediction of the amount of gravitational bending of light passing near the sun was tested and confirmed during a total eclipse of the sun in 1919. Einstein later worked out the theory of gravitational lensing in an article published in 1934. Zwicky himself pointed out a few years later that gravitational lensing, if it were ever observed, could be used to confirm his discovery of dark matter, but it was many years before telescopes were good enough. What Einstein and Zwicky understood decades earlier was that it is possible to deduce the amount of mass in the foreground object when it acts as a gravitational lens, because the gravity of that mass is what's bending the light.
27. The gravitational lensing and X-ray data on galaxies and clusters shows that dark matter does not surround them spherically. Instead the dark matter "halos" are typically elongated—just as predicted from cold dark matter simulations based on the Double Dark theory. Measurements of the motions of satellites around galaxies show that their gravity also falls off as predicted by

the Double Dark theory. These are examples of observations that appear to be impossible to explain with a modified-gravity theory.

28. Arthur Conan Doyle, *The Sign of Four* (1890), chapter 6.
29. The L stands for lambda (Λ), the Greek capital letter that Einstein used to represent the cosmological constant, the simplest form of dark energy.
30. The key sources of this data have been NASA's Wilkinson Microwave Anisotropy Probe satellite, which released its first data in March 2003, and ground-based observatories at the Atacama Plateau in Chile and at the South Pole.
31. The Two-degree Field (2dF) survey, which finished in June 2003, obtained redshifts for about 250,000 galaxies visible from Australia. The ongoing Sloan Digital Sky Survey (SDSS) is determining redshifts for nearly a million galaxies visible from the Apache Point observatory in New Mexico, and digitally mapping a large patch of sky.
32. It can't be the usual elementary particles and it also can't be the hard-to-see or actually invisible macro-objects astronomers know about, such as white dwarfs, neutron stars, or black holes. (Some of these white dwarfs may explain the massive compact halo objects, or MACHOs, that have been detected by their gravitational lensing effects.) Recall that the total quantity of hydrogen and the other material of which stars are made is less than 5 percent of critical density, but there is at least five times as much non-atomic dark matter. Thus stars and their remnants can account for only a small part of the total cosmic matter. Moreover, when a star becomes a white dwarf, a neutron star, or a black hole at the end of its life, heavy elements are ejected, so the amount of heavy elements astronomers see gives another limit on the possible quantity of stellar remnants. There are also supermassive black holes at the centers of elliptical galaxies and the bulges of spiral galaxies, but although these are as much as several billion times as massive as the sun, their masses are only about 0.1 percent of the mass of the stars in these objects. So there is no way these could account for the dark matter.
33. Astrophysicists Gary Steigman and Michael Turner coined the term "WIMPs" in 1985. The development of modern cosmology has been chronicled by Dennis Overbye, *Lonely Hearts of the Cosmos: The Story of the Scientific Quest for the Secret of the Universe* (HarperPerennial, 1992), Michael D. Lemonick, *The Light at the Edge of the Universe: Dispatches from the Front Lines of Cosmology* (Princeton University Press, 1993), and Tom Yulsman, *Origins: The Quest for Our Cosmic Roots* (Institute of Physics, 2003).

34. The exchanged particle is called the "W-boson." Another sort of weak interaction involves exchange of a "Z-boson," which has a mass ninety-seven times that of a proton.
35. A large, sensitive neutrino detector in Japan called Super-Kamiokande has been able to see the sun during the night by observing such neutrinos that have traveled through the earth. The heart of Super-Kamiokande is 50,000 tons of ultrapure water watched by thousands of phototubes in a tank in a deep mine near Kamioka, Japan. Particles produced when the rare neutrino interacts produce light signals that are observed by these phototubes. The neutrinos from the 1987 supernova that we mentioned earlier were detected by two smaller detectors after going through the earth.
36. Such detectors are operated in deep underground laboratories to shield them from cosmic rays. There are increasingly sensitive WIMP detectors operating in the United States, England, France, and Italy. Researchers in the United States and several European countries are now looking for WIMPs with supersensitive apparatus located in deep underground laboratories to protect the experiment from cosmic rays, powerful radiation from space from which our atmosphere mostly protects us. Other researchers are looking for the indirect consequences if the dark matter is supersymmetric WIMPs, such as gamma rays or neutrinos from WIMP annihilation in our galaxy.
37. Supersymmetry also solves several other problems of particle physics. For example, with supersymmetry the electromagnetic, strong, and weak forces all have the same strength and "Grand Unify" at a high energy, about 10^{16} times the mass of the proton. Supersymmetry is also the basis of string theory. String theory (also called superstring theory) is discussed further in Chapter 6.
38. Benjamin Franklin discovered this. See I. Bernard Cohen, *Benjamin Franklin's Science* (Harvard University Press, 1990). In quantum theory, however, two signs of charge are not the only possibility. All the possibilities were classified by mathematicians working on "group theory," and several have subsequently been found by physicists to occur in nature. The strong interactions are based on the quarks carrying three types of charge, called "color." The simplest Grand Unified Theory is based on five types of charge, the three colors associated with the strong interactions plus two more associated with the weak and electromagnetic interactions.
39. When Paul Dirac first successfully combined special relativity with quantum theory in 1928, the resulting theory predicted that the electron had to have a mirror particle with the same mass but the opposite charge. This particle—

the anti-electron, or positron—was discovered in cosmic rays in 1932, and Dirac shared the 1933 Nobel Prize with Erwin Schrödinger, one of the main inventors of quantum theory. Physicists later realized that all particles must have anti-particles (although some particles with no electric charge, like the photon, are their own anti-particles), and that if a particle meets its anti-particle, the two can "annihilate"—that is, disappear and turn into other particles with the same total energy.

40. According to a convention invented by the British physicist John Ellis, who for many years has worked at the European Center for Nuclear Physics (CERN) in Geneva, the hypothetical superpartners of bosons are called *-inos* and those of fermions are called *s*-particles. For example the *photino* (spin ½) is the superpartner of the photon (spin 1), and *selectrons* and *squarks* (both of spin 0) are the superpartners of electrons and quarks (both spin ½).
41. In most versions of supersymmetry the lightest partner particle is predicted to be stable, and even if it's very massive there was plenty of energy in the early universe to make a great many of them. Furthermore, when we calculate theoretically how many of them would have survived the conditions of the early universe, the answer is in the same ballpark as the amount of dark matter that astronomers have actually measured to exist in the universe today.
42. Examples of such invisible dark matter candidates come from the "shadow matter" theory and other speculative theories.
43. Saul Perlmutter, a young physicist at Lawrence Berkeley National Laboratory, doggedly pursued the idea that one could measure the history of the expansion of the universe using Type Ia supernovas and a new research strategy modeled on the actuarial approach used by life insurance companies. Stars die at a steady rate, and a small but roughly constant proportion of them are Type Ia supernovas, like Tycho Brahe's supernova of 1572. A Type Ia supernova can outshine a whole galaxy for the few weeks when it is brightest. If one photographs a sufficiently large region of the sky through a telescope, and then takes comparison images of the same region two weeks later, on the average about ten (give or take about three) bright supernovas will be visible, and time can thus be scheduled in advance at large telescopes to allow spectra of them to be taken to determine the redshift of the galaxy and confirm that they are Type Ia supernovas. Since all Type Ia supernovas come from exploding white dwarfs of approximately the same critical mass, they should be about the same brightness. Since the intrinsically brighter ones also stay bright slightly longer, it is possible to use these "standard candles" to measure the distance accurately

to supernovas at different redshifts, and thus determine whether the expansion of the universe has been slowing down as expected. Once Perlmutter's team started to do this, a rival group organized by Australian astronomer Brian Schmidt and Robert Kirshner of Harvard began to compete with them. When both groups independently reached the same conclusion in 1998 that the expansion of the universe had changed from decelerating to accelerating, *Science* magazine declared it the Breakthrough of the Year. A great deal of evidence has accumulated since then confirming this discovery.

44. Actually, it was Einstein's friend Willem de Sitter (1872–1934), a Dutch astronomer, who first showed that Einstein's cosmological constant could lead to an ever-accelerating universe. See Pierre Kerszberg, *The Invented Universe: The Einstein–de Sitter Controversy* (Oxford University Press, 1989), and Helge Kragh, *Cosmology and Controversy* (Princeton University Press, 1996).
45. Einstein realized that it was not necessary to assume that the theory had to reduce exactly to Newtonian gravity in the limit of small size regions of space—it only needed to do so within the limits of measurement available then.
46. "Much later, when I was discussing cosmological problems with Einstein, he remarked that the introduction of the cosmological term was the biggest blunder he had ever made in his life," according to George Gamow, *My World Line* (Viking, 1970), p. 44.
47. If the dark energy is really just Einstein's cosmological constant, then the horizon around us empties out but our Local Group of galaxies, which is gravitationally bound, is unaffected. But if the dark energy is an extreme version called "phantom energy," then this contraction will be catastrophic and everything will be torn apart. If, on the other hand, the dark energy eventually decays away, then our cosmic horizon could eventually expand to infinity.
48. "Thou shalt not make unto thee any graven image, or any likeness of any thing that is in heaven above, or that is in the earth beneath, or that is in the water under the earth" (Exodus 20:4, Deuteronomy 5:8, King James Version).
49. The dollar bill with the Great Seal of the United States was introduced in 1935. For more on the history of the symbols, see David Ovason, *The Secret Symbols of the Dollar Bill* (HarperCollins, 2004).
50. Several different experiments have shown that neutrinos oscillate into each other, which implies that they have mass. But these experiments only measure the difference of the squared masses of the neutrinos, not the masses themselves; they are responsible for the lower limit of 0.01 percent on the contribution of neutrinos to the cosmic density. The 2 percent upper bound

on this contribution comes from cosmology, and it can be improved with better measurements.

51. Alan Dressler, *Voyage to the Great Attractor: Exploring Intergalactic Space* (Knopf, 1994), p. 335.
52. Jeremy Naydler, *Temple of the Cosmos: The Ancient Egyptian Experience of the Sacred* (Inner Traditions International, 1996), p. 285.

CHAPTER 5

1. The speed of light is unchanging even if the galaxy that emitted it is rushing away from us, as all distant galaxies are. Closer to home, if an imaginary car shines its headlights at us while backing away at half the speed of light, the light from the headlights travels here at the same speed as if the car were at rest. But if a bullet (which goes only one millionth the speed of light) were shot straight at us from the imaginary car, the speed of the car would have to be subtracted from the speed of the bullet, with the end result that the bullet would rush *away* from us at nearly the same speed as the car).
2. The definition of "cosmic horizon" we are using here corresponds to what is called the "particle horizon" by astronomers and physicists. The sphere of invisibility around a black hole is a different but related concept called an "event horizon." For more on horizons see, for example, Edward R. Harrison, *Cosmology: The Science of the Universe*, 2nd ed. (Cambridge University Press, 2000), chapter 21.
3. Saint Augustine, *Confessions* (397 C.E.), XI, 14.
4. The figure below illustrates how the universe expands, and why it looks from every galaxy as though that galaxy is the center of the expansion. Imagine we are observers in galaxy A (top bar, second from left). We see galaxy B

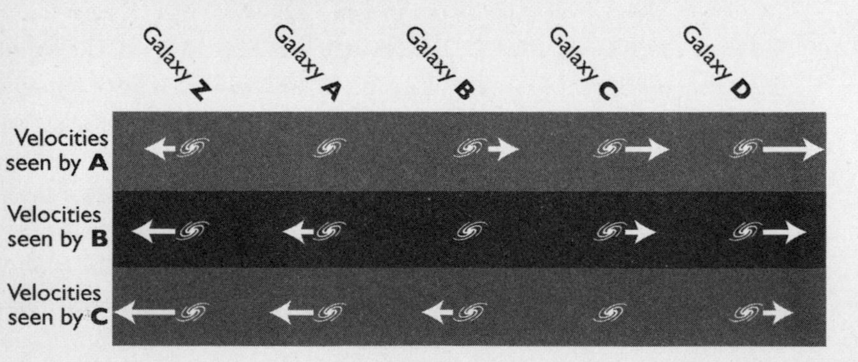

moving away at a velocity that is represented by the length of the arrow: longer arrow means higher speed. Galaxy Z, equally far away in the opposite direction, is moving at the same speed in that direction. We see galaxy C, which is twice as far away from us as galaxy B, moving away twice as fast. We see galaxy D, three times as far away as galaxy B, moving away three times as fast. So it appears that we on galaxy A are at the center of the expansion. But observers in galaxy B (second bar, center) see exactly the same pattern of velocities, as do those in galaxy C. Thus all observers appear to be at the center of a uniformly expanding universe. (Redrawn from Figure 1 of Steven Weinberg, *The First Three Minutes: A Modern View of the Origin of the Universe* [Basic Books, 1977], p. 22.)

5. If galaxies are moving *toward* us, their light is squeezed to shorter wavelengths ("blueshifted"), but only a few relatively nearby galaxies—including our neighboring giant spiral galaxy, Andromeda—are moving toward us. The only wavelengths human eyes can see are those between red and blue. This visible spectrum, however, lies in the middle of a vastly larger spectrum of invisible radiation. There are longer wavelengths than red (infrared, microwaves, radio waves) and shorter wavelengths than blue (ultraviolet, X-rays, gamma rays). With instruments we can detect all of these, although most infrared, ultraviolet, X-ray, and gamma-ray telescopes have to be in space, because most of these wavelengths don't penetrate the atmosphere.

   Each chemical element when hot emits light with certain characteristic patterns of colors, or wavelengths. For example, the brilliant red glow of the Orion Nebula, as well as other star-forming regions of our Galaxy, is a characteristic wavelength emitted by hydrogen, and it is always accompanied by certain aqua and blue wavelengths. The pattern of wavelengths becomes visible when astronomers spread the colors out by a prism or a grating.

   When light is traveling toward us from a distant source, all the wavelengths expand by the same factor as the universe has expanded. The cosmic background radiation was emitted about 400,000 years after the Big Bang; in the time since then the universe has expanded by a factor of about 1,000, and the wavelength of the cosmic background radiation has therefore also expanded by the same factor. Most of the light from the earliest galaxies we can see has redshifted into infrared wavelengths. New infrared telescopes in space such as the Spitzer Space Telescope, launched in 2003, and the James Webb Space Telescope, which NASA hopes to launch in about 2016, should allow us to see even earlier and smaller galaxies.

6. The mechanical clock was the greatest achievement of medieval mechanical ingenuity. It was originally invented to wake monks so that their before-dawn prayers could be synchronized with those at other monasteries. David S. Landes, *Revolution in Time: Clocks and the Making of the Modern World* (Harvard University Press, 1983), pp. 62–63, explains that this is what the song "Frère Jacques" is about. Soon cities competed to build elaborate mechanical clocks, and "no European community felt able to hold up its head unless in its midst the planets wheeled in cycles and epicycles, while angels trumpeted, cocks crew, and apostles, kings, and prophets marched and countermarched at the booming of the hours," according to Lynn White, Jr., *Medieval Technology and Social Change* (Oxford University Press, 1962), p. 124. These public clocks eventually were used to regulate work hours. The cosmos that had been a "great chain of being" in the Middle Ages began to be seen as a "clockwork universe." Mechanical clocks plus the steam engine were the inventions that led to the Industrial Revolution, which was a revolution not only in power technology but in social organization, according to Lewis Mumford, *Technics and Civilization* (Harcourt, Brace & World, 1934), p. 15. Thus although the mechanical clock was invented as a religious tool, it ended up secularizing our metaphorical understanding of time.
7. The use of equal-length hours we owe to Greek astronomy, and our division of hours into 60 minutes and the minute into 60 seconds we owe to the Mesopotamians. "Thus our present division of the day into 24 hours of 60 minutes each is the result of a Hellenistic modification of an Egyptian practice combined with Babylonian numerical procedures." Otto Neugebauer, *The Exact Sciences in Antiquity* (Harper Torchbook, 1962), p. 81.
8. Our modern calendar is descended from a modified Egyptian one that Julius Caesar adopted for the Roman Empire in 46 B.C.E. The ancient Egyptian civil calendar had 12 months of 30 days each, with 5 additional days at the end of each year. Although it did not keep pace with the seasons because the actual length of the year is close to 365¼ days, its simplicity recommended it for astronomical calculations, for which it is useful to have a year of fixed length without leap years. Astronomers still use "Julian dates," a continuous count of days since noon Greenwich Mean Time on January 1, 4713 B.C.E. (on the Julian calendar). The first moment of January 1, 2001, was Julian day 2451910.5.
9. Bishop Ussher's chronology became famous because it was included in popular English editions of the Bible. Ussher had Jesus born exactly 4,000 years

after the creation, and he also believed the common superstition that the world would end after 6,000 years—in 1997 (taking into account that there was no year 0). For a lively account, see Martin Gorst, *Measuring Eternity* (Broadway, 2001), chapter 2. Many earlier authors, including Saint Augustine, had arrived at a similar date for the creation. Reflecting this common view, Shakespeare has Rosalind say that "the poor world is almost six thousand years old" (*As You Like It*, IV, 1).

10. The English geologist James Hutton, in his book *Theory of the Earth* (1788), showed how to read in the layers of the earth the history of sedimentation, uplift, and then subsequent sedimentation, but he explained these processes as part of recurring geological cycles. He had no need for recourse to catastrophes or supernatural explanations such as Noah's flood. He concluded that, concerning the history of the earth, "we find no vestige of a beginning,—no prospect of an end." Charles Lyell, whose *Principles of Geology* helped inspire Charles Darwin, vigorously championed the idea that geological changes were gradual, and that the earth had changed little over time. Thus these early geologists were still believers in the ultimately cyclical nature of time—they just saw longer cycles. As the evidence grew that fossils had changed a great deal over time, the steady-state, cyclical geological theory gave way to a linear one. Stephen J. Gould, *Time's Arrow, Time's Cycle: Myth and Metaphor in the Discovery of Geological Time* (Harvard University Press, 1987), tells this story in detail.
11. The great Muslim scholar Ibn Sina (980–1037), also known by his Latin name, Avicenna, was perhaps the first to write that mountains are formed by earthquakes and erosion, and he noted that such processes would require very long periods of time. It was the philosopher Immanuel Kant who dared put a number to it, speaking seriously of a past comprising "myriads of millions of centuries." Kant, *Universal Natural History and Theory of the Heavens* (1755; see note 36 of Chapter 3); Stephen Toulmin and Jane Goodfield, *The Discovery of Time* (University of Chicago Press, 1965), pp. 64 and 132–133.
12. The only heat source the eminent physicist Lord Kelvin could imagine to power the sun was the energy released by its gravitational contraction, which couldn't last long. By the 1890s, Kelvin and others calculated an age for the sun of less than 40 million years, which didn't seem long enough for evolution to have shaped all the fossil and living species. The laws of thermodynamics proclaim that although energy cannot be lost, processes like friction inevitably reduce useful energy to waste heat. It predicts the "heat death" of

the universe. However, according to modern cosmology heat death is not the main concern in our future, as we discuss in Chapter 7.

13. In 1896 the French physicist Henri Becquerel discovered that invisible rays coming from uranium, the heaviest natural element (number 92), could expose photographic plates (the precursor to film) as if they had been exposed to light. X-rays had been discovered to do the same a year earlier. Soon afterward, Marie Curie found that another element, thorium (number 90), also emitted invisible rays, and that this occurred independently of the chemical state of the thorium—so the rays must be somehow coming from the atoms themselves. At the time, many physicists believed that matter was a continuous substance and atoms were merely a convenient fiction. Marie and Pierre Curie called the strange property of emitting invisible rays "radioactivity," and in 1898 they identified an element a million times more radioactive than uranium, which they called "radium." They became the only husband-and-wife scientific team ever to win the Nobel Prize, which they shared in 1903 with Becquerel. Pierre Curie subsequently showed that the rays from radium were so powerful they could heat its own weight of water from freezing to boiling in an hour, that the amount of heat was independent of the temperature of the radium, and that the radium hardly diminished its prodigious heat output over a long time period; again, these were indications that the rays were coming from somewhere deep inside the atoms. Ernest Rutherford and his collaborators subsequently named the three kinds of radioactivity alpha, beta, and gamma rays, and showed using alpha rays (helium nuclei) that almost all the mass of an atom is concentrated in its tiny nucleus.
14. We can tell how old rocks are by comparing the quantity of a radioactive element found in the rock, for example thorium or uranium, with the quantity of its decay products, for example lead. The best method available uses the ratios of the various lead isotopes (i.e., having the same number of protons but different numbers of neutrons in the nucleus), which were produced by different decay chains starting from thorium and uranium. Since thorium and uranium have half-lives of many billion years, their decay is suitable for dating rocks. Other radioactive decays are also used, for example potassium-40 (half-life 1.25 billion years). For pieces of wood and other plant and animal products, it is possible using radioactive carbon-14 (half-life 5,700 years) to determine the time since their carbon was drawn from the carbon dioxide in the atmosphere by a plant that may have subsequently been eaten.
15. Clair Patterson (1922–1995) was working on his dissertation at the Univer-

sity of Chicago when his advisor pointed out that the Canyon Diablo iron meteorite, which had created the Meteor Crater in Arizona, contained no uranium, so any lead in it was primordial. Assuming that the earth had formed with the same sort of lead, Patterson and his colleagues obtained in 1953 an age of 4.5 billion years, give or take about 0.3 billion. In 1956, Patterson dated the oldest meteorites, and got an age of 4.55 billion years, plus or minus 0.07 billion years. The same data would give an age of 4.48 billion years using today's more accurate decay rates, according to G. Brent Dalrymple, *Ancient Earth, Ancient Skies: The Age of the Earth and Its Cosmic Surroundings* (Stanford University Press, 2004), p. 166.

Some of the matter from which the solar system formed was radioactive elements from supernovas that had recently exploded nearby, and by dating meteorites and tracing the radioactive decay products in them, scientists have been able to produce a detailed chronology of the formation of the solar system. The best current data indicate that the first objects in the solar system formed 4.566 billion years ago, with an uncertainty of only about 2 million years.

There was also a socially beneficial result from Patterson's work. In order to measure lead decay products with the needed precision, Patterson had to eliminate contamination from lead in the environment. Using the sensitive techniques developed for geochronology, he traced the human use of lead for the past 6,000 years. He found that the concentration of lead in fresh arctic snow was hundreds of times higher than in ice from before Roman times, and that the levels of lead in human bones were now at least a thousand times higher than any "natural" level. He also found that the main recent cause was the use of tetraethyl lead as a gasoline additive, and his work provided essential data for the successful environmental campaign to require cars to burn unleaded gasoline and to remove lead from paint. See Shirley Cohen, "Duck Soup and Lead: Oral History—Clair C. Patterson," *Engineering & Science* (Caltech alumni magazine) **60** (1), 21 (1997); Stephen G. Brush, *Transmuted Past: The Age of the Earth and the Evolution of the Elements from Lyell to Patterson* (Cambridge University Press, 1996), pp. 85–86.

16. Perhaps the easiest way to understand this is to do the kind of "thought experiment" that Einstein liked to do. Imagine that we on Earth and travelers in a spaceship moving at close to the speed of light both use the same kind of clock: simply a light beam that bounces back and forth between two parallel surfaces. We look at our clock and see the light beam go up and down in a certain amount of time.

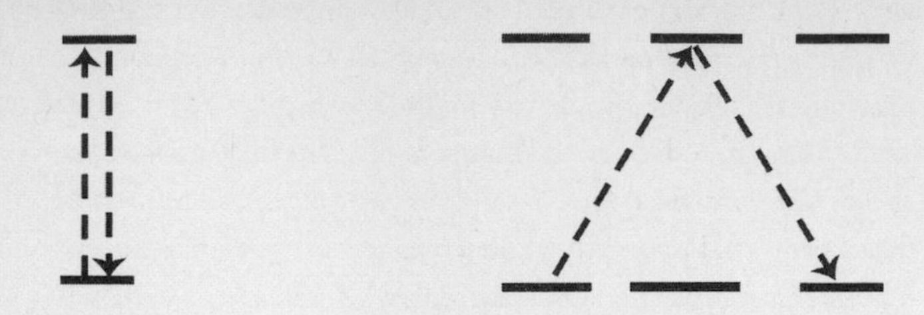

Light clock at rest. Light clock moving to the right.

From our point of view, however, the travelers' light beam doesn't go straight up and down, but instead diagonally because as it moves up the whole clock is moving to the right with the motion of the spaceship. From our point of view, therefore, their light beam crosses a longer distance on each tick, so it has to be taking a longer time; and that is all it means to say their clock is running slower. From their point of view, their clock is running normally and ours is slow. Nevertheless, careful analysis of thought experiments such as the "twin paradox" (see note 17 below) shows that there is no contradiction. (Readers are invited to run this and other Einsteinian thought experiments, and also to play relativity video games that Joel and his students created, at http://physics.ucsc.edu/relativity.)

That moving clocks run slow is constantly demonstrated at physics accelerator laboratories, at which the lifetimes of unstable elementary particles are sometimes lengthened by huge factors when they are moving close to the speed of light. Both this effect, and the effect of gravity on the speed of clocks described by Einstein's general theory of relativity, must be taken into account in the operation of the Global Positioning System (GPS). If they were neglected, the GPS errors would amount to many kilometers per day! See Peter L. Galison, *Einstein's Clocks, Poincaré's Maps: Empires of Time* (Norton, 2003), pp. 285–289. Galison also explains one way that Einstein was inspired to think about moving clocks: he was surrounded by newly installed synchronized clocks (whose minute hands all move in sync) in the Swiss capital city of Bern, and his work at the Swiss patent office included examining patent applications for synchronized clocks.

17. Since the traveler saw her homebody sister's time slow down too, what determines which one looks younger when they get back together? The answer is that the one who turns around will be younger. People haven't traveled like this yet because our fastest rockets move far more slowly than light, but na-

ture does this sort of experiment with elementary particles all the time. Unstable elementary particles called muons are the main component of the cosmic rays that reach low altitudes where most people live, but they would have decayed high in the atmosphere if their lifetimes were not greatly lengthened by this relativistic effect. These muons, which have a half-life of only 1.52 microseconds (1.52×10^{-6} s), live much longer when they move at nearly the speed of light, exactly as predicted by relativity.

18. There is a preferred reference frame (the cosmic clock reference frame) at every point in space and time, in which the cosmic background radiation has the same temperature in all directions (except for the tiny fluctuations discussed later in this chapter). In special relativity, which applies to motion in straight lines at constant speed in empty space, there cannot be any special reference frame. But our universe isn't empty space—it is filled with the cosmic background radiation and the light from all the galaxies. If one is moving with respect to the reference frame of the cosmic background radiation, the radiation will appear blueshifted in the spaceship's front window and redshifted looking out the rear. Measurements of the cosmic background radiation using the COBE satellite showed just these sorts of redshifts and blueshifts as the earth moved around the sun. But this was not the first proof that Copernicus was right that the earth orbits the sun: a slight seasonal change in the apparent direction of stars ("stellar aberration") caused by the earth's motion was discovered in 1726 by astronomer James Bradley.

19. The age that astronomers had deduced for these stars was based on the now standard theory of how stars shine and evolve as they fuse hydrogen to helium, and also on the very reasonable assumption that all the stars in dense star clusters called "globular clusters" were born at about the same time. In a given cluster, stars that are more massive than the sun are brighter and burn the hydrogen in their centers faster, and thus have shorter lives than the sun; less massive stars are dimmer and take longer to use up the hydrogen in their centers. The age of the star cluster can thus be determined by measuring the brightness of the brightest stars that are still burning hydrogen in their centers—the ones that are just about to run out.

    The disagreement between the age of the oldest star clusters and the age of the universe was partly resolved when better data showed that the distances to the clusters with the oldest stars had been underestimated, so these stars were actually brighter and therefore younger—about 13 billion years old, as expected in a 14-billion-year-old universe. The revised distances and ages were first published in 1997 from analysis of data from the European Space Agency's Hipparcos satellite, which was launched in 1989. Hipparcos was the first astrometric satellite, designed to measure the positions of stars with unprecedented accuracy. It determined the distances to more than 100,000 stars using stellar parallax, and thereby allowed astronomers to calibrate methods that could be used to determine the distance to the oldest globular star clusters in the Milky Way whose ages were in question.

20. The age of the universe has been confirmed recently with radioactivity. Thorium and uranium, the same elements used to date the formation of the earth, have now been measured in some of the oldest stars in our galaxy, revealing that they are about 13 billion years old.
21. In the November 2004 Gallup poll on creationism, evolution, and public education, 45 percent of respondents chose "God created human beings pretty much in their present form at one time within the last 10,000 years or so"; 38 percent selected "Human beings have developed over millions of years from less advanced forms of life, but God guided this process"; and 13 percent chose "Human beings have developed over millions of years from less advanced forms of life, but God had no part in this process." The American public has not notably changed its opinion on this question since Gallup started asking it in 1982.
22. Figure 1 in Chapter 3 contains two such cross-section drawings, showing the Ptolemaic and Copernican planetary pictures.
23. This figure shows where the objects whose light we see are located today, not where they were when they emitted that light. The reason that these are different is that the universe has been expanding. As we mentioned, we now have a tested cosmological theory—the Double Dark theory—that tells us where on such a diagram to place every galaxy that we see, back to the beginning of galaxies and before. When the cosmic background radiation was emitted, the material that emitted it was actually only about 13 million light-years away from the material that would become our galaxy, but it is now about 44 billion light-years away.
24. When a lot of stars form, most of the light is emitted by the few really massive ones. But these bright and massive stars have short lifetimes of only a few million years before they become supernovas, and after they are gone the rest of the stars shine more dimly. Only galaxies that are forming stars rapidly will have many massive stars still shining brightly.

25. This term was introduced by Sir Martin Rees, Astronomer Royal and now president of the Royal Society.
26. The cosmic background radiation is perfectly uniform on average, once we have removed the effects of the motion of the sun and the Milky Way with respect to the average location of a large number of nearby galaxies. But on small scales, the cosmic background radiation has small differences in temperature in different directions reflecting the small differences in density from region to region in the Big Bang. The large denser regions evolved to form cosmic structures such as galaxies, clusters, and superclusters.
27. The discovery of the small details in the cosmic background radiation by the Cosmic Background Explorer (COBE) satellite in 1992 was the first confirmation of the Cold Dark Matter (CDM) theory—and the greatest thrill in Joel's scientific career. The detailed mapping of these small details by the Wilkinson Microwave Anisotropy Probe (WMAP) satellite in 2003 agreed in such detail with the predictions of the Double Dark version of CDM that it helped convince astronomers that the modern picture explained in this book is right. The following graph shows an example of this agreement. The curve is the *prediction* of the Double Dark theory regarding the frequency of spots of various angular sizes in the temperature of the cosmic background radiation, and the points are the *data* from WMAP and ground-based experiments.

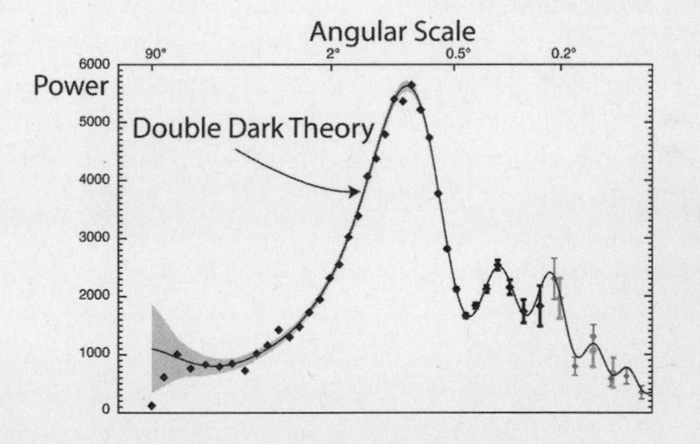

28. Only relatively high-energy neutrinos have been detected thus far; we do not yet know how to detect the much lower-energy neutrinos from the early universe. Experiments are planned to attempt to detect gravity waves from the Big Bang—for example NASA's LISA project, which aims to put gravity wave detectors into orbit around the sun.

29. For example, recent measurements show that one of the most fundamental quantities in physics is the same in distant galaxies as it is in laboratories on earth, to at least five decimal places. This quantity is the "fine structure constant" (the square of the charge on an electron, divided by Planck's constant and the speed of light), which governs how tightly atoms are bound. (Earlier measurements had suggested that there might be a change in the fifth decimal digit.)
30. In another way of representing the expanding universe the lightcone is similar, except that the backward lightcone narrows toward the Big Bang, reflecting the fact that space was smaller in the early universe. Such diagrams are included in textbooks such as Harrison, *Cosmology* (see note 2 above).
31. The Russian physicist Andrei Sakharov (who was the leader of the Soviet H-bomb project, and later a leader in ending the Cold War) showed in 1967 that there are three requirements for this tiny difference between matter and antimatter to occur, and according to currently favored particle-physics theories, all three are likely to have happened. Sakharov's three requirements are violation of (1) time reversal invariance, (2) baryon conservation, and (3) thermal equilibrium.

    (1) Although it is obvious that a movie is being played backward when billiard balls appear to assemble into a triangular pattern with the cue ball flying off, physicists thought that individual elementary particle interactions are symmetric in time, like individual billiard ball collisions. That expectation was exploded by an experiment led by James Cronin and Val Fitch in 1964, so the first of Sakharov's requirements is indeed consistent with behavior seen in the laboratory.

    (2) The proton is the least massive particle with "baryon number" B. If B is conserved, we could immediately understand why the proton does not decay into (for example) a positron (anti-electron), a much lighter particle with the same electric charge. But theories such as "Grand Unification" of the electromagnetic, weak, and strong forces predict that B will be violated at the high temperatures expected in the very early universe—thus satisfying the second Sakharov requirement.

    (3) Our present theories of the weak and strong interactions, as well as various extrapolations of these theories, predict that there will be phase transitions in the very early universe in which it goes out of thermal equilibrium—the third requirement.

Thus all three of Sakharov's conditions are likely to have been fulfilled in the very early universe. This could happen in various different ways, but there is not yet enough data and theoretical understanding to allow us to figure out which way it actually happened. Experiments have meanwhile attempted to see protons decay in the laboratory, since that would be both an important confirmation of the theory and a significant clue about how the excess of particles over antiparticles came about in the early universe.

32. The reverse process—two photons turning into an electron and positron—also occurs, but only if the photons have enough energy. Photons redshift as the universe expands; that is, their wavelengths stretch and their energies decrease. By about one second after the beginning, two photons no longer had enough energy to make an electron-positron pair. The electrons and positrons continued to annihilate into photons, but without the reverse reaction any longer occurring, their numbers decreased extremely rapidly. Pretty soon no positrons were left.
33. Reminder: a hydrogen atom is just one proton plus one electron. Deuterium ("heavy" hydrogen) has an additional neutron in its nucleus, and is the starting point for further fusion producing heaver nuclei, in particular helium. A helium atom has two protons and two neutrons in its nucleus, plus two electrons outside the nucleus.
34. Radio telescopes now being built will be larger than any existing ones, and they will be tuned to the characteristic wavelengths of radio waves emitted by neutral hydrogen. Two of the leading projects are LOFAR in the Netherlands and Mileura Wide-Field Array in Australia. The big new infrared telescope in space is the James Webb Space Telescope, mentioned in note 5 above.
35. When a system assumes a stable configuration through random processes, physicists say that the system has "relaxed." Since in gravitational collapse this occurs as rapidly as the collapse itself, it is appropriate to say that such a fast process is "violent." Lynden-Bell's paper on this was published in 1967.
36. These are frames from an early computer simulation of gravitational collapse: P. J. E. Peebles, *Astrophysical Journal* 75, 13 (1970). There would also be dark matter particles outside the spherical region pictured here, of course. But a fundamental theorem of general relativity insures that they will have no effect on the inner particles as long as they, too, are distributed in a spherically symmetric way.
37. The sizes of galaxies are determined by an interplay between the gravitation of the dark matter and the heating and cooling of the gaseous ordinary mat-

ter. Observations by the WMAP satellite indicate that when the universe was only about 100 million years old and had expanded to only about 1/20 of its present dimensions, this process of ionization must have been underway; it finished when the universe was almost a billion years old, and had expanded to about 1/7 of its present size.

38. For more details, see http://cosmicweb.uchicago.edu/sims.html. These simulations were run and analyzed by Anatoly Klypin, Andrey Kravtsov, and Joel, and their students and colleagues.
39. T. J. Cox and Patrik Jonsson, who recently finished their Ph.D.s with Joel, and other astrophysicists including Lars Hernquist and Volker Springel, have run many supercomputer simulations of collisions of spiral galaxies, taking into account the dark matter, stars, and gas. These simulations appear to agree pretty well with what astronomers have observed, which gives us confidence that we are beginning to understand what happens when galaxies collide.
40. Most of these quasars were shining around 10 to 12 billion years ago. And when quasars were imaged with the Hubble Space Telescope (and more recently with ground-based telescopes using adaptive optics to get very high-resolution images), it turned out that they are usually at the centers of merging galaxies. According to the Double Dark theory, it is precisely during this period when the majority of the major galaxy mergers should have occurred. As we look back before about 12 billion years ago, we predict and find that big galaxies were increasingly rare; and if we look closer to the present than about 10 billion years ago, we find that with expansion the universe was thinning out and big galaxies less and less frequently came near enough to merge. The energy we see streaming out of quasars is radiated by the infalling gas, not the black hole itself. Much more of the mass of the gas is converted to energy in this process than by the fusion of gas in a star.
41. This result is produced in computer simulations. The amount of gas that reaches the central black holes before it gets blown out by these winds is controlled by gravity, and thus by the amount of matter (mostly stars) near the galaxy centers, and this leads to the final mass of the black holes being a fixed fraction—about 1/1000—of the total central stellar mass, as observed.
42. We can date the formation of the disk of the Milky Way to about 10 billion years ago, soon after the largest progenitor of our galaxy probably merged with another large protogalaxy. But the Milky Way has a relatively small central stellar bulge, and correspondingly a relatively small central black hole with a mass about 3 million times that of the sun. There are stars in the Milky Way

that clearly formed not here but in other small galaxies with far fewer heavy atoms, galaxies that the Milky Way absorbed long ago just as it is now eating the Sagittarius Dwarf galaxy. Arcturus, the fourth-brightest star in the sky, is probably one of the stars that have come from another galaxy.

CHAPTER 6

1. "The left hemisphere is more associated with specific mathematical functions, while the right appears better equipped for comparing numbers." Andrew B. Newberg, Eugene G. D'Aquili, and Vince Rause, *Why God Won't Go Away* (Ballantine, 2001), p. 169.
2. Astronomers have actually observed two kinds of black holes. One kind forms when giant stars reach the ends of their lives and become "core-collapse" supernovas with massive cores. These supernova-remnant black holes have masses a few times larger than that of the sun. The other kind of black holes are found at the centers of galaxies and have much larger masses, as much as billions of times that of the sun. The masses of these "supermassive black holes" are observed to be tightly correlated with the properties of their host galaxies, and this shows that they are produced as part of the galaxy formation process.

   Physicists don't actually know how objects smaller than stars can be compressed so much that they can become black holes. Stephen Hawking suggested that black holes with much smaller masses than stars might have formed during the Big Bang at the beginning of the universe, and he showed that such small-mass black holes would eventually evaporate by emitting "Hawking radiation." But such black holes have not been found and may not in fact exist.
3. It takes more energy to localize an object in a smaller region. (For example, we have to look with shorter wavelength radiation, and the shorter the wavelength, the higher the frequency and therefore the greater the energy, according to quantum theory.) Mass is related to energy by Einstein's famous formula $E = mc^2$, so more energy means more mass. This leads to the counterintuitive result that as particles get smaller, they must get more massive, since it takes more energy to localize them. Thus an electron, which weighs only $\frac{1}{2000}$ as much as a proton, is considered "bigger" than a proton in size (not mass) because it's harder to pinpoint. It does not occupy a fixed location but is within a cloud of that size. The Planck length is named after the physicist

Max Planck, one of the inventors of quantum mechanics, who first calculated it. The Planck time (about 10^{-43} seconds) is the tiny amount of time it would take light to cross the Planck length. General relativity and quantum mechanics do not permit scales below the Planck limit. To observe a particle on the Planck scale, one would have to bombard it with an amount of energy equivalent to the Planck mass (about 2.2×10^{-5} grams), which would cause the region to collapse into a black hole, which would then evaporate in a Planck time. This is why it makes no sense to think of a region the Planck size as if it were ordinary space. A comprehensive theory of "quantum gravity," which would encompass and thus supersede both quantum mechanics and general relativity, is still a dream of physics. Superstring theory is our current best hope of such a theory.

These ideas are represented graphically in the plot of size versus mass at the top of page 349. Our modern theory of gravity, Einstein's general relativity, rules out the upper region of the figure, above the line from A to B; any object in this region is a black hole of negligible size. Quantum uncertainty rules out the lower region, below the line from A to C. Thus all possible physical objects must lie in the wedge-shaped region where various representative objects are plotted. The point of the wedge A is at the Planck length. Note that the sloping lines on which most objects lie are lines of constant density, and that living organisms, planets, and normal stars like our sun all have approximately the same density as water, while neutron stars have the same density as atomic nuclei.

4. Sheldon Glashow was one of the originators of the idea of a Grand Unified Theory that would combine the weak and electromagnetic interactions with the strong interaction. Glashow's first version of the Cosmic Uroboros is reproduced in Tim Ferris, *The New York Times Magazine*, September 26, 1982, p. 38; see also Sheldon Glashow with Ben Bova, *Interactions* (Warner, 1988), chapter 14.
5. Old Norse *gard* meant earth, place, or home—in modern English, *yard*.
6. See Victor F. Weisskopf, *Science* **187**, 605 (1975), and John D. Barrow and Frank J. Tipler, *The Anthropic Cosmological Principle* (Oxford University Press, 1986), pp. 307–308.
7. The nucleus of an atom is made of positively charged protons and uncharged (electrically neutral) neutrons, as we explained in Chapter 4. The strong force is what holds the nucleus together. It far overpowers the electrical force, which would on its own cause the protons to repel each other and fly apart. But the

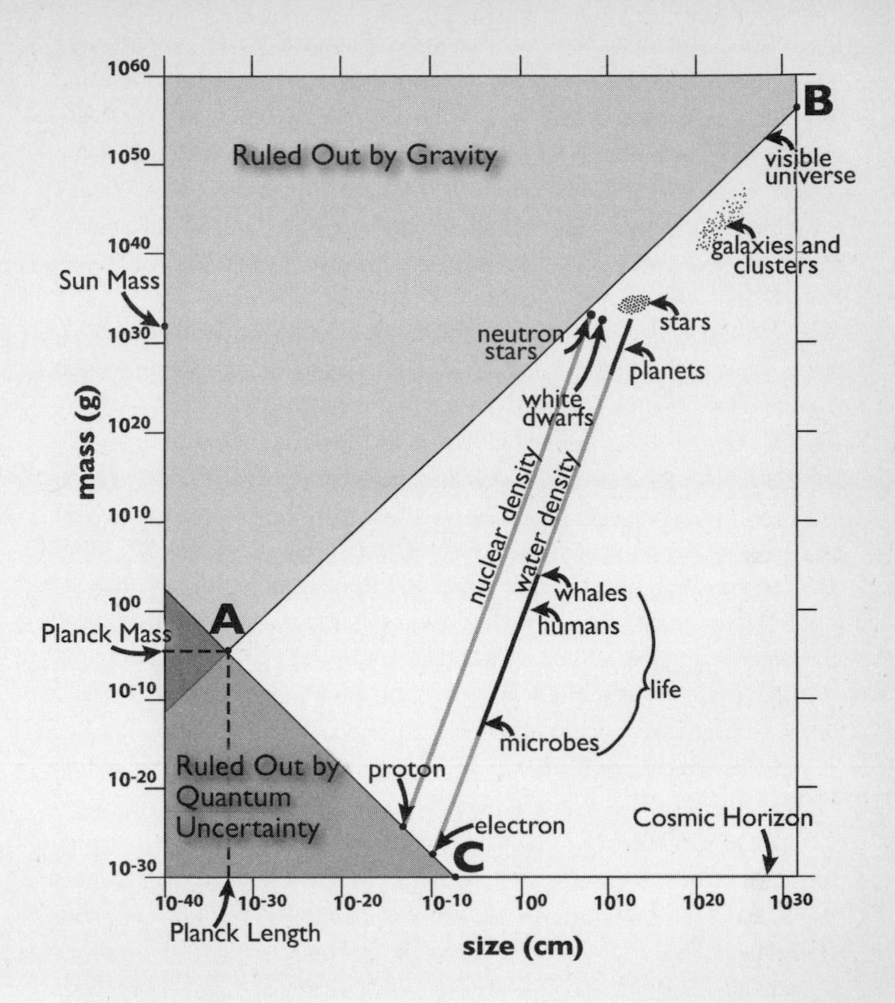

strong force has a short range that does not extend beyond the atomic nucleus. The weak force, which comes into play in many processes in which particles decay or are transformed into other particles, has an even shorter range.

8. This example is from J. B. S. Haldane, *On Being the Right Size and Other Essays,* ed. John Maynard Smith (Oxford University Press, 1985). In his essay "On Being the Right Size" (originally published in 1928), Haldane goes on to explain that "a rat is killed; a man is broken; a horse splashes. For the resistance presented to movement by the air is proportional to the surface of the moving object. Divide an animal's length, breadth, and height each by ten; its weight is reduced to a thousandth, but its surface only to a hundredth. So

the resistance to falling in the case of the small animal is relatively ten times greater than the driving force." For a critique, along the same lines, of the failure of creatures in science-fiction films to obey scientific laws, see Michael LaBarbera, "The Strange Laboratory of Dr. LaBarbera," *University of Chicago Magazine,* October–December 1996 (http://magazine.uchicago.edu/9612/9612LaBarbera.html). The irrelevance of gravity to bacteria is explained by R. C. Lewontin, in *Hidden Histories of Science,* ed. Robert B. Silvers (New York Review of Books, 1995).

9. This remapping, or duality, of string theory is explained in chapter 10 of Brian Greene, *The Elegant Universe: Superstrings, Hidden Dimensions, and the Quest for the Ultimate Theory* (Norton, 1999), especially pp. 252–254.
10. Einstein's remark to his assistant Ernst Straus, as quoted in Gerald Holton, *The Scientific Imagination: Case Studies* (Cambridge University Press, 1978), p. xii. Max Jammer, in *Einstein and Religion* (Princeton University Press, 1999), p. 124, gives the quotation in German from E. Straus, "Assistent bei Albert Einstein," in Carl Seelig, *Helle Zeit—Dunkle Zeit* (Europa, Zurich, 1956), p. 72.
11. See Erich Neumann, *The Origins and History of Consciousness* (Harper, 1962) for pictures of uroboros symbols from many cultures, and E. O. Wilson, *Consilience: The Unity of Knowledge* (Knopf, 1998), pp. 85–88 and 138–139, on the universality of the fear of snakes among primates and the psychological effects of serpents. "A snake is merely the zoological entity, but 'serpent,' as we will see, opens up vast metaphorical possibilities . . . the bearded serpents of ancient Egyptian and Greek religion; the partly human-bodied cobras with multiple fused hoods, the nagas of Hindu mythology; the horned, winged, hairy, feathered, or fire-spitting species of fable and legend; the basilisk, the dragon," according to Balai Mundkur, *The Cult of the Serpent: An Interdisciplinary Survey of Its Manifestations and Origins* (State University of New York Press, 1983), pp. 2–5. For example, the Hindu god Vishnu sleeps on the coils of Ananta, the serpent of infinity; see Richard Cavendish, ed., *Encyclopedia of Mythology* (Little, Brown, 1992), pp. 25f. In Dahomey, West Africa, the creative force controlling all life and motion is Da, or serpent. A Tibetan myth of origin has a female serpent born from the void; the crown of her head becomes the sky, her eyes are the sun and moon, her tongue becomes lightning, and so on creating the world. In many of the oldest creation stories not only of Europe, the Middle East, and Asia but of the Americas, there is a Goddess-Mother in the form of a serpent. In Aztec mythology, Quetzalcoatl (the Plumed Serpent) and Tezcatlipoca pulled Coatlicue, the Earth Goddess (Lady of the Serpent Skirt)

down from the heavens. Taking the form of two serpents, they ripped her into two pieces to form the earth and sky. Her hair became plants, her eyes and mouth were caves and sources of water, other parts of her body became mountains and valleys. In the Maya creation story, the Popol Vuh, in the beginning there was nothing. "Whatever might be is simply not there: only murmurs, ripples, in the dark, in the night. Only the Maker, Modeler alone, Sovereign Plumed Serpent . . ." A godlike figure spoke to the Serpent, and together they formed the thoughts that became the earth—D. A. Leeming and M. A. Leeming, *Encyclopedia of Creation Myths* (ABC-CLIO, 1994), p. 188.

12. Galileo Galilei, *Discourses Concerning Two New Sciences* (Great Books, Encyclopaedia Britannica, 1952), vol. 28, p. 187.
13. Scale models are useful in engineering and biology when the physics of the situation shows how to apply the results from scale models to full-size prototypes. See Thomas A. McMahon and John Tyler Bonner, *On Size and Life* (Scientific American, 1983).
14. John H. Holland, *Emergence: From Chaos to Order* (Helix, 1998); Steven Johnson, *Emergence: The Connected Lives of Ants, Brains, Cities, and Software* (Scribners, 2001); Harold J. Morowitz, *The Emergence of Everything: How the World Became Complex* (Oxford University Press, 2002).
15. Roger W. Sperry, "A Search for Beliefs to Live by Consistent with Science," in C. N. Matthews and R. A. Varghese, eds., *Cosmic Beginnings* (Open Court, 1995), pp. 319f.
16. David Grinspoon, *Lonely Planets: The Natural Philosophy of Alien Life* (HarperCollins, 2003), pp. 223–227.
17. Niels Bohr, quoted in Ivan Tolstoy, *The Knowledge and the Power: Reflexions on the History of Science* (Cannongate, 1990).
18. Richard Dawkins, *The Selfish Gene* (Oxford University Press, 1976; 2nd ed., 1990).
19. James Lovelock, *Gaia: A New Look at Life on Earth* (Oxford University Press, 1979).
20. Steven Weinberg, "Two Cheers for Reductionism," *Dreams of a Final Theory* (Pantheon, 1992). P. W. Anderson argues that radical reductionism "breaks down when confronted with the twin difficulties of scale and complexity." "More Is Different: Broken Symmetry and the Nature of the Hierarchical Structure of Science," *Science* **177,** 393–396; revised version in N. Ong and R. Bhatt, eds., *More Is Different: Fifty Years of Condensed Matter Physics* (Princeton University Press, 2001). Ernst Mayr argues that reductionism has not

been a productive approach in biology, in *What Makes Biology Unique* (Cambridge University Press, 2004), pp. 67–82.

21. The best example we know of is the reduction of the laws of thermodynamics to the statistical properties of large numbers of particles—subsequently discovered to be what we now call atoms and molecules. But like the reduction of electricity and magnetism to quantum electrodynamics by Feynman and others, it is better to view this as an encompassing revolution rather than an example of radical reductionism. Karl Popper, *Unended Quest* (Open Court, 1974), pp. 269–281, surveys the limited success of reductionism in the physical sciences and concludes that "as a philosophy, reductionism is a failure . . . we live in a universe of emergent novelty."
22. For more on this, see, for example, John D. Barrow and Frank J. Tipler, *The Anthropic Cosmological Principle* (Oxford University Press, 1986); John Gribbin and Martin Rees, *Cosmic Coincidences* (Bantam, 1989), pp. 11f; and Martin Rees, *Just Six Numbers: The Deep Forces That Shape the Universe* (Basic Books, 2001).
23. Strong and weak forces govern both atomic nuclei, on the left of the Uroboros, and stars, directly across on the right. Dark matter is probably made of particles, such as WIMPs or axions, which are associated with very small scales—but the gravity of the dark matter holds together galaxies and clusters of galaxies. Grand Unified Theories (GUT) and the hoped-for superstring theory of everything may connect phenomena on even smaller scales with the whole cosmic horizon and beyond. A drawing of the Cosmic Uroboros that Joel used in papers published in 1983–84 showed these connections across the diagram between medium small and medium large scales, between even smaller and even larger scales, and so on.
24. *Hamlet*, III, ii.
25. From an address by Chief Oren Lyons to the Non-Governmental Organizations of the UN, Geneva, Switzerland, 1977. ©1990 Steve Wall and Harvey Arden. From the book *Wisdomkeepers: Meetings with Native American Spiritual Elders* (Hillsboro, Oregon: Beyond Words Publishing, 1990), p. 71. Used by permission.

CHAPTER 7

1. See Eugene Wigner, "The Unreasonable Effectiveness of Mathematics in the Natural Sciences," in *Symmetries and Reflections* (MIT Press, 1970), pp.

222–237; and John D. Barrow, *Pi in the Sky: Counting, Thinking, and Being* (Little, Brown, 1992), pp. 270–272.

2. The problem concerned magnetic monopoles. Grand Unified Theory (GUT), in which the electromagnetic, weak, and strong forces are unified into a single force at very high energy, predicts that so many extremely massive particles called magnetic monopoles are produced in the early universe that the universe would promptly collapse on itself. It was to solve that problem that Guth originally proposed the idea of cosmic inflation. But he pointed out that it would also solve two other fundamental problems of cosmology: why the universe has critical density (discussed in Chapter 4), and why the temperature of the cosmic background radiation is the same all across the visible universe (except for tiny fluctuations) even though the parts across which light could have traveled at the time the radiation was emitted were quite small compared to the horizon that we can see today. He and others soon realized that inflation would also naturally produce the tiny fluctuations needed to explain the origin of galaxies. The main versions of inflation theory that are used now are based on work by Andrei Linde. See Alan H. Guth, *The Inflationary Universe: The Quest for a New Theory of Cosmic Origins* (Perseus, 1997), for a personal account, and for more general introductions, see, for example, Timothy Ferris, *The Whole Shebang: A State-of-the-Universes(s) Report* (Simon & Schuster, 1997); and Martin Rees, *Before the Beginning: Our Universe and Others* (Perseus, 1998).
3. George Gamow, *One, Two, Three . . . Infinity* (Viking, 1947), pp. 7–9.
4. Ten doublings corresponds to an increase by a factor of 1,000 ($2^{10} = 1024 \approx 10^3$). To fill the chessboard would require $2^{64} = 2^4 \times 2^{60} \approx 16 \times (10^3)^6 \approx 2 \times 10^{19}$ grains of wheat, and each grain has a mass of about 0.05 gram.
5. See Guth, *The Inflationary Universe,* especially Appendix A, for an explanation of how negative gravitational energy arises. Guth also explains that the idea that the universe could have arisen as a quantum fluctuation was first proposed by physicist Ed Tryon in 1973, but it was not until cosmic inflation was understood that physicists realized that such a quantum fluctuation could naturally grow into a gigantic universe. Inflation can be caused by a quantum field that fluctuates around a particular numerical value at every point in space. In physics, fields can be thought of as particles. In the particle metaphor, inflation was caused by a sea of particles called "inflatons."
6. These unobserved particles are called "virtual particles." Including them in calculations leads to readily observable effects—for example, changing the elec-

tric force between the electron and proton in a hydrogen atom to one that is stronger than the classical $1/r^2$ force at small distances r. If such virtual processes were left out, the theory would completely disagree with experiment; when they are included, the predictions of the quantum theory are in agreement with observations to amazing accuracy. See Richard P. Feynman, *QED: The Strange Theory of Light and Matter* (Princeton University Press, 1988), p. 7.

7. For example, the rate of decay of radioactive materials is governed by this kind of probabilistic law. Every radioactive element has a "half-life," which means that after its half-life, half of its atoms will have decayed, no matter how many there were to start with; but which of its atoms will be the ones to decay and in what order is unpredictable in principle.
8. Quoted in George Johnson, *Strange Beauty: Murray Gell-Mann and the Revolution in Twentieth-Century Physics* (Vintage, 2000), p. 224. This principle was also enunciated in works of fiction: T. H. White, *The Once and Future King* (1939; repr. G. P. Putnam's Sons, 1965), p. 114; Robert Heinlein, *The Moon Is a Harsh Mistress* (Orb, 1966), p. 374. It is the opposite of the totalitarian principle that "everything that is not compulsory is forbidden"—John T. Whittaker, "Italy's Seven Secrets," *The Saturday Evening Post*, December 23, 1939, p. 53.
9. Einstein said, "I can, if the worst comes to the worst, still realize that God may have created a world in which there were no natural laws. In short, a chaos. But that there should be statistical laws with definite solutions, i.e., laws that compel God to throw dice in each individual case, I find highly disagreeable." *The Quotable Einstein*, ed. Alice Calaprice (Princeton University Press, 1996), p. 183. However, John Bell and others subsequently showed that experiment can test whether nature is essentially non-causal as quantum theory predicts, and experiments always confirm the quantum predictions. See Heinz Pagels, *The Cosmic Code: Quantum Physics as the Language of Nature* (Simon & Schuster, 1992); Bruce Rosenblum and Fred Kuttner, *The Quantum Enigma: Consciousness in the Physical World* (Oxford University Press, 2005).
10. Erwin Schrödinger, *What Is Life?* (1958; repr. Cambridge University Press, 1967).
11. The earlier events naturally got to inflate many more times. And yes, forty orders of magnitude means that many of these irregularities are bigger than our visible universe, and correspond to structures many times the size of our cosmic horizon.
12. The quantum fluctuations can be thought of as causing inflation to last

slightly less long or slightly longer in different places, which results in these regions being a little more or less dense than average. From the cosmic background radiation temperature differences (first seen by NASA's COBE satellite in 1992), astronomers measured that the differences in density from place to place were typically about thirty parts per million (i.e., about 0.003 percent). This value, often called Q, is just what the Cold Dark Matter theory had predicted in 1984 based on the amplitude needed for the dark matter to grow into structures like galaxy clusters. These differences in density are initially tiny in physical size, but their amplitude is preserved as they are inflated to macroscopic sizes. (Think of the intensity of a sound wave as reflecting the amplitude, but the wavelength as the tone. During cosmic inflation, the amplitude of density fluctuations stays constant as their wavelengths inflate.) Cosmic inflation can be arranged to produce fluctuations of this tiny amplitude Q, but that is not something that cosmic inflation theory predicted naturally. Like many other features of the observed universe, such as the masses of the elementary particles, Q is an input to theory that we someday hope to understand on the basis of an encompassing theory.

13. The closer in size the sparkpoints were to the Planck length, the stronger the gravity waves that are predicted to be generated during inflation. Experiments are attempting to measure these gravity waves, both through their effects on the polarization of the cosmic background radiation and also directly.
14. This is how the physicist John A. Wheeler summarizes Einstein's general relativity.
15. The earth intercepts about 2×10^{14} kilowatts of power from the sun, but this enormous amount corresponds to only about 2 kilograms of mass per second. Where does the mass that provides the sun's energy come from? The mass of the four hydrogens that are fused in the center of the sun is slightly greater than that of the resulting helium. The sun converts about 4 million (4×10^{6}) metric tons of matter to energy per second. (This is a small fraction of the sun's mass of 2×10^{27} metric tons.)
16. Why is the cosmic background such a big source of entropy? The energy of cosmic inflation (dark energy) went into producing vast numbers of particles and antiparticles, almost all of which annihilated each other in the first few seconds (as we explained in Chapter 4), so that most of this energy was converted into about a billion photons and neutrinos for every proton or electron in the universe. These photons and neutrinos are now the cosmic background radiation.

17. Even if dark energy keeps the universe inflating faster and faster forever on the large scale, inside galaxies things could still run down. Fred C. Adams and Gregory Laughlin, *The Five Ages of the Universe: Inside the Physics of Eternity* (Free Press, 2000), discusses the long-term future of the universe. See also Fred Adams, *Our Living Multiverse* (Pi Press, 2003).
18. The key predictions of inflation are as follows: (1) The universe is flat or Euclidean (that is, on large scales, parallel lines never meet, which follows from general relativity if the total matter-energy density is critical, i.e., $\Omega_{total} = 1$). The fluctuations are (2) "adiabatic" (that is, all components of the universe fluctuate together) and (3) "Gaussian" (the most random possible, as predicted by quantum mechanics). (4) The fluctuations have a nearly "flat" spectrum (that is, they are approximately the same amplitude on different size-scales). (5) The fluctuations of different angular sizes in the temperature and the polarization of the heat radiation from the Big Bang have definite relations. (6) There are definite relations between the gravity waves and the density fluctuations from inflation. Predictions 1 through 5 have been tested and confirmed (some of these are also tests of the Cold Dark Matter theory). Tests of greater precision are in progress, and (as mentioned in the text) we hope to be able to detect the gravity waves from the Big Bang and test prediction 6. Various theories have been proposed as alternatives to cosmic inflation, such as "ekpyrotic" and "cyclic" scenarios, but these either do not solve the problems that cosmic inflation does or else depend on finding a solution to the cosmological singularity problem which the usual approach to cosmic inflation sidesteps.
19. The large-scale homogeneity of the cosmic background radiation plus an argument based on general relativity imply in the context of eternal inflation that our bubble is at least 100 times the size of our horizon. In most versions of cosmic inflation theory, our bubble is many orders of magnitude larger than our horizon. The argument mentioned in the text is due to Leonid Grischuk and Jakob B. Zel'dovich (1978). For the application to inflation, see Andrew R. Liddle and David Lyth, *Cosmological Inflation and Large Scale Structure* (Cambridge University Press, 2000), pp. 121–122.
20. Eternal inflation is thus eternal forward in time, but we naturally want to know whether it is also eternal in the backward direction—that is, did inflation start a finite or an infinite length of time ago? We do not have any reason now to believe that there was a single beginning of the evolution of the

whole universe; for details and references to recent work on this question, see Andrei Linde, "Prospects for Inflation," *Physica Scripta* **T117**, 40 (2005). Saint Thomas Aquinas had distinguished ordinary time, eternity, and *aevum*, which has a beginning but no end; see G. J. Whitrow, *Time in History* (Oxford University Press, 1988), p. 130. Kant's *Universal Natural History* (1755) said, "The creation . . . has indeed once begun, but it will never cease." Stephen Toulmin and June Goodfield, *The Discovery of Time* (University of Chicago Press, 1965), p. 133.

21. Fred Hoyle and the other inventors of the "Steady-State Universe" in the 1940s and 1950s get the last laugh after all!
22. Andrei Linde tells us that his own favorite metaphor for eternal inflation is based on the Hindu concept of an eternally branching Cosmic Tree (Asvattha) of Eternity, which in the Upanishads is identified with Brahman, pure spirit. For an illustration, see Andrei Linde, "The Self-Reproducing Inflationary Universe," *Scientific American* **271** (5), 48 (November 1994).
23. See note 10 of Chapter 6.
24. See John D. Barrow and Frank J. Tipler, *The Anthropic Cosmological Principle* (Oxford University Press, 1986); Ian G. Barbour, *When Science Meets Religion* (HarperSanFrancisco, 2000); John D. Barrow, *The Book of Nothing* (Vintage, 2002); Martin Rees, *Just Six Numbers: The Deep Forces That Shape the Universe* (Basic Books, 2001); and the works cited in note 2 above.
25. Paley's book *Natural Theology* (1802) argued that if you found a watch in the woods, you would know it had not just sprouted there but had been made by an intelligent watchmaker; and similarly animals fit so perfectly into nature that they too must have a Maker. Richard Dawkins, *The Blind Watchmaker: Why the Evidence of Evolution Reveals a Universe Without Design* (Norton, 1996), explains what's wrong with Paley's argument.
26. For example, Heinz Pagels said that "physicists and cosmologists who appeal to anthropic reasoning seemed to me to be gratuitously abandoning the successful program of conventional physical science of understanding the quantitative properties of our universe on the basis of universal physical laws. Perhaps their exasperation and frustration in attempting to find a complete, quantitative account of the cosmic parameters that characterize our actual universe has gotten the better of them." *Perfect Symmetry: The Search for the Beginning of Time* (Simon & Schuster, 1985), p. 359. The main success claimed for anthropic reasoning in cosmology is the argument, due to Steven Wein-

berg, that if Dark Energy made a much larger contribution to the cosmic density than it is observed to do, then galaxies could not have formed, and thus we could not be here, either. But Anthony Aguirre showed in 2001 that this argument is greatly weakened if other cosmological parameters are also allowed to vary from their observed values.

27. The idea that the Creator followed a blueprint that existed before the universe is seen in Plato's *Timaeus,* according to which the Ideas were a kind of plan that preceded the creation. In Jewish Midrashic literature the idea also occurs; *Genesis Rabbah* 1:1 says: "A ruler building a palace consults an architect's plans. The Blessed Holy One, in creating the universe, also worked from a plan—the Torah." And according to the Gospel of John, "In the beginning was the Word"; one could interpret the Word as preexisting instructions or a blueprint. But in none of these traditions is there the idea that the awakening of creativity itself was a precondition for the blueprint. See also Dorothy L. Sayers, *The Mind of the Maker* (1941; repr. HarperSanFrancisco, 1987).
28. At the end of cosmic inflation, the grand unity between the strong, weak, and electromagnetic forces may have broken, and then the symmetry between particles and antiparticles. As the universe cooled further, the symmetry between the weak and electromagnetic interactions broke as the W and Z particles, carriers of the weak interaction, acquired large masses but the photon stayed massless. Still later, about 10^{-4} second into the Big Bang, as the temperature fell further and the strong interaction became stronger, triples of quarks were bound together in small regions where they have been confined forever since, in the interiors of particles such as protons and neutrons. The similarity between the breaking of physical symmetries in the early universe and the Kabbalistic idea of shevirah is discussed in Daniel C. Matt, *God and the Big Bang: Discovering Harmony Between Science and Spirituality* (Jewish Lights, 1996), especially pp. 79–90.
29. However, Kabbalists also believed that God's emanation goes on and creation is continual.
30. This interpretation of Lurianic Kabbalah is due to the great scholar Gershom Scholem; see especially the first chapter of his *Sabbatai Sevi: The Mystical Messiah, 1626–1676* (Princeton University Press, 1973). For a different view, see Moshe Idel, *Kabbalah: New Perspectives* (Yale University Press, 1988).
31. For a Christian view of tzimtzum and its connection with *kenosis,* see Jurgen Moltmann, *God in Creation: Gifford Lectures 1984–1985* (Fortress, 1993).
32. When performed in awareness of the cosmology, even eating became ele-

vated: "When you eat and drink, . . . arouse yourself every moment to ask in wonder, 'What is this enjoyment and pleasure? What is it that I am tasting?' Answer yourself, 'This is nothing but the holy sparks from the sublime, holy worlds that are within the food and drink.' " Daniel C. Matt, *The Essential Kabbalah: The Heart of Jewish Mysticism* (HarperSanFrancisco, 1996), p. 150.

33. By Jews, that is, except for Chassidism, a joyous mystical movement begun by Rabbi Yisrael (1698–1762), the *Ba'al Shem Tov* (Master of the Good Name), which grew partly out of Kabbalah. There was also a Christian version of Kabbalah, usually spelled Cabala or Qabala, which became an esoteric, mystical school of thought. Even Newton studied it.
34. The creation story from ancient Egypt, told early in this book, could also be used as a provocative and fertile metaphor system for helping visualize the key transitions from eternal inflation to cosmic inflation and then to expanding spacetime. To recap that story, the Egyptians called the Primeval Watery Chaos (of which nothing could be known) Nun, and Nun was the father of the gods. Nun's first tingling of *desire* to create was called Atum, the Creative Principle; Atum was the crucial transition from eternal nothing to the beginning of something. Atum gave birth to Order/Harmony/Truth (his daughter Ma'at), and this Order/Harmony/Truth included the plan for the future heaven and earth *before* they were created. When Atum breathed Ma'at back into himself—when Creativity began to follow the plan—there appeared Shu, the space for the world to come. So the four parallel pairs are Ein Sof/Nun, Keter/Atum, Hokhmah/Ma'at, Binah/Shu. Interestingly, Egypt reversed the Kabbalistic genders: Wisdom was female (the goddess Ma'at) and space was male (Shu). The personification of the Creative Principle, of Order/Harmony/Truth, and of Space should not distract us from seeing that together these ideas were a theory of creation strikingly similar to eternal inflation, cosmic inflation, and expansion of spacetime. There are of course differences between the Egyptian mythology and modern science: for example, Order/Harmony/Truth represented not only physical but also moral order, while the scientific theory contains no shred of moral content. Nevertheless, this ancient Egyptian story is imaginatively very close to our new story.
35. However, we understand so little about the nature of dark energy that we cannot rule contraction out. The most we can do now is to estimate a minimum lifetime for the universe. According to a recent article, the *minimum* time until cosmic doomsday in a large class of cosmic inflation theories is a comfortable 24 billion years.

CHAPTER 8

1. T. S. Eliot, *Four Quartets,* "Little Gidding," V.
2. The first extra-solar planets were found in 1992 orbiting a neutron star. The first detection of a planet around a sunlike star was reported by Michel Mayor and Didier Queloz in October 1995, and confirmed a few days later by Geoffrey Marcy and Paul Butler. See Michael D. Lemonick, *Other Worlds: The Search for Life in the Universe* (Touchstone, 1999). Many additional extra-solar planets have been found since then; up-to-date lists of extrasolar planets are at www.obspm.fr/planets and exoplanets.org/.
3. Paul Davies ("Goodbye Mars, Hello Earth," *The New York Times,* OpEd page, April 10, 2005) suggests that we search for unrelated life on Earth—since if we found it, we would know that life started at least twice on our planet. For example, it may be just a historical accident that our kind of life exclusively uses right-handed sugar molecules such as dextrose, so Davies suggests that experiments test whether there are any anomalous life-forms on Earth that can use left-handed sugars.
4. In Egypt in 1911, a dog was hit and killed by a meteorite, the only meteorite canine fatality ever recorded. In the 1990s scientists realized that this particular meteorite was from Mars. About thirty such meteorites from Mars have now been discovered. One (known as ALH84001, discovered near the Allan Hills of Antarctica in 1984) may even include fossils of bacteria-like organisms from Mars, although this is controversial. These objects were blasted off of Mars by impacts, and they struck Earth after orbiting the sun for a few million years. The relatively low gravity and thin atmosphere of Mars makes such ejection relatively easy. See Paul Davies, *The Fifth Miracle: The Search for the Origin and Meaning of Life* (Simon & Schuster, 1999), especially chapter 8.
5. Stuart Ross Taylor, *Destiny or Chance: Our Solar System and Its Place in the Cosmos* (Cambridge University Press, 1998). Peter D. Ward and Donald Brownlee, *Rare Earth: Why Complex Life Is Uncommon in the Universe* (Copernicus/Springer-Verlag, 2000); for a critique, see James F. Kasting, *Perspectives in Biology and Medicine* **44** (1), 117 (Winter 2001). Jonathan I. Lunine, *Astrobiology: A Multidisciplinary Approach* (Pearson/Addison-Wesley, 2005).
6. This is partly because the methods astronomers use to make these discoveries favors finding such close-orbiting massive planets. One such method is to look for small motions of the host star responding to the gravity of the planet

as it orbits the star, and these motions are larger and thus more noticeable for massive planets close to their stars. With special instruments on telescopes, it is relatively easy to detect motion of the star along our line of sight by looking for small redshifts and blueshifts in characteristic wavelengths of light. That the motions are really caused by giant planets has been checked by finding a few cases in which the planet is actually seen to transit across the face of its star, or where the planet itself has been observed in infrared light.

7. If the inward migration of a giant planet is caused by material orbiting the star farther away, earthlike planets could subsequently form from this material. It may also be possible for some planets to survive the inward migration of giant planets.
8. Jupiter also created the asteroid belt by preventing the formation of a planet there. Since the impact event that wiped out the dinosaurs 65 million years ago was probably caused by an asteroid rather than a comet (as indicated by the abundance of iridium and other elements common in asteroids in the layer just above the latest dinosaur fossils), Jupiter may be partly responsible for this particular extinction event.
9. Pluto is a big comet, a member of the Kuiper Belt of comets in the outskirts of the solar system, with an average distance from the sun about 40 times that of the earth. In 2004 another Pluto-like object in the Kuiper Belt was discovered; this frigid planet, three times as far from the sun as Pluto, was named Sedna, after the Inuit goddess of the sea who is thought to live at the bottom of the Arctic Ocean. In 2005 the discovery was announced of three other large Pluto-like objects. One, provisionally named 2004 UB313, is larger than Pluto. It is currently about 97 times as far from the sun as is the earth, which makes it the most distant object known to be in orbit around the sun. The other two, provisionally named 2003 EL61 and 2005 FY9, are both currently about 52 times as far from the sun as is the earth, with diameters about three-fourths Pluto's; 2003 EL61 has its own moon. These objects were discovered only recently because their highly elliptical orbits take them far outside the ecliptic plane in which the inner planets orbit.
10. If too much water vapor is evaporated into the atmosphere, water in the stratosphere is split by ultraviolet light into oxygen and hydrogen, and the lighter hydrogen escapes to space. This would cause the oceans to disappear in a few hundred million years. It is this phenomenon that determines the inner edge of the habitable zone.

11. But since the early earth had an atmosphere rich in the greenhouse gases carbon dioxide (CO_2) and methane (CH_4), the surface temperature could have remained above freezing even though the early sun is calculated to have emitted 30 percent less energy than today. See James F. Kasting and David Catling, "Evolution of a Habitable Planet," *Annual Review of Astronomy & Astrophysics* **41,** 429 (2003).
12. David H. Grinspoon, *Venus Revealed: A New Look Below the Clouds of Our Mysterious Twin Planet* (Addison-Wesley, 1997).
13. The origin of Earth's water remains uncertain. It was once thought that comets were the main source, but the comets that have been studied have twice as much deuterium (heavy hydrogen) relative to ordinary hydrogen as water on Earth does.
14. The Moon's mass is about 1.2 percent of the mass of Earth, and this ratio is the largest in the solar system except for Pluto, whose moon Charon is about one-seventh its mass. But Pluto itself is only about one-sixth the mass of our Moon, and as we already mentioned, these small frozen worlds on the outskirts of the solar system are more like giant comets than planets.
15. Although the *tilt* (obliquity) is nearly constant, the *direction* of the earth's axis rotates slowly like the wobbling of a spinning top, going around once every 26,000 years. (See note 17 of Chapter 2.) Oscillation of the tilt between about 21.5 and 24.5 degrees occurs with a period of about 41,000 years, which may be related to the occurrence of ice ages. The tilt was likely caused by the impact event that created the Moon, which also caused the early Earth to rotate rapidly. We don't know what the spin period of Earth would have been if the impact event had not occurred; if it were less than about twelve hours, Earth's tilt would be stable without the Moon.
16. Charles H. Lineweaver, Yeshe Fenner, and Brad K. Gibson, *Nature* **303,** 59 (January 2, 2004).
17. Prokaryotes may also possess small circular pieces of DNA called plasmids.
18. There are still *living* stromatolites in Shark Bay (western Australia) and a few other locations. The surface is a mat of living cyanobacteria and other prokaryotes, which very slowly add limestone to the underlying rock.
19. In rocks from Greenland there is molecular evidence for life as early as 3.8 billion years ago, when the bombardment may still have been going on. The evidence for early life is reviewed by Andrew H. Knoll, *Life on a Young Planet: The First Three Billion Years of Evolution on Earth* (Princeton Univer-

sity Press, 2003), chapter 4. The interpretation that the 3.5-billion-year-old layered limestone stromatolites were created by cyanobacteria is uncertain.

20. According to geochemist Everett L. Shock, "Life continues the release of energy begun by planetary differentiation but stalled by sluggish oxidation/reduction reactions. In this sense, life is inevitable, as it greatly enhances the efficiency of reactions that release energy, which would otherwise not proceed in its absence."
21. John Maynard Smith and Eörs Szathmáry, *The Origins of Life: From the Birth of Life to the Origin of Language* (Oxford University Press, 1999); Christian de Duve, *Life Evolving: Molecules, Mind, and Meaning* (Oxford University Press, 2002); and Lynn Margulis and Michael F. Dolan, *Early Life: Evolution on the PreCambrian Earth*, 2nd ed. (Jones and Bartlett, 2002), are excellent summaries of the evolution of life on earth. See also Sean B. Carroll, "Chance and Necessity: The evolution of Morphological Complexity and Diversity," *Nature* **409,** 1102–1109 (February 22, 2001), and Knoll, *Life on a Young Planet* (see note 19 above).
22. Some primitive organisms intermediate between prokaryotes and eukaryotes have survived, and they give us an idea how eukaryotes may have evolved. An example is Giardia lamblia, discussed in this context by Christian de Duve, *Vital Dust: Life as a Cosmic Imperative* (Basic Books, 1995), pp. 138–146. Giardia was one of the first microscopic organisms identified by the inventor of the microscope, Anton van Leeuwenhoek. Despite being so primitive, Giardia is an enormously successful organism. It has recently and unfortunately spread widely as a parasite among wild animals in the United States. While it was possible to drink directly from wild streams in places like the high Sierra mountains of California until the 1960s, now it is necessary to treat or filter the water before drinking it. Giardia has become the most common cause of water-borne parasitic illness in the United States, and perhaps 20 percent of the world's population is chronically infected.
23. For example, chloroplasts, which turn light into chemically stored energy, and mitochondria, which process energy. These little factories may have originated as separately living prokaryotes that either invaded or were engulfed by primitive eukaryotes. This endosymbiosis (Greek for "living inside together") theory was championed by biologist Lynn Margulis, and has now been widely accepted.
24. David J. Bottjer, "The Early Evolution of Animals," *Scientific American* **293** (2), 42 (August 2005).

25. Simon Conway Morris, *Crucible of Creation: The Burgess Shale and the Rise of Animals* (Oxford University Press, 1998).
26. The "Snowball Earth" theory was proposed by the paleomagnetism expert and biogeologist Joseph Kirschvink, and strong evidence for it was discovered by geologists Paul Hoffman and Daniel Schrag; see their article on this in *Scientific American* **282** (1), 68–75 (January 2000). Gabrielle Walker, *Snowball Earth: The Story of the Great Global Catastrophe That Spawned Life As We Know It* (Crown, 2003), tells some of the human story of this scientific revolution. That the Snowball Earth periods really did last millions of years each is supported by the recent discovery of layers of iridium and other meteoritic material just above the snowball layers; see Richard Kerr, "Cosmic Dust Supports a Snowball Earth," *Science* **308,** 181 (April 8, 2005). Kirschvink and others have found geological evidence for a previous Snowball Earth period about 2.3 billion years ago. This appears to have been associated with the first increase in the oxygen content of the atmosphere, which may have decreased the amount of the greenhouse gas methane. See L. R. Kump, J. F. Kasting, and R. G. Krane, *The Earth System*, 2nd ed. (Pearson/Prentice Hall, 2004), pp. 236–244.
27. In the 1800s there were 280 parts per million (ppm) of CO_2 in the atmosphere; by 1900, about 295 ppm; now it's 365 ppm. If we continue with business as usual, CO_2 will increase to about 600 ppm by 2050. Strenuous efforts could lower that to only 500 ppm, but that would still be nearly twice what it's been for the last several hundred thousand years. If we remain addicted to burning fossil fuels, CO_2 could go over 1,000 ppm by 2100. That's serious greenhouse territory.
28. High oxygen levels may have been required for the evolution of large animals. Knoll, *Life on a Young Planet*, especially pp. 217–219, discusses the likely role of the increase in atmospheric oxygen in the evolution of the Ediacarian animals and in the Cambrian explosion.
29. Early geologists like Lyell assumed that the earth had always been the same in the past, or at least that the same forces had always been acting. This helped them develop their understanding of mountain building, sedimentation, etc., but it led later geologists to resist the importance of continental drift and catastrophic impacts.
30. David M. Raup, *Extinction: Bad Genes or Bad Luck* (Norton, 1991).
31. Walter Alvarez, *T. Rex and the Crater of Doom* (Princeton University Press, 1997).

32. Stephen J. Gould, *Wonderful Life: The Burgess Shale and the Nature of History* (Norton, 1990), pp. 289, 283. Gould devoted another book to arguing that evolution has no direction and does not lead to progress except by chance: *Full House: The Spread of Excellence from Plato to Darwin* (Random House, 1996).
33. Loren C. Eiseley, *The Immense Journey* (Random House, 1957), pp. 160–162.
34. Jared M. Diamond, "Alone in a Crowded Universe," *Natural History* (January 1990), reprinted as chapter 12 of Diamond's *The Third Chimpanzee: The Evolution and Future of the Human Animal* (HarperCollins, 1992), and in Ben Zuckerman and Michael H. Hart, eds., *Extraterrestrials—Where Are They?* (Cambridge University Press, 1995), pp. 157–164.
35. Ernst Mayr, "Are We Alone in This Vast Universe?" in *What Makes Biology Unique* (Cambridge University Press, 2004), pp. 209–217. Hominids are the primates that evolved into humans, and no other kind of hominid besides us remains.
36. Simon Conway Morris, *Life's Solution: Inevitable Humans in a Lonely Universe* (Cambridge University Press, 2003).
37. Ibid., p. 215. After playing with a responsive octopus in an aquarium tank, we can attest that it is hard to avoid the impression that it is intelligent.
38. Ibid., p. 258, citing D. Reiss and L. Marino, *Proceedings of the National Academy of Sciences* **98**, 5937 (2001), who say: "The emergence of self-recognition is not a by-product of factors specific to great apes and humans but instead may be attributable to more general characteristics such as a high degree of encephalization and cognitive ability. . . . More generally, these results represent a striking case of cognitive convergence in the face of profound differences in neuroanatomical characteristics and evolutionary history."
39. See, for example, Julian Paul Keenan with Gordon G. Gallup, Jr., and Dean Falk, *The Face in the Mirror: How We Know Who We Are* (HarperCollins, 2003); also David Premack and Ann J. Premack, *The Mind of an Ape* (Norton, 1983).
40. Conway Morris, *Life's Solution*, pp. 269–70.
41. Rob Foley, *Trends in Evolution & Ecology* **8**, 196–197 (1993), quoted ibid., pp. 272–273.
42. Charles Darwin, *The Descent of Man; And Selection in Relation to Sex* (1871; repr. Penguin, 2004).
43. Geoffrey Miller, *The Mating Mind: How Sexual Choice Shaped the Evolution of Human Nature* (Random House, 2000); Matt Ridley, *The Red Queen: Sex and the Evolution of Human Nature* (Macmillan, 1994).

44. Richard Dawkins, *The Ancestor's Tale: A Pilgrimage to the Dawn of Evolution* (Houghton Mifflin, 2004), pp. 263–273.
45. Thomas A. McMahon and John Tyler Bonner, *On Size and Life* (Scientific American, 1983), p. 8. Thus you have about 10^{14} (100 trillion) cells in your body. As we explained above, cells of all plants and animals have nuclei, and are said to be eukaryotic. The more primitive cells without nuclei, called prokaryotic cells, often have diameters as small as about 3×10^{-4} cm. But single-celled eukaryotes, known as protists, can be much larger, 10^{-2} cm or more in diameter.
46. From Ary L. Goldberger, David R. Rigney, and Bruce J. West, "Chaos and Fractals in Human Physiology," *Scientific American* **262** (2), pp. 42–49 (February 1990). The illustration, by Carol Donner, appears on p. 47 of the article; it is reproduced here by permission of the illustrator and authors.
47. Benoit B. Mandelbrot, *The Fractal Geometry of Nature* (Freeman, 1982); John Briggs, *Fractals: The Patterns of Chaos* (Simon & Schuster, 1992).
48. G. B. West, J. H. Brown, and B. J. Enquist, "The Fourth Dimension of Life: Fractal Geometry and the Allometric Scaling of Organisms," *Science* **284,** 1677–1679 (1999). See James H. Brown and Geoffrey B. West, eds., *Scaling in Biology* (Oxford University Press, 2000). J. H. Brown et al., "Toward a Metabolic Theory of Ecology," *Ecology* **85** (7), p. 1771 (2004), summarizes more recent work and discusses the implications of this theory for ecological communities; this article is followed by several articles critically discussing this work, with a reply from the authors.
49. In English units, the tree shrew is 0.15 foot long and weighs 0.01 pound, and the whale is 100 feet long and weighs 220,000 pounds.
50. In other words, the heartbeat scales as $(\text{mass})^{-1/4}$ and lifetime as $(\text{mass})^{1/4}$.
51. The measurement of the scaling of properties by organisms with their size and mass was pioneered by the biologist Julian S. Huxley in the 1920s, who named it allometry. That metabolic rate scales as $(\text{mass})^{3/4}$ was discovered by the veterinary biologist Max Kleiber in 1932. This implies the scaling of metabolic rate per unit mass as $(\text{mass})^{-1/4}$. One might naively expect that metabolic rate would scale as surface area of organisms. This would imply that metabolic rate would scale as $(\text{mass})^{2/3}$, which would in turn imply a scaling of metabolic rate per unit mass as $(\text{mass})^{-1/3}$. But the fractal nature of the circulatory systems for air and blood allows a more efficient scaling.
52. This may reflect a continuation of the fractal scaling into the microtubules that form part of the inner structure of cells. Geoffrey B. West, William H.

Woodruff, and James H. Brown, "Allometric Scaling of Metabolic Rate from Molecules and Mitochondria to Cells and Mammals," *Proceedings of the National Academy of Sciences* **99,** suppl. 1, pp. 2473–2478 (2002).

53. McMahon and Bonner, *Size and Life,* p. 66. Since the infant is about one-twentieth of its mother's weight, scaling with $(\text{mass})^{-1/4}$ predicts that the metabolic rate of its cells should be $(1/20)^{-1/4} \approx 2$ times greater.
54. See ibid., p. 116. Since a typical human pulse rate is about 80 beats per minute, this corresponds to about 36 years—but the typical human lifetime in modern industrial societies is about twice this. Some biologists have pointed out lifespans of animals scale better with brain size than with body mass.
55. Kurt Schmidt-Nielsen, *Scaling: Why Is Animal Size So Important?* (Cambridge University Press, 1984), p. 147, says that lifespan in years scales with body mass M in kilograms as lifespan = $11.8\ M^{0.20}$ for mammals in captivity and as $28.3\ M^{0.19}$ for birds in captivity, so birds tend to live about 2.5 times as long as mammals of the same mass. (It does not appear that difference of the scaling exponent 0.20 from the 0.25 for metabolic rate is significant.)
56. The great evolutionary biologist George Gaylord Simpson concluded that humanoids on alien worlds probably do not exist and that, even if they do exist, we could not communicate with them in a meaningful way. "The Non-prevalence of Humanoids," *Science* **143,** 769 (1964), reprinted in Donald Goldsmith, ed., *The Search for Extraterrestrial Life* (University Science, 1980), pp. 214–221.
57. The website is http://setiathome.ssl.berkeley.edu/.
58. The SETI Institute employs about 130 people, primarily scientists, in a variety of fields including astronomy and the planetary sciences, chemical evolution, the origin of life, biological evolution, and cultural evolution. See the Institute's website, www.seti.org.
59. That lasers could be used for interstellar communication was pointed out by Charles Townes, who won the Nobel Prize for inventing the maser (which led to the laser). The most ambitious SETI search for laser signals is led by Harvard physicist Paul Horowitz (see his website, http://seti.harvard.edu/oseti/), a SETI pioneer who is also the coauthor of a leading textbook on electronics.
60. SETI pioneer Frank Drake and others have speculated that the first extraterrestrial signals may well be from such immortal civilizations. See David Grinspoon, *Lonely Planets: The Natural Philosophy of Alien Life* (HarperCollins, 2003), chapter 23.
61. Giuseppe Cocconi and Philip Morrison, "Searching for Interstellar Commu-

nications," *Nature* **184,** 844 (September 19, 1959); reprinted in Goldsmith, *The Search for Extraterrestrial Life,* pp. 102–104.

62. This figure assumes the most efficient matter-antimatter rockets or laser-propelled light-sail spaceships. See, for example, Eugene Mallove and Gregory Matlock, *The Starship Handbook: A Pioneer's Guide to Interstellar Travel* (Wiley, 1989).
63. This assumes interstellar spacecraft that travel only at the speed of our current rockets, about 10^{-4} of the speed of light. At that speed even a trip to the nearest star would take 40,000 years, so the spacecraft would have to accommodate many generations of voyagers—who would therefore need to have a very stable culture—or else very reliable robots. Such spacecraft could be created by using rockets to adjust the orbit of a small asteroid so that the gravity of one of the planets ejects it from the solar system in the desired direction. (This sort of "gravitational slingshot" has been used repeatedly to accelerate spacecraft toward the outer planets.) Spacecraft that can travel at higher speeds would require much higher inputs of energy, but they would allow much faster colonization of the galaxy. At 1 percent of the speed of light, it would take about ten million years—only a thousandth of the age of the Galaxy.
64. "Fermi's Question," in B. R. Finney and E. M. Jones, *Interstellar Migration and the Human Experience* (University of California Press, 1985), pp. 298–300.
65. Our friend is Józef Życiński, who subsequently became the archbishop of Lublin, Poland, and chancellor of the Catholic University of Lublin. Nancy was inspired by this conversation to write the title song on her CD *Alien Wisdom*—see http://expandinguniverse.org.
66. The astrophysicist Brandon Carter argued in 1983 that the sun's lifetime and the evolution of intelligent technological life are unconnected, so that it would be merely a coincidence if both time-scales were the same. Mario Livio and others have proposed that there are plausible connections between the lifetimes of stars (and the buildup of elements like carbon and oxygen in the galaxy) and the time required for intelligence to evolve. See David Koerner and Simon LeVay, *Here Be Dragons: The Scientific Quest for Extraterrestrial Life* (Oxford University Press, 2000), especially chapter 10.

CHAPTER 9

1. Donald Brownlee and Peter D. Ward, *The Life and Death of Planet Earth: How the New Science of Astrobiology Charts the Ultimate Fate of Our World* (Times Books, 2003).

2. This could be done by altering the orbit of a large comet so that on its rare passages through the inner solar system it passes repeatedly close to both the earth and Jupiter, as proposed by Don Korycansky, Gregory Laughlin, and Fred C. Adams, "Astronomical Engineering: A Strategy for Modifying Planetary Orbits," *Astrophysics and Space Science* **275**, 349 (2001). To do this would require perturbing the comet's orbit using large rockets. It would of course be crucial to prevent this comet from crashing into the earth during these maneuvers. Meanwhile, astronomical catastrophes like the one that destroyed the dinosaurs 65 million years ago can also be averted in the future by changing the orbits of asteroids and comets. These tasks would require improvements in our astronomical and rocketry capabilities that should be possible relatively soon. Such efforts are far easier than transporting many people to another planetary system.
3. For the few stars that collide it will be cataclysmic, but the probability that our sun will be involved is vanishingly small. In such a galactic collision, however, stars could pass close enough to other stars to dislodge their outer planets or comets—or send them inward to bombard their inner planets.
4. Simon Mitten, *The Crab Nebula* (Charles Scribner's Sons, 1978). The debris from the supernova of 1054, known as the Crab Nebula, is the most famous and conspicuous supernova remnant, number 1 in Messier's catalog of "nebulae" (Latin for "clouds," i.e., things that looked in a telescope like fuzzy luminous objects).
5. Source: *Statistical Abstract of the United States: 2004–2005*, especially tables 822, 861, 862. Inspired by Figure 2.1 of Marc H. Ross and Robert H. Williams, *Our Energy: Regaining Control* (McGraw-Hill, 1981).
6. Edward O. Wilson, "The Bottleneck," *Scientific American* **286** (2), 82–91 (February 2002); also Edward O. Wilson, *The Future of Life* (Knopf, 2002), from which this article was excerpted.
7. Steven Schneider, *Laboratory Earth: The Planetary Gamble We Can't Afford to Lose* (Basic Books, 1997); Spencer R. Weart, *The Discovery of Global Warming* (Harvard University Press, 2003); Intergovernmental Panel on Climate Change website, http://www.ipcc.ch/.
8. Peter H. Gleick, "Making Every Drop Count," *Scientific American* **284** (2), 40 (February 2001); Sandra Postel, "Growing More Food with Less Water," ibid., 46; Lester R. Brown, *Plan B: Rescuing a Planet Under Stress and a Civilization in Trouble* (Norton, 2003).
9. Humans now use 72 percent of the total biological productivity in east and

south central Asia, where almost half of the world's people live, and the same fraction in Western Europe, according to Marc L. Imhoff, et al., "Global Patterns in Human Consumption of Net Primary Production," *Nature* **429,** 870–873 (June 24, 2004). Net primary production is the net amount of solar energy converted to plant organic matter through photosynthesis. It can be measured in units of elemental carbon, and it represents the primary food energy source for the world's ecosystems. Human appropriation of net primary production, apart from leaving less for other species to use, alters the composition of the atmosphere, levels of biodiversity, energy flows within food webs, and the provision of important ecosystem services. See also Millennium Ecosystem Assessment, *Ecosystems and Human Well-being: Synthesis* (Island, 2005).

10. George Lakoff and Mark Johnson, *Metaphors We Live By*, 2nd ed. (University of Chicago Press, 2003), pp. 255–257. This chapter's discussion of metaphor and the brain is largely based on a groundbreaking series of books over the last quarter-century by linguist George Lakoff of UC Berkeley with several collaborators in different fields, who have investigated the nature of metaphor and its impact on our understanding of ordinary life, poetry, politics, psychology, philosophy, and mathematics. In addition to the book just mentioned, see Lakoff and Mark Turner, *More Than Cool Reason: A Field Guide to Poetic Metaphor* (University of Chicago Press, 1989); and Lakoff, *Moral Politics: How Liberals and Conservatives Think* (University of Chicago Press, 1996, 2002).
11. Steven Pinker, *How the Mind Works* (Norton, 1999), p. 357.
12. Lakoff and Johnson, *Metaphors We Live By*, pp. 8–9. See also Robert Levine, *A Geography of Time: The Temporal Misadventures of a Social Psychologist, or How Every Culture Keeps Time Just a Little Bit Differently* (Basic Books, 1997).
13. Cognitive psychologist Steven Pinker speculates that early humans started out with the same brain circuitry as animals for reasoning about moving through space and exercising force and resistance; the human brain evolved by, among other things, copying these structures into new brain domains devoted to higher kinds of thinking than just moving the body. The new structures thus had no connection to eyes and muscles; instead they are like "scaffolding whose slots are filled with symbols for more abstract concerns like states, possessions, ideas, and desires." Pinker, *How the Mind Works*, p. 355.
14. George Lakoff and Mark Johnson, *Philosophy in the Flesh: The Embodied Mind and Its Challenge to Western Thought* (Basic Books, 1999), chapter 2.

15. This was explained in Chapter 5.
16. There is no friction in empty space. Angular momentum is conserved on the earth, too, but a moving object will slow down because, through friction, its angular momentum is transferred to the earth.
17. Aaron Bernstein, "Waking Up from the American Dream," *BusinessWeek*, December 1, 2003, p. 54; Paul Krugman, "The Death of Horatio Alger," *The Nation*, January 5, 2004, at http://www.thenation.com/doc/20040105/krugman.
18. Jeffrey D. Sachs, "Can Extreme Poverty Be Eliminated?" *Scientific American* **293** (3), 56 (September 2005).
19. On a still smaller scale, our own planet is held up by chemical (electromagnetic) forces between atoms.
20. For better or worse, gambling is a de facto way that this happens in many human cultures. In the United States, legal gambling accounts for more than 10 percent of all leisure expenditures, and gambling travel accounts for more than a third of all "destination leisure" expenditures (a category that includes cruises, theme parks, spectator sports, and concerts). See the National Gambling Impact Study (1999), pp. 3–4, available at http://govinfo.library.unt.edu/ngisc/.
21. Robin Dunbar, *Grooming, Gossip, and the Evolution of Language* (Harvard University Press, 1996); Malcolm Gladwell, *The Tipping Point* (Little, Brown, 2000), pp. 176–181.
22. Lewis Thomas, *The Lives of a Cell* (Viking, 1974), p. 127.
23. Samuel P. Huntington, "The Clash of Civilizations," *Foreign Affairs* 72 (3), 22 (1993). Huntington also expanded his essay into a book, *The Clash of Civilizations: The Remaking of World Order* (Touchstone, 1997).
24. "Think Globally, but Act Locally" is the title of chapter 3 of René Dubos, *Celebrations of Life* (McGraw-Hill, 1981).
25. The *rate* of population growth reached its highest point in the 1960s and has been slowing since then, but even at these lower rates of growth, the actual numbers of people are inflating. See Joel E. Cohen, *How Many People Can the Earth Support?* (Norton, 1995); Joel E. Cohen, "Human Population Grows Up," *Scientific American* **293** (3), 48 (September 2005).
26. Donella H. Meadows, Jorgen Randers, and Dennis L. Meadows, *Limits to Growth: The 30–Year Update* (Chelsea Green, 2004).
27. Jared Diamond, *Collapse: How Societies Choose to Fail or Succeed* (Viking, 2005). The collapse of the Easter Island culture as a result of destruction of forests and other biodiversity is a particularly sobering example, also dis-

cussed by Paul G. Bahn and John Flenley, *Easter Island, Earth Island* (Thames & Hudson, 1992); and David Christian, *Maps of Time: An Introduction to Big History* (University of California Press, 2004), pp. 472–475.

28. Statement of Sheldon Presser, Senior Vice President, Warner Bros., Testimony before the House Committee on Ways and Means, April 21, 2005.
29. Genesis 25:29.
30. Exponential amplification, known as the "butterfly effect," refers to the possibility that a butterfly flapping its wings on one side of the planet can set off turbulence that gets amplified by complicated weather patterns until it's a hurricane on the other side of the planet. When effects are proportional to causes, a system is called *linear*. But the weather and many other systems are known to be *nonlinear*, in which tiny changes can cause huge effects. A system as simple as a double pendulum exhibits such nonlinear behavior—illustrated at the website http://www.myphysicslab.com/dbl_pendulum.html.
31. End of tractate Kedushin in the Jerusalem Talmud, which also asks: "It is not sufficient what the Torah has forbidden you, so you seek to prohibit upon yourself other things?" The Talmud is a collection of the writings of leading Jewish rabbis, with editions edited in Babylon and Jerusalem in the fourth century.
32. Gregory Benford, *Deep Time: How Humanity Communicates Across Millennia* (Avon, 1999). Nancy E. Abrams and Joel R. Primack, "Helping the Public Decide: The Case of Radioactive Waste Management," *Environment* **22** (3), pp. 14–20, 39–40 (April 1980); reprinted in Robert W. Lake, ed., *Resolving Locational Conflict* (Center for Urban Policy Research, Rutgers University, 1987), pp. 75–91.
33. Deuteronomy 30:19–20.
34. The refrigerator is always "on," and in many homes it is the appliance that consumes the most energy. In the mid-1950s the average refrigerator used 500 kilowatt-hours per year of electricity. Electricity was cheap, so there were no incentives for efficiency. Every year the new refrigerators burned almost 7 percent more energy than in the year before until by 1975, the average electricity use per full-size refrigerator in the United States was 1,800 kilowatt-hours per year. To get a sleeker look, some manufacturers went so far as to use less insulation, and then prevented accumulation of moisture on cold outer surfaces by including heating coils under them—which of course forced the refrigerator to work that much harder. But starting with U.S. appliance efficiency labels in 1975 and California standards in 1977, the average elec-

tricity consumption of new refrigerators began to drop dramatically, by about 5 percent per year, even though the refrigerators themselves have increased in size and dropped in purchase price. New refrigerators now use as little energy as 1955 models did, even though they are much larger and better. The net saving to Americans just from these more efficient refrigerators amounts to more than $10 billion per year. Arthur H. Rosenfeld, "The Art of Energy Efficiency," *Annual Reviews of Energy and Environment* **24,** 33–82 (1999), and private communication.

35. In September 2005, officials from the California Public Utilities Commission, the Energy Commission, and Pacific Gas and Electric signed a pact with officials from China to provide expertise and training to Chinese regulators and utility companies, according to the *San Francisco Chronicle,* October 7, 2005, p. 1. The Chinese were impressed that because of the success of decades of "demand-side management" of energy use, carbon dioxide emissions per capita in California are only about half those for the United States as a whole.
36. S. Pacala and R. Socolow, "Stabilization Wedges: Solving the Climate Problem for the Next 50 Years with Current Technologies," *Science* **305,** 968–972 (August 13, 2004). New technology could help us actually reduce emissions and solve the climate problem.
37. Dubos, *Celebrations of Life,* p. 92.
38. David Sloan Wilson, *Darwin's Cathedral: Evolution, Religion, and the Nature of Society* (University of Chicago Press, 2002), p. 22.
39. Christopher Stone, *Should Trees Have Standing? Toward Legal Rights for Natural Objects* (Kauffmann, 1974).

CHAPTER 10

1. Bertrand Russell, from his essay "The Free Man's Worship," in Stefan Andersson, ed., *Russell on Religion* (Routledge, 1999), p. 32. For a scholarly edition, see volume 12 of *The Collected Papers of Bertrand Russell,* entitled *Contemplation and Action, 1902–14* (Routledge, 1985).
2. "Humans are not the end result of predictable evolutionary progress, but rather a fortuitous cosmic afterthought, a tiny little twig on the enormously arborescent bush of life." Stephen Jay Gould, *Dinosaur in a Haystack: Reflections in Natural History* (Random House, 1995), p. 327.
3. Joseph Campbell, *The Inner Reaches of Outer Space: Metaphor as Myth and as Religion* (Harper & Row, 1988), p. 18.

4. Sagan's speech at his sixtieth-birthday fest, in Yervant Terzian and Elizabeth Bilson, eds., *Carl Sagan's Universe* (Cambridge University Press, 1997), p. 148.
5. Carl Sagan, *Pale Blue Dot: A Vision of the Human Future in Space* (Random House, 1994), p. 37.
6. Joel R. Primack, "Dark Matter, Galaxies, and Large Scale Structure in the Universe," SLAC-PUB-3387 (July 1984), in N. Cabibbo, ed., *Proceedings of the International School of Physics "Enrico Fermi" XCII* (North-Holland, 1987), pp. 137–241.
7. Neil Postman, *Amusing Ourselves to Death: Public Discourse in the Age of Show Business* (Methuen, 1985).
8. Actually, Earth's axis did not point toward any north star in Caesar's time, although it did by Shakespeare's. (See note 17 of Chapter 2.)
9. But unlike the Norse giants, the giants of the expanding universe have no connotation of evil.
10. "Midgard on the Cosmic Uroboros" is our *cosmological* address—a place not on a grid but in our context of meaning. In contrast, "Our Cosmic Address" in Chapter 4 is our *postal* address in the large-scale structure of the visible universe.
11. D. H. Lawrence, *Apocalypse and the Writings on Revelation* (1931; Penguin, 1996), p. 44.
12. This is because oxygen and silicon are produced together before a star explodes in a supernova; free-floating in space, they combine with the hydrogen that is everywhere, making H_2O and SiO_2—water and sand—that travel together and become incorporated into new worlds.
13. Utterance 587, R. T. Rundle-Clark, *Myth and Symbol in Ancient Egypt* (London: Thames & Hudson, 1959), p. 37.
14. Exodus 3:14.
15. Bryan Sykes, *The Seven Daughters of Eve: The Science That Reveals Our Genetic Ancestry* (Norton, 2001); see http://www.oxfordancestors.com.
16. There is, of course, that part of human beings not controlled by conscious identity that is usually described as beyond identity, what Christians call "losing oneself to find oneself," the Hindus call *atman*, the twelve-step programs might call "anonymity." However, this self, central as it may be to *individual* experience, is nevertheless trapped below a spiritual glass ceiling and does not contribute anything new to the *human race* if it is always expressed in traditional metaphors.
17. At the beginning of Genesis, "the deep" (*tehom* in Hebrew) is the term for the

primeval waters over which God is hovering in darkness before beginning to create. The Deep with a capital D, however, was one of the names of the old Mesopotamian goddess of the primeval waters, Tiamat. ("That *tehom* never takes the definite article in Hebrew proves it to have once been a proper name, like *Tiamat*." Robert Graves and Raphael Patai, *Hebrew Myths: Stories of Cosmic Forces, Deities, Angels, Demons, Monsters, Giants, and Heroes—Interpreted in the Light of Modern Anthropology and Mythology* [Anchor, 1964], p. 31.) In Mesopotamian cosmology, the oceanic Tiamat existed before all else and gave birth to generations of concepts (called gods) so fundamental they had to be in place before the creator god could be born. "The Deep" for us is Lakoff's "cognitive unconscious," the primeval wellspring of imagery still present within our minds right here and now, in some way controlling our thinking from the distant past. According to the myth, the world was created from the parts of Tiamat's body; for us too, oceans of unconscious metaphors are the raw materials from which we create our mental world. The Deep is biologically within us today, and this is why ancient imagery can still work, even to describe the modern universe.

18. H. G. Wells, *Outline of History* (Macmillan, 1921).
19. See Thorkhild Jacobsen, "Mesopotamia," in *The Intellectual Adventure of Ancient Man* (note 5 of Chapter 2), pp. 125–219.

Index